T0214856

Communications in Computer and Information Science 1381

More information about this series at http://www.springer.com/series/7899

K. C. Santosh · Bharti Gawali (Eds.)

Recent Trends in Image Processing and Pattern Recognition

Third International Conference, RTIP2R 2020
Aurangabad, India, January 3–4, 2020
Revised Selected Papers, Part II

 Springer

Editors
K. C. Santosh 🅾
University of South Dakota
Vermillion, SD, USA

Bharti Gawali
Dr. Babasaheb Ambedkar
Marathwada University
Aurangabad, India

ISSN 1865-0929   ISSN 1865-0937 (electronic)
Communications in Computer and Information Science
ISBN 978-981-16-0492-8   ISBN 978-981-16-0493-5 (eBook)
https://doi.org/10.1007/978-981-16-0493-5

This Springer imprint is published by the registered company Springer Nature Singapore Pte Ltd.
The registered company address is: 152 Beach Road, #21-01/04 Gateway East, Singapore 189721, Singapore

Preface

It is our great pleasure to introduce the collection of research papers in the Communications in Computer and Information Science (CCIS) Springer series from the third Biennial International Conference on Recent Trends in Image Processing and Pattern Recognition (RTIP2R). The RTIP2R conference event took place at Dr. B.A.M. University, Aurangabad, Maharashtra, India, during January 03–04, 2020, in collaboration with the Department of Computer Science, University of South Dakota (USA). Further, as in 2018, the conference had a very successful workshop titled Pattern Analysis and Machine Intelligence (PAMI), with more than 100 participants.

As announced in the call for papers, RTIP2R attracted current and/or recent research on image processing, pattern recognition, and computer vision with several different applications, such as document understanding, biometrics, medical imaging, and image analysis in agriculture. Altogether, we received 329 submissions and accepted 106 papers for conference presentations. Unlike in the past, conference chairs' reports were also considered to decide on publication. Based on thorough review reports, the conference chairs decided to move forward with 78 papers for publication. As a result, the acceptance rate was 23.70%. As before, we followed a double-blind submission policy and therefore the review process was extremely solid. On average, for a conference presentation, there were at least two reviews per paper except the few that had desk rejections. We also made the authors aware of plagiarism and rejected some of them even after conference presentations.

In brief, the event was found to be a great platform bringing together research scientists, academics, and industry practitioners. Following those review reports, we categorized the papers into five different tracks: a) computer vision and applications; b) data science and machine learning; c) image analysis and recognition; d) healthcare informatics and medical imaging; e) image and signal processing in agriculture.

The conference event (with more than 150 participants) was full of new ideas, including those presented by the primary keynote speaker Prof. Umapada Pal, Indian Statistical Institute (ISI), Kolkata, India.

October 2020

K. C. Santosh
Bharti Gawali

Organization

Conference website: rtip2r-conference.org

Patron

Pramod Yeole (Hon'sble Vice Chancellor)	Dr. B A M Univ., India
Pravin Wakte (Hon'ble Pro-vice Chancellor)	Dr. B A M Univ., India
Sadhana Pande (Registrar)	Dr. B A M Univ., India
Suresh Chandra Mehrotra	Dr. B A M Univ., India
Karbhari Kale	Dr. B A M Univ., India

Honorary Chairs

P. Nagabhushan	IIIT, Allahabad, India
P. S. Hiremath	KLE Technological Univ., India
B. V. Dhandra	Symbiosis International Univ., India

General Chairs

Jean-Marc Ogier	La Rochelle Université, France
D. S. Guru (Conference Steering Committee)	Univ. of Mysore, India
Sameer Antani	National Library of Medicine, USA

Conference Chairs

Bharti Gawali	Dr. B A M Univ., India
K.C. Santosh	Univ. of South Dakota, USA

Organizing Secretary

Pravin Yannawar	Dr. B A M Univ., India

Area Chairs

Szilárd Vajda	Central Washington Univ., USA
Mickaël Coustaty	La Rochelle Université, France
Nibaran Das (Conference Steering Committee)	Jadavpur Univ., India
Nilanjan Dey	Techno International New Town, India

Publicity Chairs

Hubert Cecotti	California State Univ., Fresno, USA
Alba García Seco de Herrera	Univ. of Essex, UK
Alireza Alaei	Southern Cross Univ., Australia
Sabine Barrat	Univ. de Tours., France
Do Thanh Ha	VNU Univ. of Science, Vietnam
B. Uyyanonvara	Thammasat Univ., Thailand
Sk. Md. Obaidullah	Univ. de Évora, Portugal
V. Bevilacqua (Conference Steering Committee)	Polytechnic Univ. of Bari, Italy
R. S. Mente	Solapur Univ., India
Partha Pratim Roy	Indian Inst. of Technology (IIT) Roorkee, India
Manjunath T. N.	BMSIT, India

Finance Chairs

Ramesh Manza	Dr. B A M Univ., India
Ashok Gaikwad	Institute of Management Studies and Information Technology, India

Advisory Committee

Daniel P. Lopresti	Lehigh Univ., USA
Rangachar Kasturi	Univ. of South Florida, USA
Sargur N. Srihari	Univ. at Buffalo, USA
K. R. Rao	Univ. of Texas at Arlington, USA
Ishwar K. Sethi	Oakland Univ., USA
G. K. Ravikumar	CVS Health/Wipro, Texas, USA
Jose Flores	Univ. of South Dakota, USA
Rajkumar Buyya	Univ. of Melbourne, Australia
Arcot Sowmya	UNSW Sydney, Australia
Antanas Verikas	Halmstad Univ., Sweden
B. B. Chaudhuri	Indian Statistical Institute, India
Umapada Pal	ISI, India

Atul Negi (Conference Steering Committee) — Univ. of Hyderabad, India

Arun Agarwal — Univ. of Hyderabad, India

Hemanth Kumar — Univ. of Mysore, India

K. V. Kale — Dr. B A M Univ., India

B. V. Pawar — NMU Jalgaon, India

R. R. Deshmukh — Dr. B A M Univ., India

Basavaraj Anami — KLEIT, India

Karunakar A. K. — Manipal Inst. Of Technology, India

Suryakanth Gangashetty — IIIT Hyderabad, India

Kaushik Roy (Conference Steering Committee) — West Bengal State Univ., India

Mallikajrun Hangarge (Conference Steering Committee) — KASCC, India

T. Devi — Bharathiar Univ., India

Hanumanthappa M. — Bangalore Univ., India

G. R. Sinha — IIIT Bangalore, India

U. P. Kulkarni — SDMCET, India

Rajendra Hegadi — IIIT Dharwad, India

S. Basavarajappa — IIIT Dharwad, India

G. S. Lehal — Punjabi University, India

Yumnam Jayanta Singh — NIELIT Kolkata, India

S. K. Gupta — NIELIT Aurangabad, India

Contents – Part II

Image Analysis and Recognition

Image and Signal Processing in Agriculture

Signal Processing and Pattern Recognition

Contents – Part I

Computer Vision and Applications

Data Science and Machine Learning

Document Understanding and Recognition

Healthcare Informatics and Medical Imaging

Design New Wavelet Filter for Detection and Grading of Non-proliferative Diabetic Retinopathy Lesions

Yogesh Rajput[1]($\boxtimes$), Shaikh Abdul Hannan[2], Dnyaneshwari Patil[1], and Ramesh Manza[3]

[1] Dr. G. Y. Pathrikar College of Computer Science and IT, MGM University, Aurangabad, India
yogeshrajput128@gmail.com, dnyaneshwari03patil@gmail.com
[2] Department of Computer Science, Albaha University, Albaha, Saudi Arabia
abdulhannan05@gmail.com
[3] Department of Computer Science and IT, Dr. B. A. M. University, Aurangabad, India
manzaramesh@gmail.com

Abstract. WHO projects that diabetes will be the 7th major cause leading death in 2030. Diabetic Retinopathy caused by leakage of blood or fluid from the retinal blood vessels and it will damage the retina. For detection and extraction of non-proliferative diabetic retinopathy lesion we have invent the new wavelet filter. The proposed filter give the good extraction result as compare to exiting wavelet filter. In proposed algorithm, we have extract the microaneurysms, hemorrhages, exudates and retinal blood vessels. After extraction of lesions, grading is done by using feed forward neural network. The proposed algorithm achieves sensitivity of 98%, specificity of 92% and accuracy of 98%.

Keywords: Wavelet filter · Microaneurysms · Hemorrhages · Exudates · Retinal blood vessels

1 Introduction

Non-proliferative diabetic retinopathy is initially characterized by microaneurysms which may burst and leak to backside of the eye. Small spots or dots of blood may gather in the retina, but they frequently do not produce noticeable symptoms in the early stages of the disease. As the disease progresses, hard exudates, abnormalities in the growth of microscopic blood vessels in the retina, and bleeding from the veins that feed the retina may occur [1].

Non-proliferative diabetic retinopathy is classified in three categories like Mild: indicate by the being there of at least 1 microaneurysms, Moderate: Includes the presence of hemorrhages, microaneurysms, and hard exudates, Severe: Many more blood vessels are blocked, depriving several areas of the retina with their blood supply. These areas of the retina send signals to the body to grow new blood vessels for nourishment [2].

© Springer Nature Singapore Pte Ltd. 2021
K. C. Santosh and B. Gawali (Eds.): RTIP2R 2020, CCIS 1381, pp. 3–13, 2021.
https://doi.org/10.1007/978-981-16-0493-5_1

Mother wavelet produces all wavelet functions used in the transformation through translation and scaling, it determines the characteristics of the resulting Wavelet Transform. Therefore, the details of the particular application should be taken into account and the appropriate mother wavelet should be chosen in order to use the Wavelet Transform effectively. The wavelets are chosen based on their shape and their ability to analyze the signal in a particular application. An important property of wavelet analysis is perfect reconstruction, which is the process of reassembling a decomposed signal or image into its original form without loss of information [3].

Arturo Aquino and et al. have been detected optic disk in retinal image of diabetic patient with retinopathy and risk of macular edema. They have used image contrast analysis [4–6] and structure filtering techniques. The location methodology obtained 98.83% success rate [7].

Clara I. Sánchez and et al. have developed an automatic image processing algorithm to detect hard exudates. They have prospectively assessed the algorithm performance using a database containing 58 retinal images with variable color, brightness, and quality. They have obtained a sensitivity of 88% with a mean number of 4.83 ± 4.64 false positives per image using the lesion-based performance evaluation criterion, and achieved an image-based classification accuracy of 100%, sensitivity of 100% and specificity of 100%) [8].

Usman M. Akram and et al. have proposed a computer aided system for the early detection of DR. Blood vessels are enhanced and segmented by using Gabor wavelet and multilayered thresholding respectively. Then they localized optic disk using average filter and thresholding and detected the optic disk boundary using Hough transform and edge detection. Dark and bright lesions are detected by fuzzy classifier. They have used four public databases DRIVE, STARE, DiaretDB0 and DiaretDB1; According to them results show that proposed system gives comparable results and can be used in a computer aided system for accurate and early detection of diabetic retinopathy [9].

Rangaraj M. and et al. has detected optic nerve head in fundus image of the retina with Gabor filters and phase portrait analysis. They have used free databases and for testing and analysis they have indicated 88.9% sensitivity at 4.6 false positive per image [10].

S. Kavitha and et al. has detected hard and soft exudates in fundus image using color histogram thresholding. They have got 89.78% sensitivity, 99.12% specificity and 99.07 accuracy [11].

V. Vijaykumari and et al. has developed a method for exudates detection in retinal image using image processing techniques. Here few methods are used for the detection and the performance of all techniques was compared [12].

Neera singh and et al. have used image analysis techniques like grey mathematical morphology,top-hat transform, fuzzy clustering etc. for early detection of diabetic retinopathy [13].

P.N. Jebrani sargunar and et al. detected and classified exudates in diabetic retinopathy images by texture segmentation methods. They proposed a tool for the early detection using fuzzy c-means clustering, fractal techniques and morphological transformations. Here a accuracy of 85% is achieved [14].

Tjandrasa, H. and et al. classify the hard exudates in retinal fundus images which are employed to the moderate and severe non-proliferative diabetic retinopathy. The hard exudates are segmented using mathematical morphology and the extracted features are classified by using soft margin SVM [15]. The classification model achieves accuracy of 90.54% for 75 training data and 74 testing data of retinal images [16].

2 Methodology

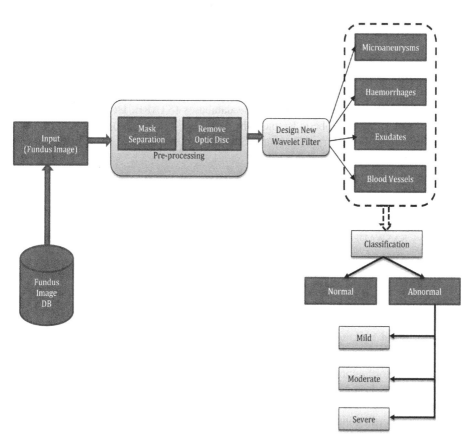

Fig. 1. Workflow for detection and grading of non-proliferative diabetic retinopathy lesions

The above block diagram is the workflow for detection and grading of non-proliferative diabetic retinopathy lesions. Initially in preprocessing mask separation and optic disc removal is done. In preprocessing, green channel extraction is done and then apply intensity transformation function. Aftrerward design new wavelet filter for extraction of non-proliferative diabetic retinopathy lesions. Afterwards grading is done on extracted features by using feed formward neural network [16] (Fig. 1).

2.1 Design New Biorthogonal Wavelet Filter

Stage 1: Generate a biorthogonal wavelet of type 2.
Stage 2: Produce the two filters associated with the biorthogonal wavelet and save them in a MAT-file.

Rf = [1/2 1/2];
Df = [7/8 9/8 1/8 -1/8]/2;

Stage 3: Insert the new wavelet family to the stack of wavelet families.
Stage 4: Show the two pairs of scaling and wavelet functions.
Stage 5: Added new biorthogonal wavelet to analyze a signal/image.

3 Result

For evaluation of this algorithm use online databases (STARE, DRIVE, Diarect DB 0, Diarect DB 1, HRF (Diabetic Retinopathy) and HRF (Glaucoma)) and local fundus image database (collected from Dr. Manoj Saswade) following table show the details of databases (Table 1).

Table 1. Fundus image database

Sr. no	Name of fundus database	Image dimension	FOV	Total images
1	SASWADE	1504 × 1000	50	200
2	STARE	605 × 700	35	402
3	DRIVE	565 × 584	45	40
4	Diarect DB 0	1150 × 1153	50	130
5	Diarect DB 1	1150 × 1153	50	89
	Total			**856**

Proposed biorthogonal wavelet filter is design for extraction of non-proliferative diabetic retinopathy lesions such as microaneurysms (MN), hemorrhages (HM), exudates (EX) and blood vessels (BV). Following table shows the features of retinal blood vessels by proposed biorthogonal wavelet filter.

Table 2. Comparison result of symlet wavelet verses proposed wavelet (rrm) of MN, HM and EX

Sr no	Image	MN (rrm)	MN (sym4)	HM (rrm)	HM (sym4)	EX (rrm)	EX (sym4)
1	im0001	1652	465	10506	8581	717	404
2	im0002	1270	321	12117	9941	1547	529
3	im0003	1901	525	10969	8994	5730	4378
4	im0004	637	169	12304	10236	0	0

(continued)

Table 2. (*continued*)

Sr no	Image	MN (rrm)	MN (sym4)	HM (rrm)	HM (sym4)	EX (rrm)	EX (sym4)
5	im0005	1000	268	12946	11035	4830	3571
6	im0006	1506	408	4771	3160	0	0
7	im0007	1738	453	22239	20651	8	0
8	im0008	1070	294	14527	12534	677	291
9	im0009	1922	507	22830	21218	14	0
10	im0010	2255	597	22832	21419	629	399
11	im0011	1000	255	29841	26487	503	128
12	im0012	1338	367	13713	11634	0	0
13	im0013	1490	392	36188	33144	0	0
14	im0014	1752	516	22581	21091	49	0
15	im0015	2267	623	22312	20663	2723	651
16	im0016	2064	576	22730	21313	2172	1119
17	im0017	1395	377	25780	24151	4203	3116
18	im0018	487	129	20182	18759	1882	1475
19	im0019	2047	500	21691	20293	15580	16312
20	im0020	945	260	22862	21510	19	0
21	im0021	1282	338	36229	34240	815	61
22	im0022	2456	676	23445	21956	4749	4172
23	im0023	1159	283	22947	21492	0	0
24	im0024	1794	494	21962	20561	98	0
25	im0025	2071	592	23076	21643	2312	1692
26	im0026	1642	426	22785	21366	4654	4048
27	im0027	1159	310	31397	27831	649	120
28	im0028	1703	461	22529	21141	144	0
29	im0029	1251	350	31181	29076	0	0
30	im0030	1016 ara>	290	27845	25622	69	0
31	im0031	2912	809	23179	21770	204	61
32	im0032	1328	387	22877	21476	16	0
33	im0033	1887	535	21999	19650	21	0
34	im0034	1633	449	23899	21855	367	30
35	im0035	1691	505	22967	21549	0	0
36	im0036	2876	736	23302	21543	0	0
37	im0037	1804	494	22616	21250	521	121
38	im0038	2016	544	23976	21888	978	312
39	im0039	2373	614	22536	20840	413	145
40	im0040	673	175	51721	46311	1509	1439

Table 3. Features of blood vessels (BV) by proposed biorthogonal wavelet filter

Sr. no	Area	Diameter	Length	Thickness	Mean	Tortuosity	Bifurcation points
1	20	14	9.95	2	20	2	651
2	33	18	4.7	2	19	4	1434
3	27	17	8.5	2	19	2	677
4	31	18	6.42	2	19	4	148
5	22	15	5.66	2	20	3	309
6	40	20	5.71	2	20	2	619
7	74	27	8.26	2	20	1	509
8	37	19	6.32	2	19	2	186
9	38	20	5.86	2	20	3	404
10	42	21	6.44	2	20	3	933
11	37	19	7.55	2	19	1	205
12	78	28	8.16	2	20	3	426
13	26	16	5.5	2	20	5	907
14	48	22	6.24	2	20	3	418
15	104	32	9.95	2	20	2	651

Table 4. Features of blood vessels (BV) by existing symlet wavelet filter

Sr. no	Area	Diameter	Length	Thickness	Mean	Tortuosity	Bifurcation points
1	17	13	6.71	2	19	2	649
2	15	12	7.67	2	19	4	1431
3	18	14	5.8	2	20	2	675
4	20	14	10.59	2	20	2	147
5	19	14	9.37	2	19	3	306
6	18	14	9.47	2	20	2	616
7	33	18	12.63	2	19	1	500
8	25	16	10.56	2	19	2	187
9	20	14	9.83	2	20	3	405
10	27	17	10.42	2	19	3	930
11	19	14	9.04	2	19	1	203
12	42	21	12.71	2	20	3	421
13	18	14	9.33	2	20	5	903
14	27	17	10.03	2	19	3	415
15	45	21	14.75	2	20	6	649

In above Table 2, 3 and 4 statistical features of non-proliferative diabetic retinopathy lesions such as microaneurysms, hemorrhages, exudates and blood vessels. After extraction of retinal blood vessels we have calculated the area, diameter, length, thickness, mean, tortuosity, bifurcation points (Fig. 2, 3, and 4).

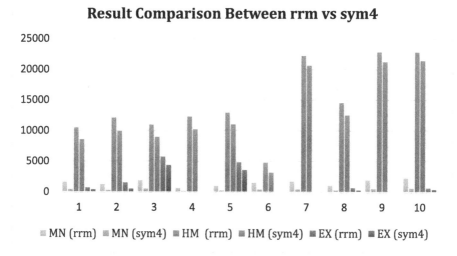

Fig. 2. Result comparison between proposed wavelet (rrm) verses symlet wavelet (sym4) for MN, HM, EX and BV

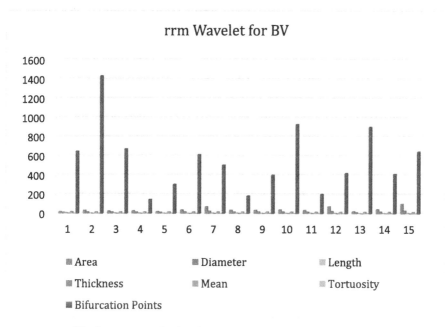

Fig. 3. Features of retinal blood vessels by proposed rrm wavelet

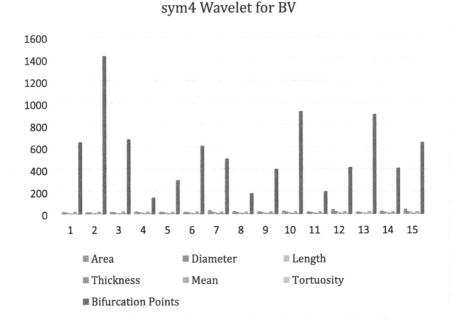

Fig. 4. Features of retinal blood vessels by existing sym4 wavelet

3.1 Grading of Non-proliferative Diabetic Retinopathy Lesions Using Multilayer Perceptron (Feed Forward Neural Network)

Artificial neural networks are generally presented as systems of interconnected "neurons" which can compute values from inputs. Examinations of the human's central nervous system inspired the concept of neural networks. In an Artificial Neural Network, simple artificial nodes, known as "neurons", "neurodes", "processing elements" or "units", are connected together to form a network which simulators a biological neural network [18].

A multilayer perceptron (MLP) is a feedforward artificial neural network model that maps sets of input data onto a set of suitable outputs. An MLP consists of multiple layers of nodes in a directed graph, with each layer fully connected to the next one. Except for the input nodes, each node is a neuron with a nonlinear activation function. MLP utilizes a supervised learning technique called backpropagation for training the network. MLP is a modification of the standard linear perceptron and can distinguish data that are not linearly separable [19–22]. For detection and grading of non-proliferative diabetic retinopathy lesions, design one graphical user interface using MATLAB 2013a.

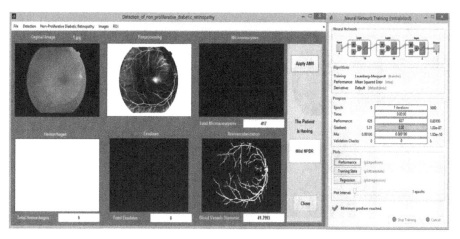

Fig. 5. Mild non-proliferative diabetic retinopathy lesions

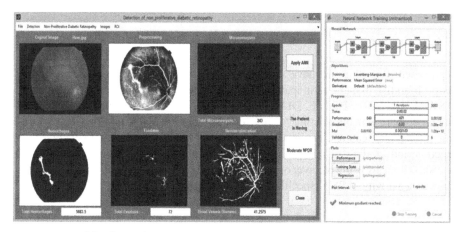

Fig. 6. Moderate non-proliferative diabetic retinopathy lesions

Above figures (Fig. 5, 6 and 7) shows the graphical user interface for the non-proliferative diabetic retinopathy lesions followed by the model of artificial neural network (ANN) for grading the lesions in the categories of mild, moderate and serve NPDR.

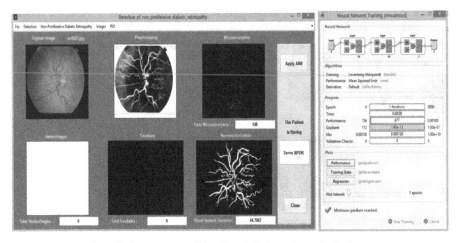

Fig. 7. Serve non-proliferative diabetic retinopathy lesions

4 Conclusion

In proposed algorithm we have design new biorthogonal wavelet for detection of non-proliferative diabetic retinopathy lesions. The proposed algorithm is evaluated on SASWADE, STARE, DRIVE, DIARECT DB0, DIARECT DB1. The proposed "rrm" filter is compare with the existing "sym4" filter. After extraction of non-proliferative diabetic retinopathy features, grading is done by using feed forward neural network.After detection and grading of non-proliferative diabetic retinopathy lesions, performance analysis is done by using receiver operating characteristic curve. Theproposed algorithm achieves sensitivity of 98%, specificity of 92% and accuracy of 98%.

References

1. Grading of Non-Proliferative Diabetic Retinopathy. https://www.nei.nih.gov/health
2. Diabetic Retinopathy. https://www.galloways.org.uk/eyeinformation/eyetab02.html
3. Kumari, S., Vijay, R.: Effect of symlet filter order on denoising of still images. Adv. Comput. Int. J. (ACIJ) **3**(1) (2012)
4. Ruikar, D.D., Santosh, K.C., Hegadi, R.S.: Contrast stretching-based unwanted artifacts removal from CT images. In: Santosh, K.C., Hegadi, R.S. (eds.) RTIP2R 2018. CCIS, vol. 1036, pp. 3–14. Springer, Singapore (2019). https://doi.org/10.1007/978-981-13-9184-2_1
5. Ruikar, D.D., Santosh, K.C., Hegadi, R.S.: Automated fractured bone segmentation and labeling from CT images. J. Med. Syst. **43**(3), 1–13 (2019). https://doi.org/10.1007/s10916-019-1176-x
6. Ruikar, D.D., Santosh, K.C., Hegadi, R.S.: Segmentation and analysis of CT images for bone fracture detection and labeling. In: Medical Imaging: Artificial Intelligence, Image Recognition, and Machine Learning Techniques, vol. 131 (2019)
7. Aquino, A., Gegúndez, M.E., Marín, D.: Automated optic disc detection in retinal images of patients with diabetic retinopathy and risk of macular edema. Int. J. Biol. Life Sci. **8**, 2 (2012)

8. Sánchez, C.I., Hornero, R., López, M.I., Aboy, M., Poza, J., Abásolo, D.: A novel automatic image processing algorithm for detection of hard exudates based on retinal image analysis. Elsevier Med. Eng. Phys. **30**, 350–357 (2008). Received 30 October 2006; received in revised form 26 March 2007. Accepted 7 April 2007

9. Akram, U.M., Khan, S.A.: Automated detection of dark and bright lesions in retinal images for early detection of diabetic retinopathy. J. Med. Sys. **36**, 3151–3162 (2011). https://doi.org/10.1007/s10916-011-9802-2

10. Rangayyan, R.M., Zhu, X., Ayres, F.J., Ells, A.L.: Detection of the optic nerve head in fundus images of the retina with gabor filters and phase portrait analysis. J. Digit. Imaging **23**(4), 438–453 (2010)

11. Kavitha, S., Duraiswamy, K.: Automatic detection of hard and soft exudates in fundus images using color histogram thresholding. Eur. J. Sci. Res. **48**(3), 493–504 (2011). ISSN 1450-216X

12. Vijayakumari, V., Suriyanarayanan, N.: Exudates detection methods in retinal images using image processing techniques. Int. J. Sci. Eng. Res. **1**(2), 1 (2010). ISSN 2229-5518

13. Singh, N., Tripathi, R.C.: Automated early detection of diabetic retinopathy using image analysis techniques. Int. J. Comput. Appl. **8**(2), 18 (2010). (0975-8887)

14. Sargunar, P.N.J., Sukanesh, R. Dr.: Exudates detection and classification in diabetic retinopathy images by texture segmentation methods. Int. J. Recent Trends Eng. **2**(4), 148 (2009)

15. Hegadi, R.S., Navale, D.I., Pawar, T.D., Ruikar, D.D.: Multi feature-based classification of osteoarthritis in knee joint X-ray images. In: Medical Imaging: Artificial Intelligence, Image Recognition, and Machine Learning Techniques, vol. 75 (2019)

16. Tjandrasa, H., et al.: Classification of non-proliferative diabetic retinopathy based on hard exudates using soft margin SVM. In: IEEE International Conference Control System, Computing and Engineering (ICCSCE) (2013). https://doi.org/10.1109/ICCSCE.2013.6719993.

17. Hegadi, R.S., Navale, D.I., Pawar, T.D., Ruikar, D.D.: Osteoarthritis detection and classification from knee x-ray images based on artificial neural network. In: Santosh, K.C., Hegadi, R.S. (eds.) RTIP2R 2018. CCIS, vol. 1036, pp. 97–105. Springer, Singapore (2019). https://doi.org/10.1007/978-981-13-9184-2_8

18. Rosenblatt, F.: Perceptron's and the Theory of Brain Mechanisms. Spartan Books, Washington DC (1961)

19. Rumelhart, D.E., Hinton, G.E., Williams, R.J.: Learning internal representations by error propagation. In: Parallel distributed processing: Explorations in the microstructure of cognition, vol. 1: Foundations. MIT Press (1986)

20. Cybenko, G.: Approximation by superpositions of a sigmoidal function. Math. Control Sig. Syst. **2**(4), 303–314 (1989)

21. Swets, J.A.: Signal Detection Theory and ROC Analysis in Psychology and Diagnostics: Collected Papers. Lawrence Erlbaum Associates, Mahwah (1996)

22. Santosh, K.C., Antani, S., Guru, D.S., Dey, N. (eds.): Medical Imaging: Artificial Intelligence, Image Recognition, and Machine Learning Techniques. CRC Press, Boca Raton (2019)

Techniques for the Detection of Skin Lesions in PH^2 Dermoscopy Images Using Local Binary Pattern (LBP)

Ebrahim Mohammed Senan[1]([⊠]) and Mukti E. Jadhav[2]([⊠])

[1] Department of Computer Science and Information Technology, Dr. Babasaheb Ambedkar Marathwada University, Aurangabad, India
Senan1710@gmail.com
[2] Marathwada Institute of Technology, Aurangabad, India
muktijadhav@gmail.com

Abstract. Skin lesion is the most deadly skin disease in humans, it arises as a result of disorders in the pigment cells, which is produced by a pigment known as melanin. This disease can be prevented and treated if there is an early diagnosis of the disease. Computer-Aided Diagnosis (CAD) has played a key role in helping dermatologists to diagnose the disease. In this proposed system, the model for diagnosis and classification of lesions consists of several stages beginning with pre-processing for the purpose of enhancing images, and identify the area of the lesion by isolating it from the healthy body, and extract features from an region of interest using the LBP method. Classification of the lesion into any of the three classes belong, benign or atypical or malignant according to database PH^2. The results obtained for both SVM and K-NN classification techniques were 93.42% and 96.05% respectively.

Keywords: Gaussian filter · Active contour technique · Morphological method · LBP method · Skin cancer · Support Vector Machine (SVM) · K-Nearest Neighbour (K-NN)

1 Introduction

Human skin contains pigment cells such as melanin, which produced from melanocytes, these pigment cells absorb some radiation that is dangerous for human health such as radioactive Ultraviolet (UV) radiation from the sun. The skin also contains DNA repair enzymes, as people who lack these enzymes have skin cancer [1]. Skin cancer is one of the most deadly types of diseases if there is no early diagnosis. It grows from abnormal cells and has the ability to move to the rest of the body if there is no early diagnosis. People with skin cancer also have a great chance of healing from the disease early and taking appropriate treatment. The risk of exposure to skin lesions can be reduced by reducing and avoiding exposure to ultraviolet rays from sunlight. There are many types of skin lesions namely Melanoma, Nevus, Squamous cell carcinoma and Basal cell carcinoma of the skin [2–4]. Melanoma is the most dangerous and deadly in skin

K. C. Santosh and B. Gawali (Eds.): RTIP2R 2020, CCIS 1381, pp. 14–25, 2021.
https://doi.org/10.1007/978-981-16-0493-5_2

cancer. It can appear in any area of the body, including the sun-covered area and any type of skin. Melanoma markers show irregular, multicolor, asymmetric lesions and their boundaries are constantly growing [5]. There are some databases such as PH^2 and ISIC are available that contain three types of disease: Malignant, Benign and Atypical, in this proposed system, the PH^2 database has been used.

2 Related Work

In this paper presented by Rohini S. Mahagaonkar, et al., a proposed system for the detection of skin cancer, by several steps Pre-processing to enhance the image and remove undesirable effects through the average filter. followed by segmentation to separate the lesion area from the rest of the skin using the global thresholding method. Extract features from the region of interest are two main methods to extract features are GLCM features and CSLBP features. These features are used as an inputted to the KNN and SVM the classifiers to classify the dermoscopy image as normal or abnormal. Results for 300 images were 79.7315% using KNN and 84.7615% using SVM [6]. Other than average filter, contrast stretching [7–10], and fuzzy binarization [11] methods can be used to reduce unwanted artifacts and to improve subjective image quality.

In the study presented by Şerban-Radu-Ştefan Jianu and others, how to detect skin cancer using seven of the features extracted from each image in the lesion area (histogram of oriented gradients, perimeter, area, fractal dimension, local binary patterns, diameter, and lacunarity) These features are used to classify lesions using the classifier. The accuracy obtained from the proposed system is 85% [12].

In the paper presented by Adria Romero Lopez, et al., suggesting a solution to assist the diagnosis of infected skin diseases as benign or malignant, building a model by deep convolutional neural networks. The images input in this model are predictive as benign or malignant. In this proposal, a pre-processing method is used to remove artifacts and noise. The ISBI 2016 database consists of 900 images for training and 379 images for testing. Results obtained from the proposed system are a sensitivity is 78.66% and a precision is 79.74% [13].

In the paper presented by Fonseca-Pinto, R., et al. a new approach to the classification of skin lesion from dermatoscopy images, by Local Binary Pattern Variance (LBPV) histograms after the Bidimensional Empirical Mode Decomposition (BEMD) scale-based decomposition methodology. They used the Derm101 database consisting of 200 training images and 240 images to evaluate. All images were enhanced by removing artifacts and converted from the RGB system to the gray-level system. Results obtained from the proposed system SE = 97.83, SP = 94.44, ACC = 96.00 [12].

In the study presented by Richa Sharma, et al., a new method for detecting skin lesion is proposed by the integration of features in the MRILPP (median robust extended local binary pattern) and DRLBP (Dominant rotated local binary pattern) methods for features extraction from the area of the lesion. ese features are used as an input to SVM classifier to classify images as malignant or benign based on features. The ISIC database contains 367 images. Results obtained the best results in accuracy and sensitivity [14].

This paper presents by González-Castro, V., et al. a description of the dermoscopy images and their classification as benign or malignant using the features extracted from the area of interest by the Local Binary Patterns (LBP) method, calculated from the geometrical features of each color component. Both the adaptive and classical approaches were better than the dermatologists' expectations and the accuracy of the proposed system has reached 0.792 [15].

System to a pattern recognition, that includes three stages to analyze the images of dermoscopy, segmentation to separate the lesion area from the rest of the skin, this step is performed using active contours. Feature extraction is the next step to extract the most important features from the area of the lesion, using the local binary pattern (LBP) to extract the features from the area of interest and put each features image in a vector. Classification is the next stage after selecting the features of each image, where feature vectors are considered input to an SVM classifier, which classifies each image as malignant or benign. Results reached by the proposed system are sensitivity 91.6% and specificity 78.7% by automatic segmented masks [16].

In the paper presented by Faouzi Adjed, et al., for classification of skin cancer and non-skin cancer, using the features extracted from the lesion area by the local binary patterns method. LBP method calculates the local texture features from each skin cancer image. The LBP then uses the local information to calculate statistical features that have the ability to distinguish between malignant and benign lesions. Support vector machine (SVM) has been applied and receives a feature matrix to classify dermascope images into malignant and benign. The proposed system results are as follows: sensitivity of 75.6%, accuracy of 76.1% and specificity of 76.7% [17].

In the paper presented by Farhan Riaz, et al., diagnosis System to detect the skin lesion of patients. Color and texture are two properties of skin cancer. In this paper, they suggested a combination of the color and texture properties extracted from the PH^2 database. The texture features consist of a variety of local features extracted by the local binary pattern (LBP), which are used to extract features that adapt to each pixel in the lesion, followed by extraction of the features of the color using standard HSV histograms. The extracted features are stored in vectors features to form a matrix of features of each vector representing image data. Then the classification method using support vector machines to classify each image as malignant or benign. Results reached by the proposed system Sensitivity 84% and Specificity 94% [18].

In the paper presented by POORNIMA M S, et al., method of detection of skin cancer by methods of image processing, dermoscopy images is input to system, and the images were enhanced by removing the negative effects, and the images were segmented to isolate the lesion area from the rest of the image using an active contour model. Extract features from the area of the lesion by checking the color, diameter and perimeter of the area by texture, and storing the features in the vector features that are inserted into the SVM classifier to classify the lesion as benign or malignant [19].

3   Methodology

Image processing is technology for extracting information from images. our proposed system as shown in Fig. 1. The dataset used in the system is PH2 considered as input to the system. The proposed system consists of several steps: pre-processing using the Gaussin filter to enhance images and delete the negative features that negatively affect in the next steps of image processing, segmentation method is to identify the lesion area and isolate the infected area from the healthy body. Morphological method for enhancing digital images and producing enhanced digital images. Feature extraction from the Region of Interest (RoI) using local binary patterns (LBP). For classification technology in the proposed system, we used two types of classification techniques SVM and K-NN.

3.1 Data Set

The increased incidence of skin cancer encouraged computer-aided diagnostic systems to pay attention to the disease. The PH² database was created to be used by researchers to early diagnose disease and to classify dermoscopy images. PH² is a database obtained from at the Dermatology Service of a hospital, Pedro Hispano, Matosinhos, Portugal. The PH² contains 200 dermoscopy images, divided to three types: 40 Melanomas, 80 Benign and 80 Atypical Nevi. All images in PH² were obtained in the same conditions and contain high resolution of 768 * 560 pixels in RGB colour system.

3.2 Pre-processing (Gaussian Filter Method)

Filters are used to suppress high or low frequencies. Effects and noise are present in all images, including medical images caused by factors either natural or external. Gaussian filter belongs to linear filters to pass low frequencies and reduce noise and image details of unwanted pixels. The proposed system used a Gaussian filter to enhance images. Removing noise, air bubbles, hair, and all unwanted pixels. Images used in the proposed system All PH² images, all in RGB color system and contain high resolution of 768 * 560 pixels so we resized the images to a resolution of 200 * 160 with preserving the important pixels and for the purpose of fast computation. The Gaussian filter is used to remove unwanted pixels by replacing it by averaging the pixels of the adjacent pixels. Figure 2(a) shows image enhancement in the proposed system using the Gaussian filter for three types of skin lesions in the PH2 database.

$$h(x, y) = \frac{1}{2\pi\sigma^2} e^{\frac{x^2 - y^2}{2\sigma^2}} \tag{1}$$

Where x represents the distance from the central pixel on the horizontal axis, y represents the distance from the central pixel on the vertical axis, and σ is the standard deviation of the Gaussian distribution.

3.3 Segmentation Method

The next step is to segment the area of the lesion from the input image, and to identify the region of interest (RoI) and isolate it from the healthy body. The Segmentation step is one of the important steps in image processing to ensure a high diagnosis in the following steps. In this paper Active Contour Technique (ACT) is the method used in the proposed system, which is an effective technology in segmentation of medical images, and one of the techniques of segmentation methods is known to use of energy forces

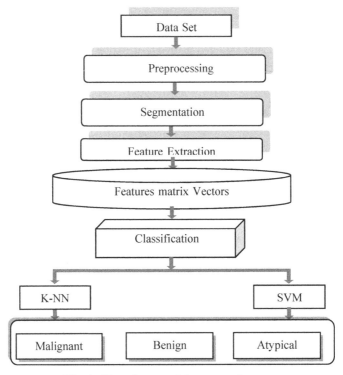

Fig. 1. Flow diagram of the proposed system

to separate the pixels to be analyzed from the rest of the body. It is a set of points that undergoes to the process of interpolation, which is linear and multidimensional, which describes the boundaries of the object or any features to form a parametric curve or contour using internal or external forces applied [20]. Figure 2(b) shows the process of image segmentation in the proposed system using Active Contour Technique (ACT) for three types in the PH^2 database.

3.4 Morphological Method

Morphological method follows after the process of segmentation, digital images produced by the process of segmentation may contain defects such as noise and holes, the morphological method aims to remove defects by calculating the shape and structure of the image. The morphological method is a nonlinear process that does not depend on numerical values but is based on the relative order of pixels, so also suitable for grayscale images. Each location in the image is compared to adjacent neighborhood pixels. Morphological method produces a digital image that does not have a value of zero. Where each pixel of the structure element is associated with the digital image, and each element of the structure fits the image, if for each of pixels set to 1. Erosion and dilation are fundamental operations in the morphological method. In the proposed system, we used a morphological method to enhance digital images and produce enhanced digital images

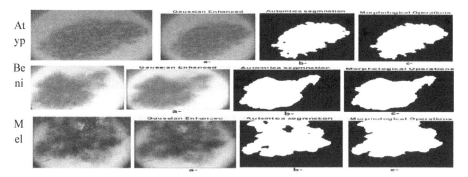

At
yp

Be
ni

M
el

Fig. 2. Results of the proposed system performance **a-** Pre-processing **b-**Segmentation **c-** Morphological

to ensure that features are clearly extracted in the next step. Figure 2(c) Morphological Method shows the images in the proposed system for three types in the PH² database.

3.5 Feature Extraction Method

Extraction of features is the next step after morphological method. It is one of the most important Stage of image processing, which extract useful features of images such as color, texture and shape, and stores the features of each image in a vector of features, and stores the features of many images in a matrix of features. There are many techniques to extract features in image processing. In the proposed system we used the local binary pattern to extract features from images. The method transfers the image into an array, which describes the image's texture. In the proposed system, we focused on 8 properties extracted from the images as follows: Mean Red, Mean Green, Mean Blue, Area, Major Axis Length, Minor Axis Length, Orientation and Perimeter. LBP compares the gray level between one pixel called the central pixel with the neighborhood pixels.

How Does the Local Binary Pattern Work?
Suppose that we have an image $I(x, y)$ and g_c indicates the gray level of arbitrary pixel. g_p indicates the gray level of the P pixels in the neighborhood with radius R around the point (x, y):

$$g_P = I(x_P, y_P), p + 0, 1, \ldots \ldots p \quad (1)$$

$$\quad (2)$$

$$x_P = x + R \cos(2\pi/P) \quad (3)$$

$$y_P = y + R \cos(2\pi/P) \quad (4)$$

We obtain the central pixel through the neighborhood pixels as in Eq. (5)

$$T = t\left(g_c, g_0 - g_c, g_1 - g_c, \ldots \ldots g_p - g_c\right) \quad (5)$$

Calculate Local Binary Pattern (LBP) mathematically

$$LBP_{R,P} = \sum_{P=0}^{P-1} s\big((g_p - g_c)\big).2^P \tag{6}$$

As in Eq. (6) shows how to calculate LBP per pixel based on neighborhood pixels. where, neighborhood pixels g_p in each block, s threshold by its center pixel value g_c, P neighborhood pixels and R radius. Using Eq. (6) many texture features extracted from images as the standard deviation, the mean, entropy, contrast, and energy of images .

Binary threshold function S(X):

$$S(X) = \begin{cases} 0, X < 0 \\ 1, X \geq 0 \end{cases}$$

3.6 Classification Algorithms

Classification techniques for medical images, by the Computer Aided Diagnosis (CAD), it has a role of paramount importance in teaching of medicine and early diagnosis of the lesions, for treatment and prevention, and help the Dermatologist to accurate diagnosis. The classification stage comes after the extraction of features, in which the image is classified into any class. There are many Classification techniques such as Convolutional Neural Network (CNN), Support Vector Machine (SVM), Artificial Neural Networks (ANN), K Nearest Neighbors (KNN) etc. In the proposed system we used two classification techniques, the Support Vector Machine (SVM) and K Nearest Neighbors (KNN). All image features of the PH2 database are stored in matrix features where each vector represents the feature for a single image. The feature matrix is an input to the classification technique. The database used in the system consists of 200 images. The images in the classification technology were divided into 60% for training and 40% for testing.

SVM. A Support Vector Machine (SVM) is belong to a supervised learning technique that works in both regression and classification. The paper is an employee for the purpose of classifying images into three classes: either malignant, benign or atypical. A Support Vector Machine (SVM) uses a hyperplane to separate different data [21]. The best hyperplane has maximizes margin, the margin is the distance between the close points and a hyperplane, where these close points are the support vectors and it control of the hyperplane. Overall, support vector machines are great classifiers, in the proposed system we have obtained good results using SVM as we will see later.

KNN. KNN is one of the techniques used in the classification and regression. It is a non-parametric. And it is used for easy interpretation and needs less time to calculate. In the proposed system to be used for the purpose of classification, and k = 5. The purpose from KNN is to use the database and divide it into several classes, after that a new data point is predicted to any belong classe.

How to Calculate the Nearest K Neighbors Algorithm: First, the parameter K = the number of nearest neighbors is determined. Second, calculate the distance between points and all training samples to be classified. Third, sort the distances and select the nearest neighbors based on the minimum distance.

4 Experimental Results

The approach was tested in the proposed system on PH^2 database which is divided into 80 Atypical, 80 Benign images, and 40 Malignant. After all the steps of the image processing we got good results. The database has been divided 60% for training and 40% for testing. The metrics obtained from the proposed system are as follows: Accuracy, Sensitivity, Specificity, Precision, Recall, Fscore, and gmean obtained as in the following Eqs. (7), (8), (9), (10), (11), (12) and (13) respectively.

$$\text{Accuracy} = \frac{TN + TP}{TN + TP + FN + FP} * 100\% \tag{7}$$

$$\text{Sensitivity} = \frac{TP}{TP + FN} * 100\% \tag{8}$$

$$\text{Specificity} = \frac{TN}{TN + FP} * 100 \tag{9}$$

$$\text{Precision} = \frac{TP}{TP + FP} * 100\% \tag{10}$$

$$\text{Recall} = \frac{TP}{TP + FN} * 100\% \tag{11}$$

$$\text{Fscore} = 2 * \frac{\text{Precision} * \text{Recall}}{\text{Precision} + \text{Recall}} * 100 \tag{12}$$

$$\text{gmean} = \sqrt{\frac{TP}{P} * \frac{TN}{N}} \tag{13}$$

True Positive (TP) is correctly classified as malignant pixels. True Negative (TN) is correctly classified as benign pixels. False Positive (FP) is incorrectly classified as malignant pixels where as the are benign. False Negative (FN) is incorrectly classified as benign pixels where as the are malignant. Negative (N) number of benign images. Positive (P) number of malignant images [36] (Tables 1, 2 and 3).

A Comparative Study. We have presented a comparative study of our proposed system with other relevant systems. As shown in Table 4 and 5.

Table 1. Shows the results obtained using SVM classification.

	Classified as Benign	Classified as Malignant	Accuracy%	Sensitivity%	Specificity%	Precision%	Recall%	Fscore%	Gmean%
Benign and Atypical	TN = 47	FN = 1	93.42	96.3	95.92	92.86	96.3	94.55	93.3
Malignant	FP = 2	TP = 26							
	N = 48	P = 28							

Table 2. Shows the results obtained using K-NN classification techniques for training images.

	Classified as Benign	Classified as Malignant	Accuracy%	Sensitivity%	Specificity%	Precision%	Recall%	Fscore%	Gmean%
Benign and Atypical	TN = 45	FN = 1	96.05	96.55	95.74	93.33	96.55	94.92	95.55
Malignant	FP = 2	TP = 28							
	N = 46	P = 30							

Table 3. The evaluation results obtained from the proposed system are shown using the SVM and K-NN classification techniques.

Evaluation	SVM	K-NN
Accuracy%	93.42	96.05
Sensitivity%	96.3	96.55
Specificity%	95.92	95.74
Precision%	92.86	93.33
Recall%	96.3	96.55
Fscore%	94.55	94.92
Gmean%	93.3	95.55

Table 4. Comparative study using K-NN technique.

Reference paper	Accuracy %	Sensitivity%	Specificity%	Precision%	Recall%	Fscore%	Gmean%
Proposed system	96.05	96.55	95.74	93.33	96.55	94.92	95.55
[31]	92	98	86	–	–	–	–
[32]	97	96.2	97.7	–	–	–	–
[33]	76	69	83	–	–	–	–

Table 5. Comparative study using SVM technique

Reference paper	Accuracy%	Sensitivity%	Specificity%	Precision%	Recall%	Fscore%	Gmean%
Proposed system	**93.42**	**96.3**	**95.92**	**92.86**	**96.3**	**94.55**	**93.3**
[22]	93.8	93.8	93.8	–	–	–	–
[23]	91.55	92.8	90.30	–	–	–	–
[24]	66.7	60.7	80.5	–	–	–	–
[25]	82	92	72	–	–	–	–
[26]	79	74	83	–	–	–	–
[27]	77.9	79.7	76	–	–	–	–
[28]	86.7	96.8	85.7	–	–	–	–
[29]	83.6	96.6	73.5	–	–	–	–
[30]	84	98	70	–	–	–	–

5   Conclusion

In the proposed system we used the PH^2 database, which contains 200 images that are divided into 80 Benign, 80 Atypical and 40 Malignant. Data set was divided into two parts: 60% for testing and 40% for training. The aim of the proposed system is to classify, detect and treat skin lesions early. The proposed system consists of several stages: a pre-processing stage to enhance the input images using the Gaussian filter method, separation of the lesion area from the rest of the image using Active Contour Technique [34, 35], for more enhancement of digital images using the Morphological method, and extraction of features from the region of interest (RoI) using the LBP. We created an matrix of features as an input to the classification techniques SVM and K-NN. The accuracy obtained from the proposed system were 93.42% and 96.05% for both SVM and K-NN, respectively.

Acknowledgments. We thank the larger community of a hospital, Pedro Hispano, Matosinhos, Portugal (PH2) to provide this database and make it available to researchers. We thank all the collaborators for completing this paper.

References

1. Maton, A., et al.: Human Biology and Health. Prentice Hall, Englewood Cliffs, New Jersey (1893). ISBN 978-0-13-981176-0
2. Cakir, B.Ö., Adamson, P., Cingi, C.: Epidemiology and economic burden of nonmelanoma skin cancer. Facial Plast. Surg. Clin. North Am. **20**(4), 419–422 (2012). https://doi.org/10.1016/j.fsc.2012.07.004. pmid 23084294
3. Sajjad, R., Jerry, M. (eds.): ABC of skin cancer, pp. 5–6. Blackwell Pub, Malden (2008). ISBN 978-1-44-431250-8. Archived from the original on 29 April 2016
4. Dunphy, L.M.: Primary Care: The Art and Science of Advanced Practice Nursing. F.A. Davis. p. 242 (2011). ISBN 978-0-80-362647-8. Archived from the original on 20 May 2016

5. General Information About Melanoma. NCI. 17 April 2014. Archived from the original on 5 July 2014. Accessed 30 June 2014
6. Mahagaonkar, R.S., Soma, S.: A novel texture based skin melanoma detection using color GLCM and CS-LBP feature. Int. J. Comput. Appl. **975**, 8887 (2015)
7. Ruikar, D.D., Santosh, K.C., Hegadi, R.S.: Contrast stretching-based unwanted artifacts removal from CT images. In: Santosh, K.C., Hegadi, R.S. (eds.) RTIP2R 2018. CCIS, vol. 1036, pp. 3–14. Springer, Singapore (2019). https://doi.org/10.1007/978-981-13-9184-2_1
8. Ruikar, D.D., Santosh, K.C., Hegadi, R.S.: Automated fractured bone segmentation and labeling from CT images. J. Med. Syst. **43**(3), 1–13 (2019)
9. Ruikar, D.D., Santosh, K.C., Hegadi, R.S.: Segmentation and analysis of CT images for bone fracture detection and labeling. In: Medical Imaging: Artificial Intelligence, Image Recognition, and Machine Learning Techniques, vol. 131 (2019)
10. Santosh, K.C., Antani, S., Guru, D.S., Dey, N. (eds.): Medical Imaging: Artificial Intelligence, Image Recognition, and Machine Learning Techniques. CRC Press, Boca Raton (2019)
11. Santosh, K.C., Roy, P.P.: Arrow detection in biomedical images using sequential classifier. Int. J. Mach. Learn. Cybern. **9**(6), 993–1006 (2016). https://doi.org/10.1007/s13042-016-0623-y
12. Zohora, F.T., Antani, S., Santosh, K.: Circle-like foreign element detection in chest x-rays using normalized cross-correlation and unsupervised clustering. In: Medical Imaging 2018: Image Processing. vol. 10574, p. 105741V. International Society for Optics and Photonics (2018)
13. Jianu, Ş., Ichim, L., Popescu, D.: Advanced processing techniques for detection and classification of skin lesions. In: International Conference on System Theory, Control and Computing (ICSTCC). IEEE (2018)
14. Lopez, A.R., Giro-i-Nieto, X., Burdick, J., Marques, O.: Skin lesion classification from dermoscopic images using deep learning techniques. In: 2017 13th IASTED International Conference on Biomedical Engineering (BioMed), pp. 49–54. IEEE, February 2017
15. Fonseca-Pinto, R., Machado, M.: A textured scale-based approach to melanocytic skin lesions in dermoscopy. In: 2017 40th International Convention on Information and Communication Technology, Electronics and Microelectronics (MIPRO), pp. 279–282. IEEE, May 2017
16. Sharma, R., Lal, M.: Skin cancer lesion classification using LBP based hybrid classifier. Int. J. Adv. Res. Comput. Sci. **8**(7) (2017)
17. González-Castro, V., et al.: Automatic classification of skin lesions using geometrical measurements of adaptive neighborhoods and local binary patterns. In: 2015 IEEE International Conference on Image Processing (ICIP), pp. 1722–1726. IEEE, September 2015
18. Naeem, S., Riaz, F., Hassan, A., Nisar, R.: Description of visual content in dermoscopy images using joint histogram of multiresolution local binary patterns and local contrast. In: Jackowski, K., Burduk, R., Walkowiak, K., Woźniak, M., Yin, H. (eds.) IDEAL 2015. LNCS, vol. 9375, pp. 433–440. Springer, Cham (2015). https://doi.org/10.1007/978-3-319-24834-9_50
19. Adjed, F., Faye, I., Ababsa, F., Gardezi, S.J., Dass, S.C.: Classification of skin cancer images using local binary pattern and SVM classifier. In: AIP Conference Proceedings, vol. 1787, no. 1, p. 080006. AIP Publishing, November 2016
20. Riaz, F., Hassan, A., Javed, M.Y., Coimbra, M.T.: Detecting melanoma in dermoscopy images using scale adaptive local binary patterns. In: 36th Annual International Conference of the IEEE Engineering in Medicine and Biology Society, pp. 6758–6761. IEEE, August 2014
21. Hegadi, R.S., Navale, D.I., Pawar, T.D., Ruikar, D.D.: Multi feature-based classification of osteoarthritis in knee joint X-ray images.In: Medical Imaging: Artificial Intelligence, Image Recognition, and Machine Learning Techniques, vol. 75 (2019)
22. Poornima, M.S., Shailaja, K.: Detection of skin cancer using SVM. Int. Res. J. Eng. Tech. (2017). e-ISSN 2395-0056. p-ISSN 2395

23. Laurent, C.D.: On active contour models and balloons. CVGIP Image Understand. **53,** 211–218 (2004). https://doi.org/10.1016/1049-9660(91)90028-n

24. Schaefer, G., Krawczyk, B., Celebi, M.E., Iyatomi, H.: An ensemble classification approach for melanoma diagnosis. Memetic Comput. **6**(4), 233–240 (2014). https://doi.org/10.1007/s12293-014-0144-8

25. Kasmi, R., Mokrani, K.: Classification of malignant melanoma and benign skin lesions: implementation of automatic ABCD rule. IET Image Process. **10**(6), 448–455 (2016)

26. Abuzaghleh, O., Barkana, B.D., Faezipour, M.: Automated skin lesion analysis based on color and shape geometry feature set for melanoma early detection and prevention. In: IEEE Long Island Systems, Applications and Technology (LISAT) Conference 2014, pp. 1–6. IEEE, May 2014

27. Barata, C., Ruela, M., Francisco, M., Mendonça, T., Marques, J.S.: Two systems for the detection of melanomas in dermoscopy images using texture and color features. IEEE Syst. J. **8**(3), 965–979 (2013)

28. Glaister, J., Amelard, R., Wong, A., Clausi, D.A.: MSIM: Multistage illumination modeling of dermatological photographs for illumination-corrected skin lesion analysis. IEEE Trans. Biomed. Eng. **60**(7), 1873–1883 (2013)

29. Barata, C., Celebi, M.E., Marques, J.S.: Improving dermoscopy image classification using color constancy. IEEE J. Biomed. Health İnf. **19**(3), 1146–1152 (2014)

30. Mustafa, S., Kimura, A.: A SVM-based diagnosis of melanoma using only useful image features. In: 2018 International Workshop on Advanced Image Technology (IWAIT), pp. 1–4. IEEE, January 2018

31. Lee, H.D., et al.: Dermoscopic assisted diagnosis in melanoma: reviewing results, optimizing methodologies and quantifying empirical guidelines. Knowl.-Based Syst. **158**, 9–24 (2018)

32. Barata, C., Marques, J.S., Celebi, M.E.: Towards an automatic bag-of-features model for the classification of dermoscopy images: the influence of segmentation. In: 2013 8th International Symposium on Image and Signal Processing and Analysis (ISPA), pp. 274–279. IEEE, September 2013

33. Cavalcanti, P.G., Scharcanski, J., Baranoski, G.V.: A two-stage approach for discriminating melanocytic skin lesions using standard cameras. Exp. Syst. Appl. **40**(10), 4054–4064 (2013)

34. Oliveira, R.B., Pereira, A.S., Tavares, J.M.R.S.: Computational diagnosis of skin lesions from dermoscopic images using combined features. Neural Comput. Appl. **31**(10), 6091–6111 (2018). https://doi.org/10.1007/s00521-018-3439-8

35. Riaz, F., et al.: Active contours based segmentation and lesion periphery analysis for characterization of skin lesions in dermoscopy images. IEEE J. Biomed. Health Inf. (2018)

36. Hegadi, R.S., Navale, D.I., Pawar, T.D., Ruikar, D.D.: Osteoarthritis detection and classification from knee x-ray images based on artificial neural network. In: Santosh, K.C., Hegadi, R.S. (eds.) RTIP2R 2018. CCIS, vol. 1036, pp. 97–105. Springer, Singapore (2019). https://doi.org/10.1007/978-981-13-9184-2_8

Effect of Quality Enhancement Techniques on MRI Images

Deepali N. Lohare[1]([✉]), Rupali Telgad[2], and Ramesh R. Manza[3]

[1] Dr. Babasaheb Ambedkar Marathwada University, Aurangabad, India
loharedeepali@yahoo.co.in
[2] Shri Vyankatesh College, Deulgaon Raja, India
rupalitelgad@gmail.com
[3] Bio-Medical Image Processing Laboratory, Department of Computer Science and Information Technology, Dr. Babasaheb Ambedkar Marathwada University, Aurangabad, Maharashtra, India
manzaramesh@gmail.com

Abstract. Magnetic resonance image is a greatly developed medical representation system, cynical to generate prominent renowned imagery of human being and there part. It gives feature in sequence to analyze the disease. MRI play significant role to give in turn innovative scope intended such precede representation. Inventive image of MRI are usually comprise squat disparity. It is dense intended in favor of health center to analyze them. Escalating the difference of representation, it resolve exist simple designed in favor to analyze a complete in sequence. This manuscript compares special method of development of brain MRI by histogram based technique and by using different statistical measures.

Keywords: Statistical measures · Brain MRI · Histogram equalization

1 Introduction

A variety of quality measures [1, 8] are standard deviations, mean mode, variance, median, sleekness and kurtosis and covariance. These Statistical measures are used in extensive sort of logical and societal study, with: computational sociology, biostatistics, network biology, computational biology, etc. [1]. In this manuscript we cover diverse kind of statistical measure in revere to image processing and pretend all of these [2]. In this paper, different contrast enhancement techniques are compared with the help of quality measure to improve the quality or brightness of MRI of brain image. These methods are comparing on the basis of MSE, PSNR, Median, Mode, Total Mean, Total Covariance, Standard Deviation and histograms. After converting input image to histogram image and contrast enhancement to enhance hidden features of interest. Usually, at all type of magnetic resonance image examination start with an image enhancement process. The selection of enhancement technique has a straight crash on the concluding outcome, in view of the fact that the image superiority has immense impact on consequent study [3].

An idea behind writing this paper is to show enhancement of brain MRI using different effects of quality measures on brain MRI images using histogram based techniques.

© Springer Nature Singapore Pte Ltd. 2021
K. C. Santosh and B. Gawali (Eds.): RTIP2R 2020, CCIS 1381, pp. 26–35, 2021.
https://doi.org/10.1007/978-981-16-0493-5_3

Section 1 contains the introduction. Section 2 contains literature review of the research work. Section 3 contains the definitions and formulae of the different Statistical measures. Section 4 contains experimental analysis which shows the Table 1 of sample original MRI images and comparison of quality measures such as MSE, PSNR, Mean, Mode, Total Mean, Total Median, Total standard deviation Table 2 shows comparison of same quality measures after histogram equalization of original MRI images. In this paper 302 original images are taken on the way to find out various Statistical measures, MRI images are obtained from the internet (BRATS-2012, MIDAS.) database. After obtaining the statistical measures of original MRI image. The histogram equalization of original MRI image has taken, after histogram equalization again its statistical measures has taken and the comparison is shown using different graphs. Section 5 contains Result Analysis of various quality measures. Section 6 concludes a detailed review work of enhancement of brain MRI images using various statistical measures.

2 Literature Review

Brain tumor is renowned on or after the opening of the 18th century. An interlude of research on the variety of aspect of brain tumor start from it. Set of treatment are identified on or after that era beyond. A lot of research centre were set up in order to find out imaging and treatment methods. Soon after the 19th century. Charles Wilson set up societies akin to "The Brain Tumor Research Centre" are liability a variety of research on recognition of tumor and treatment. Recently "Automatic segmentation algorithms" has been developed to identify tumor inside the brain.

The images so as to obtain during MRI or the CT are in use to evaluate brain tumor. A number of special effects with the intention of practical in these images which are in use both throughout Magnetic Resonance Image or Computer Tomography might contain artifact, short contrast production for recognition of brain tumor anxious or noise present in individual image [4–7]. Consequently it is able to be believed so as to imaging at this point plays a vital role in segmentation or recognition of brain tumor. So many imaging techniques are used to identify tumor inside the image. Thresholding and region growing algorithms are used to segment brain cancer. The available technique is based on k means clustering. While these methods were useful there are a small number of disadvantage such as thresholding method unnoticed spatial uniqueness which are significant for malignant tumor discovery.

In the region based segmentation it requisite supplementary user interface for the assortment of seed. Because of these drawback this technique is not been extensively adopted, for future research in the segmentation of medical images will escort towards improving the correctness, exactness, and computational speed of segmentation approaches, in addition to minimizing the quantity of manual interface. In research region segmentation methods have proved their usefulness and are now emphasizing improved use for automated conclusion and radiotherapy. These will be mostly imperative in applications such as computer included surgery, where imagine of the structure is a major section [8, 9].

3 Statistical Measures

3.1 Peak Signal to Noise Ratio (PSNR) and Mean Square Error (MSE)

The Peak signal to Noise Ratio and Mean Square Error is defined as the two-error metrics used to evaluate picture firmness superiority. The Mean Square Error stand for collective squared fault stuck among flattened and unique image, whereas PSNR signifies a quantity of the topmost fault. If MSE have minor value, error is minor. If the value of PSNR is higher, the image superiority of recreated image copy is better.

To calculate the Peak Signal Noise Ratio, the wedge initial calculate the mean-squared inaccuracy using the equation given below:

$$\text{Mean Square Error} = \frac{1}{mn} \sum_{i=0}^{m-n} \sum_{j=0}^{n-1} \left[I(i,j) - k(i,j) \right]^2 \qquad (1)$$

$$\text{Peak Signal Noise Ratio} = 10.\log 10 \left(\frac{\text{MAX}^2}{\text{MSE}} \right) \qquad (2)$$

3.2 Mean

Mean is the average of all numbers. Add all the figures together in set to calculate mean and at that time spilt the summation by the overall count of all numbers.

$$\frac{a_1 + a_2 + a_3 + \ldots\ldots\ldots\ldots\ldots + a_n}{n} \qquad (3)$$

3.3 Median

Median is classify as, the core numeral in a place of prearranged statistics. Median is a middle number in a sorted list; the list may be ascending or descending.

$$\{(n+1) \div 2\}^{\text{th}} \qquad (4)$$

Where "n" is the integer of objects in the position and "th" is (n)th numeral.

3.4 Mode

The mode is the number that arises utmost frequently in a set of numbers. Mode helps classify the most common or frequent occurrence of an individual.

$$\frac{\text{Sum of values of a set}}{\text{Number of values}} \qquad (5)$$

3.5 Standard Deviation

Standard deviation is defined as, It is a determine to calculate the quantity of deviation or scattering of a traditional statistics value. It measures the total unpredictability of an allocation; the develop the scattering or unpredictability, the larger is the standard deviation and larger will be the magnitude of the deviation of the value from their mean.

The most commonly used quantity of unpredictability castoff in statistics. In relationship of image processing it illustrations in what way considerable deviation, happens since the standard (mean or else predictable value). A squat average variation shows with the aim of the statistics point lean to live equal to the mean, while elevated usual divergence specifies with the intention of statistics points are extensive in excess of a huge array of standards. Scientifically standard divergence is specified by [10].

$$S_X = \frac{\sqrt{\sum_{i=1}^{n} (x_i - \bar{x})^2}}{n-1} \tag{6}$$

N = the number of data points, $\bar{x}$ = the mean of the x_i, x_i = each of the value of the data.

3.6 Covariance

The covariance amongst two mutually dispersed real-valued arbitrary variables X and Y with predetermined subsequent instants is well-defined as the predictable creation of their deviations from their discrete predictable values.

$$Cov(x, y) = \sum E((X - \mu) E(Y - V)/n - 1 \tag{7}$$

X is an unsystematic variable, $E(X) = \mu$ is predictable values (the mean) of the unsystematic variable, $E(Y) = V$ expected values (mean) of the unsystematic variable, n = number of objects in the records [11].

4 Histogram Equalization

Histogram Equalization is uses to improve the divergence of the picture, it discriminate the backdrop and the entity [11]. The Histogram of an image is a string of quantity of actions of grim levels in the MRI image alongside the grim level qualities. The histogram provide appropriate sketch out of the concentrations in an image, so far it can't stretch a little information about three-dimensional relation among pixels. The Histogram Equalization can be defined as "Plotting of every pixel of participation representation into involving pixel of equipped productivity representation is defined as Histogram". HE stands as in equation [12–14].

$$p = \frac{\text{no. of pixels with intensity nk}}{\text{Total no. of pixels n}} \tag{8}$$

$$n = 0.1 \ldots \ldots L - 1$$

[O... L − 1] is range of the gray level values image [13].

5 Experimental Analysis

In this research paper total 400 online images are obtained from (BRATS-2012) database. Out of four hundred 302 images are used for this research work. The Graphical User

Interface (GUI) is constructed for the classification of MRI images [15, 16]. Statistical measure of original MRI image has taken then original image is converted in histogram equalization. After histogram equalization same statistical measure of that image has taken and the comparison is shown below by graph. Table 1 and Table 2 given below represent the various values of different statistical measures.

Table 1. Shows a quality measure of ten original MRI images.

MRI IMAGE	MSE	PSNR	Median	Mode	Total Mean	Total covariance	Standard Deviation
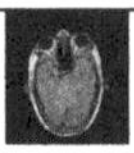	871.9632	18.3072	0	0	30.8197	2.56E+03	50.591
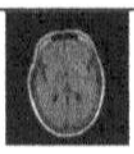	956.5911	18.3255	0	0	31.5604	2.27E+03	47.6676
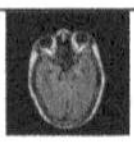	934.9801	18.4228	0	0	31.3895	2.36E+03	48.6265
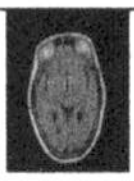	985.8834	18.1925	0	0	31.5696	2.28E+03	47.7656
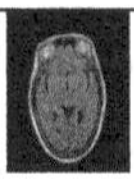	956.5911	18.3255	0	0	31.5604	2.27E+03	47.6676
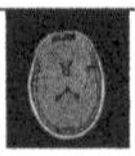	921.8665	18.4841	0	0	30.8277	2.23E+03	47.1776
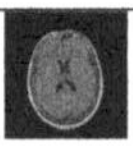	879.5344	18.6883	0	0	31.3941	2.31E+03	48.0108
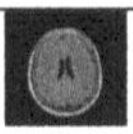	798.4677	19.1082	0	0	31.3499	2.35E+03	48.4986
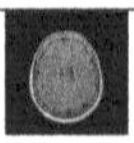	605.2014	20.3118	0	0	33.8492	2.66E+03	51.5345
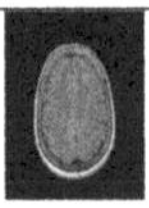	564.8112	20.6118	0	0	32.7343	2.61E+03	51.1037

Table 2. Shows a quality measure of histogram of same original MRI images.

MRI IMAGE	MSE	PSNR	Median	Mode	Total Mean	Total covariance	Standard Deviation
	581.5903	20.4846	0	0	225.1926	1.78E+03	42.1818
	567.1263	20.594	0	0	225.2051	1.78E+03	42.1846
	587.5946	20.44	0	0	225.2035	1.78E+03	42.1816
	600.5339	20.3454	0	0	226.166	1.65E+03	40.679
	593.9173	20.3935	0	0	226.166	1.65E+03	40.679
	601.5671	20.338	0	0	227.1408	1.53E+03	39.1317
	606.9791	20.2991	0	0	229.1479	1.30E+03	36.0019
	606.9791	20.2991	0	0	231.2829	1.07E+03	32.7836
	597.4663	20.3677	0	0	233.5064	868.6291	29.4725
	564.8112	20.6118	0	0	235.8073	680.682	26.0899

Table 1 and Table 2 shows difference between the quality measures of original MRI image and histogram equalization image of same MRI images.

The comparisons of statistical measures between sample original MRI image and histogram equalization of same original image has shown with the help of graph (Figs. 1, 2, 3 and 4).

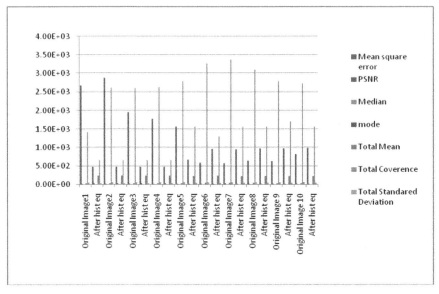

Fig. 1. Sample MRI images from 1 to 10 are shown in above graph 1.1

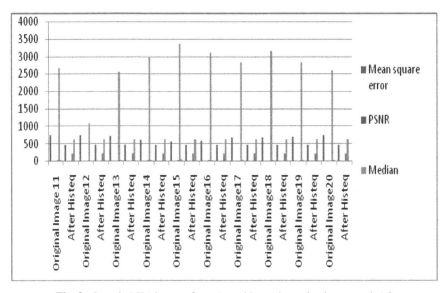

Fig. 2. Sample MRI images from 11 to 20 are shown in above graph 1.2

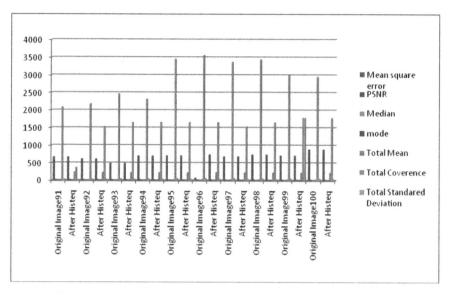

Fig. 3. Sample MRI images from 91 to 100 are shown in above graph 1.3

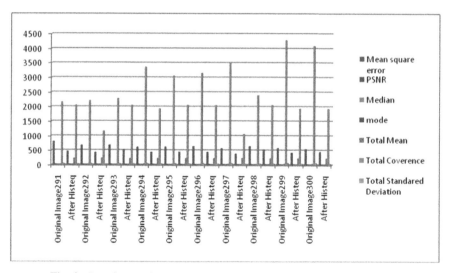

Fig. 4. Sample MRI images from 291 to 300 are shown in above graph 1.4

6   Result Analysis

This paper presents the comparison between original MRI image and histogram equalization of same MRI image with the help of different quality measures applied on that images as shown in the above table. The MRI records quantity is obtained on or after a virtual brain database (BRATS-2012).

To show this comparison we plot a graph which shows the difference between Quality measures of original image and histogram equalization of 302 images. The following figure shows the increases and decrease of statistical measures such as MSE, PSNR, MODE, MEDIAN, TOTAL COVARIANCE AND STANDARD DEVIATION in original MRI image, and histogram equalization of same MRI image. It concludes that the image is enhanced and quality of image is improved in histogram equalization.

7 Conclusion

This manuscript presents an idea of evaluation of a variety of Quality measures to improve the worth of original MRI image and comparative analysis of different enhancement Quality measure on histogram image. Depending in the lead of unique temperament of representation, suitable technique use for enrichment and improved the quality of image mentioned above. The value represent in above table shown the difference in original image of MRI and histogram images showed that the quality of image is improved in histogram images. This difference is shown by graph with 302 images. In future many filtering methods would be use by using statistical measures to enhanced the quality of image, and get more improved result.

References

1. Oak, P.V., Kamathe, R.S.: Contrast enhancement of brain MRI images using histogram based techniques. Int. J. Innov. Res. Electric. Electron. Instrum. Control Eng. **1**(3) (2013)
2. Nisa, M.U., Khawaja, A.: Contrast enhancement impact on detection of tumor in brain MRI. Sci. Int. (Lohare) **27**(3), 2161–2163 (2015). ISSN 1013-5316
3. Kauri, H., Rani, J.: MRI brain image enhancement using histogram equalization technique. Presented at the IEEE WiSPNET Conference (2016)
4. Ruikar, D.D., Santosh, K.C., Hegadi, R.S.: Contrast stretching-based unwanted artifacts removal from CT images. In: Santosh, K.C., Hegadi, R.S. (eds.) RTIP2R 2018. CCIS, vol. 1036, pp. 3–14. Springer, Singapore (2019). https://doi.org/10.1007/978-981-13-9184-2_1
5. Ruikar, D.D., Santosh, K.C., Hegadi, R.S.: Automated fractured bone segmentation and labeling from CT images. J. Med. Syst. **43**(3), 1–13 (2019). https://doi.org/10.1007/s10916-019-1176-x
6. Ruikar, D.D., Santosh, K.C., Hegadi, R.S.: Segmentation and analysis of CT images for bone fracture detection and labeling. In: Medical Imaging: Artificial Intelligence, Image Recognition, and Machine Learning Techniques, Chap 7. CRC Press (2019). ISBN 978-0-36-713961-2
7. Pandey, A., Beg, S., Yadav, A.: Image processing technique for the enhancement of brain tumor pattern. Int. J. Res. Dev. Appl. Sci. Eng. (IJRDASE) (2016). ISSN 2454-684
8. Ruikar, D.D., Hegadi, R.S., Santosh, K.C.: A systematic review on orthopedic simulators for psycho-motor skill and surgical procedure training. J. Med. Syst. **42**(9), 168 (2018)
9. Kumar, V., Gupta, P.: Importance of statistical measures in digital image processing. Int. J. Emerg. Tech. Adv. Eng. **2**(8) (2012). www.ijetae.com. ISSN 2250-2459
10. Suneetha, B., JhansiRani, A.: A survey on image processing technique for brain tumor detection using magnetic resonance imaging. In: International Conference on Innovations in Green Energy and Healthcare Technologies (IGEHT) (2017)

11. Hegadi, R.S., Navale, D.I., Pawar, T.D., Ruikar, D.D.: Osteoarthritis detection and classification from knee X-ray images based on artificial neural network. In: Santosh, K.C., Hegadi, R.S. (eds.) RTIP2R 2018. CCIS, vol. 1036, pp. 97–105. Springer, Singapore (2019). https://doi.org/10.1007/978-981-13-9184-2_8

12. Santosh, K.C., Antani, S., Guru, D.S., Dey, N. (eds.): Medical Imaging: Artificial Intelligence, Image Recognition, and Machine Learning Techniques. CRC Press, Boca Raton (2019)

13. Hegadi, R.S., Navale, D.I., Pawar, T.D., Ruikar, D.D.: Multi feature-based classification of osteoarthritis in knee joint X-ray images. In: Medical Imaging: Artificial Intelligence, Image Recognition, and Machine Learning Techniques, vol. 75 (2019)

14. Janani, P., Premaladha, J., Ravichandran, K.S.: Image enhancement techniques: a study. Indian J. Sci. Tech. 8(22) (2015)

15. Senthilkumaran, N., Thimmiaraja, J.: Histogram equalization for image enhancement using MRI brain image. In: 2014 World Congress on Computing and Communication Technologies (2014)

16. Hemanth, D.J., Smys, S.: Computational Vision and Bio Inspired Computing. Springer, Cham (2018). https://doi.org/10.1007/978-3-030-37218-7

Osteoarthritis Detection in Knee Radiographic Images Using Multiresolution Wavelet Filters

Shivanand S. Gornale[1](✉), Pooja U. Patravali[1](✉), and Prakash S. Hiremath[2](✉)

[1] Department of Computer Science, Rani Channamma University,
Belagavi 591156, Karnataka, India
shivanand_gornale@yahoo.com, pcdongare@gmail.com
[2] Department of Computer Science (MCA), KLE Technological University,
Hubballi 580031, Karnataka, India
hiremathps53@gmail.com

Abstract. Osteoarthritis (OA) is one of the chronic diseases related to joints. The joint pain caused due to Osteoarthritis is unbearable and if not treated may cause deformity and disability. It is observed that few numbers of researchers have implemented identification and grading of Osteoarthritis utilizing diverse methodologies based on their own datasets for experimentation. However, there is still need of automatic computer aided methods to detect Osteoarthritis for early recognition. In this work, cartilage region is automatically identified based on density and range of wavelet filters were used for appropriate and early analyses of radiographic Knee Osteoarthritis images. The computed wavelet features are classified using decision tree and k-nearest neighbour classifiers. The indicated experimental outcomes are challenging and viable which are examined by medical experts.

Keywords: Osteoarthritis · Knee radiography · Region of interest · Wavelet transformation · Decision tree · k-nearest neighbor

1 Introduction

Arthritis is one of the chronic diseases related to joints [1, 2]. There are different types of arthritis the most widely recognized are Rheumatoid Arthritis (RA) and Osteoarthritis (OA). Osteoarthritis is one such disease which makes the patients feel its existence all the time. The joint pain caused due to Osteoarthritis is unbearable and if not treated will cause deformity and disability. Generally any form of arthritis are not curable, the ailment can be controlled but rather can't be totally restored. The important clinical symptoms of OA in the initial stage are joint pain in knee, hip, ankle, spine etc. [12, 27]. The medical experts precisely observe the patient clinically and may ask the patient to go for a radiographic assessment. Osteoarthritis can also be caused due to some accidental joint injuries. The relationship between the joint injury and knee OA was initially depicted by Kellgren and Lawrence [5, 25]. The Kellgren and Lawrence (KL) grading system is most useful framework for systemization of singular joint into 5 grades [10]. The KL grading is shown in the Table 1.

© Springer Nature Singapore Pte Ltd. 2021
K. C. Santosh and B. Gawali (Eds.): RTIP2R 2020, CCIS 1381, pp. 36–49, 2021.
https://doi.org/10.1007/978-981-16-0493-5_4

Table 1. Grading system by Kellgren & Lawrence

KL grades	OA analysis
Grade 0 (Normal OA)	No Radiographic parameters related to OA
Grade 1 (Doubtful OA)	Diminished joint space width
Grade 2 (Mild OA)	Clear/ visible narrowing of joint space
Grade 3 (Moderate OA)	Numerous bony outgrowths, sclerosis
Grade 4 (Severe OA)	Massive bone spurs, extreme sclerosis, bone deformity

Whenever the patient is suggested for any medical imaging the probabilities of noise or distortions is high due to incorrect positioning of patient as well as camera, instability in atmosphere, blood flow intrusion etc. The complexities related to the medical images make it hard to examine them in an effective way and may cause problems in analyzing the bone structures. Due to these complexities it may be possible that the experts may delay investigating the knee x-ray and may reach some vague conclusion [12]. Thus to overcome these problems various computer aided and machine vision approaches have to be developed to effectually analyze the abnormalities & problems associated with the bone structures [29, 33]. The major idea behind this work of this proposed work is to identify Knee Osteoarthritis in radiographic images utilizing multi-resolution wavelet filters.

"In the rest of the paper, Sect. 2 contains the related work tshat incorporates wavelet transformations for the analysis of the disease. Section 3 focuses on proposed methodology. Section 4 includes experimental and medical expert result analysis and finally conclusions are drawn in Sect. 5."

2 Related Work

The earlier work reveals that the researchers have used multiple automated and semi-automated segmentation methods for the extraction of region of interest [3–5]. Various wavelet transformation methods have been proposed for the recognition and grading of Osteoarthritis utilizing X-ray and MRI images of Knee Osteoarthritis. T. Janvier et al. [6], have utilized a fractal texture analysis method for analyzing the textures of trabecular bone assessed from radiographic images to predict the progression of knee osteoarthritis. Firstly semi automated segmentation method was used to extract the region of interest and later fractal texture analysis was applied using different methods. The trabecular bone texture was accessed utilizing logistic regression and ROC curve. The authors concluded a high predictive ability to detect the progression of Knee OA using trabecular bone texture parameters. Rabia Riada et al. [7], have used new approach for texture analysis for the assessment of Osteoarthritis on Knee X-ray images. Initially pre-processing of x-ray images is carried out and later relative phases and complex coefficients of image content was extracted using wavelet decomposition. The estimated parameters are used to classify and analyze the texture of knee x-ray images thus obtaining the accuracy of 80.38%. Jian WeiKoh et al. [8, 9], "proposed the genetic algorithm optimized neural network technique for osteoarthritis recognition in Knee MRI." The important steps of the experimentation are feature computation using Discrete Wavelet Transform, training and testing of neural network and finally optimization using Genetic Algorithm (GA). The results demonstrated an accuracy of 98.5% for training and 94.67% for testing. Along with classification, 17.24% of time reduction was achieved after optimization of neural network. Lior Shamir et al. [10, 11] have intricate a computer aided system for determination of Knee Osteoarthritis in radiographic images. The authors have computed various features like Zernike features, first four moments, Tamura texture features, Haralick features, Chebyshev statistics and multiclass histograms using WND-CHRM algorithm. The weighted nearest neighbour classifier is used that obtained an accuracy of 91.5% for Moderate Osteoarthritis and 80.4% for Mild Osteoarthritis. J Christopher Buckland-Wright et al. [12], have used a method to find the joint space width using high micro focal radiography. The method compares the joint space width measured from weight bearing micro radiographic images with the sum of femur and tibia cartilage thickness of non-weight bearing lateral position images. The results revealed high correlation for medial compartment but very less correlation was seen in a lateral compartment. S. S. Gornale et al. [15–19], have proposed an algorithm for Osteoarthritis recognition in radiographic images. The semi-automated active contour segmentation method was utilized to determine the region of interest [16–18]. Various features like textural based, shape based and statistical based features were estimated using Random forest classifier, and yielded the classification accuracy of 87.92% [16]. Further, the experimentation was extended that included Zernike moments and textural based feature computation. K- Nearest neighbor classifier was utilized obtaining the accuracy of 88.88% [16]. Later, from the extracted region of interest Histogram gradients were computed and classified using multiclass SVM that yielded an accuracy of 95% [17]. The ROI extraction was further conceded using different segmentation methods like Prewitt, Sobel, Texture based and Otsu's based methods [19]. Prewitt method obtained a better accuracy of 97.55% compared to other methods. Lastly a novel approach for automated identification of ROI was developed that demonstrated an accuracy of 97.86% using various local and gradient features [20]. Lior Shamir et al. [13], have developed a system that mainly concentrates on changes in bone morphology. The system uses WND-CHRM grouping method to study the texture parameters in knee x-ray images. The attributes considered for computation were high contrast features, Haralick features, Tamura features, statistical features and polynomial decomposition of image. The results demonstrated no huge difference between KL grade 0 and KL grade 1 with respect to bone surface. The alterations in bone morphology were observed only after KL grade 2 onwards. The authors assumed that the examination focused on impartial investigation of image content and further the clinical parameters like weight, pain, BMI, previous injuries etc. can be considered for clear and early analysis of the disease.

From the literature, analysis of spatial as well as time domain is understudied for the detection and prediction of OA progression. The objective of the work is to study the frequency domain analysis using Knee x-ray images for monitoring the ailment in early stage and categorizing as per Kellgren and Lawrence grading system.

3 Proposed Methodology

The proposed methodology constitutes bone edge detection of knee X-ray, preprocessing, identification of region of interest (Osteoarthritis region) based on density, feature extraction using wavelet transformation and classification of Osteoarthritis into Kellgren and Lawrence grades using decision tree and k-nn classifier. The flow of the proposed methodology is depicted in Fig. 1.

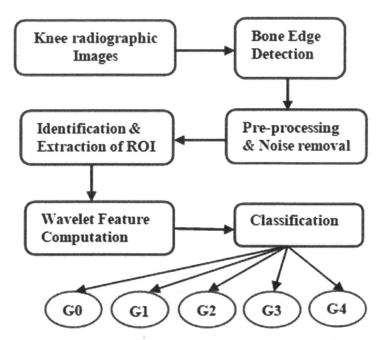

Fig. 1. Block diagram (G-0: Normal, G-1: Doubtful, G-2: Mild, G-3: Moderate, G-4: Severe)

3.1 Dataset

A dataset of 1650 patients' knee x-ray images are collected from various hospitals and diagnostic centers of rural and urban population. The fixed flexion digital knee x-ray images are acquired using PROTEC PRS 500 E X-ray machine. Original images were 8-bit 1350 × 2455 grayscale DICOM (Digital Imaging and Communications in Medicine) images. Each and every Knee x-ray is manually assigned a KL grade by two different orthopedicians.The annotations made as per KL grading framework is given in Table 2.

3.2 Bone Edge Detection

Edges are noticeable variations that are encapsulated by salient shape features of a particular image. In this proposed work, detection of bone edges is implemented by

Table 2. Kellgren and Lawrence grades assigned by two different medical experts

KL grade	Medical expert-I opinion	Medical expert-II opinion
Normal (G-0)	514	503
Doubtful (G-1)	477	488
Mild (G-2)	232	232
Moderate (G-3)	221	221
Severe (G-4)	206	206
Total	1650	1650

finding the difference with the neighboring pixels. The bone edge detection at pixel (a, b) for input image F with respect to row and column is given in (1) and (2). The output of bone edge detection is shown in Fig. 2.

$$A(m, n) = A(m, n) - A(m - 1, n) \tag{1}$$

$$A(m, n) = A(m, n) - A(m, n - 1) \tag{2}$$

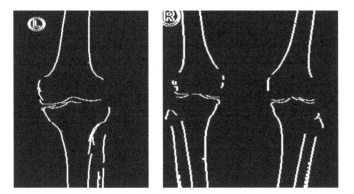

Fig. 2. Detection of bone edges

3.3 Preprocessing

Pre-processing emphasizes appropriate prominent characteristics of an image more clearly [11]. After detecting the bone edges' pre-processing of the image is conducted that includes noise removal using adaptive median filter [14]. Mainly the images acquired from x-ray detectors are eminent in salt and pepper noise. Adaptive median filtering is helpful in lessening salt and pepper noise or impulsive noise and beneficial in preserving edges of an image that result in reduction of irregular noise [26]. Filtering is important as it enhances edges that contain useful information.

3.4   Identification of Region of Interest

Identification of region of interest of an image can be implemented by partitioning the entire image into significant structures. The required or interested structures or objects from the partitioned image can be segregated from background or foreground, can be scaled, estimated or evaluated for processing [21, 23]. The identification is assigned into four steps: primarily the complete image is fragmented to numerous parts into rows and columns. Secondly, extract the region of interest (ROI) from each segmented part. Thirdly estimate the area of each segmented part to detect the bone density. The region in the image that is denser or thicker, results in high density value. Lastly, the region with high density cost is extracted and enhanced using sine adaptive filter. The steps used by sine adaptive filter are given in Eqs. (3–6).

$$FC = 0 : \frac{Pi}{2} : 0 \tag{3}$$

$$F = \frac{Pi}{2} - Filter\_Width * \frac{Pi}{2} \tag{4}$$

$$FC(FC > F) = \frac{Pi}{2} \tag{5}$$

$$I_{out} = I_{in} \sin(FC) \tag{6}$$

Where, FC is a filter co-efficient and Filter Width is calculated based on the X- ray reconstructed image characteristics. Images recorded from x-ray detectors always have the data in the centre portion of the image. Therefore by defining the filter co-efficient as a sine wave, we are adding more weight to the data in the middle of the image. Adaptive sine filter is used for allocating weights based on the geometrical axis of the reconstructed image. Thus the region of interest is accurately extracted that can be further used for the medical examination [27, 35]. The identification of ROI is shown in Fig. 3(a) and 3(b).

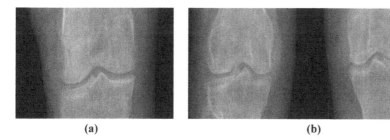

| (a) | (b) |

Fig. 3. Identification of Region of interest: **(a)** Single knee ROI (left/right) **(b)** ROI identification both left and right knee

3.5 Wavelet Transformation

Wavelets are mathematical functions which are useful for analysis of different frequency components with a resolution matched to its scale [31]. The unique characteristics of Wavelets make them useful in signal processing, Image compression and image denoising. Wavelet transform have the distinctive ability to represent the time and frequency information simultaneously and are meant to analyze the non-stationary signals. Generally, in medical imaging choosing a best wavelet filter order and its level of decomposition is a challenging task. The selection criterion of wavelets depends on some basic properties like real, complex, compactly supported, arbitrary, infinite regularity, symmetry and asymmetry, number of vanishing moments and existence of scaling functions [10]. In this work Discrete wavelet transformation is used for feature computation bearing multi-resolution wavelet filters. General decomposition of image using discrete wavelet transform is given in Fig. 4 below.

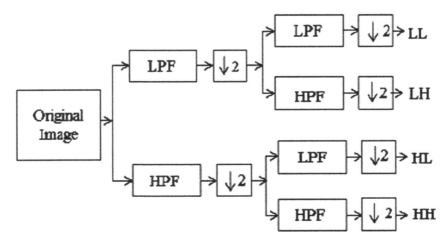

Fig. 4. Decomposition of image using wavelet transformation

The procedure of Discrete Wavelet transform includes fragmenting the entire image into approximation coefficients low pass filter and detail coefficients high pass filter. Approximate coefficients demonstrate the general pattern of pixel esteems and detail coefficients show Horizontal (H), Vertical (V) and Diagonal (D) changes of the picture [24, 32]. In Discrete Wavelet transform decomposition is always repeated on approximation band to reduce the number of low pass coefficients as it contains most original signal energy that yields good energy compaction. The four groups of information attained by DWT are named as LL (low-low), HL (high-low), LH (low-high) and HH (high-high). The LL band can be disintegrated indeed in similar behavior, for delivering more subgroups. This should be possible up to $2^n - 1$ levels, where n depends on the input image size. The pyramidal decomposition of an image is shown in Fig. 5 beneath.

Simple example of 3rd level image decomposition using discrete wavelet transforms on knee x-ray is shown in Fig. 6.

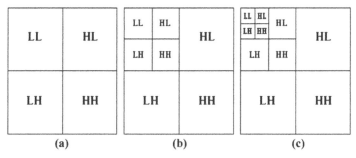

LL	HL
LH	HH

(a)

(b)

(c)

Fig. 5. Pyramidal decomposition of an image: (a) Single level decomposition (b) Two level decomposition (c) Three level decomposition

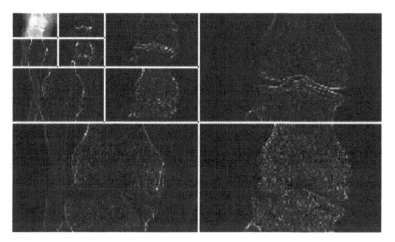

Fig. 6. 3rd level decomposition of Knee X-ray.

The multi-resolution wavelet filters with different filter order and decomposition levels are used in this work. The filter banks used are:

- Daubechies: These are represented as dbN where N represents the filter order ranging from 1–45 and having filters length of 2N.
- Symlets: Represented as symN where N represents the filter order with positive integers ranging from 1–45.
- Coifflets: Represented as coifN where N represents the filter order ranging from 1–5 and having filters length of 6N.
- Fejer Korovokin: Represented as fkN where N is the filter order fk4, fk6, fk8…..fk22.
- Discrete Meyer wavelets are the discrete form of Meyer wavelets.
- Biorthogonal wavelets: Represented as bior with order Nr for reconstruction and Nd for decomposition such as bior3.1, bior5.5 etc.

3.6 Classification

For the experimentation decision tree and K-Nearest Neighbor classifier are used to classify the features computed by wavelet transformation.

Decision Tree: It is widely used classifier for multiclass problems. In decision tree the outcome is demonstrated in terms of leaf node where as non-leaf nodes depicts the decision. Here the various decedents are encircled based on the attributes examined by non-leaf nodes. The major step while building the decision tree is to figure out which characteristic is to be investigated and which among numerous possibility tests dependent on characteristic has to be performed. In decision tree the important query is to estimate the optimal partition of m components into n sections [22]. Every leaf is allocated one single class corresponding to most suitable target value. It may so happen that the leaf node may hold the probability vector specifying target attribute comprising some specific value. Further the classification is carried out by navigating from the root to a leaf along the path.

K Nearest Neighbors (K-NN): In k-NN the affinity between the test data and the training set is virtually obtained by the classifier [20, 28]. K-Nearest Neighbour will classify the class label based on measuring the distance between testing and training data. KNN will classify by suitable K value which in turn finds the nearest neighbor and provides a class label to un-labeled images. Depending on the types of problem, a variety of different distance measures can be implemented [14, 34]. In this work, City-block distance, Cosine, Correlation and Euclidean distance is considered with K = 3 which is empirically fixed throughout the experiment [30, 35]. Basically, K-NN is non-parametric classifier which finds the minimum distance d between training sample M and testing pattern Ni and T = 3 using Eq. (7).

$$D_{Euclidean}(M, N) = \sqrt{(M - N_i)^T - (M - N_i)} \qquad (7)$$

4 Results

The dataset of 1650 radiographic knee images with DICOM standards are considered for the experimentation. The feature extraction was carried out using multi resolution wavelet filters with different filter order and decomposition level. The computed wavelet features are classified using decision tree and K-nn classifiers. The proposed algorithm is as shown below.

Input: Digital Knee X-ray image.
Output: Normal or Affected image (grade-wise).

Step-1: Bone edge detection
Step-2: Noise removal using adaptive median filter.
Step-3: Segmenting images into multiple parts based on pixel density
Step-4: Extract Region of interest from each segmented part.
Step-5: Feature computation using Wavelet Transformation

Step-6: Computed wavelet features were classified using two different classifiers namely Decision tree and K-Nearest Neighbour classifiers.
End

The experimentation is carried out to understand the robustness of the algorithm. In this experiment two fold cross validation is performed where 50% of images are considered for training and 50% for testing. The overall experiment is implemented using two different classifiers and the results are shown in Tables 3 and 4.

Table 3. Classification accuracy of different wavelets with respect to medical expert I and II opinion using decision tree

Decision classifier				
Wavelet filters and its order	Decomposition level	Computational time in seconds	Proposed method accuracies for expert-I	Proposed method accuracies for expert-II
Db2	Level 6	24.31	98.29%	97.63%
Coif5	Level 5	41.05	98.03%	97.35%
Sym2	Level 6	23.27	98.29%	97.63%
Fk22	Level 3	30.35	97.86%	97.61%
Dmey	Level 1	52.82	97.69%	97.18%
Bior1.5	**Level 4**	**23.72**	**98.54%**	**97.93%**

From Table 3, it is observed that six different wavelet filters are used for the experimentation with different filter order and decomposition level. The computational time is calculated for every filter and the average accuracy is demonstrated. The highest accuracy of 98.54% and 97.93% with respect to expert I and II opinion is obtained for Biorthogonal 1.5 wavelet filter at 4th decomposition level using decision tree classifier.

Table 4. Classification accuracy of different wavelets with respect to medical expert I and II opinion using K-nn classifier

K-NN classifier				
Wavelet filters and its order	Decomposition level	Computational time in seconds	Proposed method accuracies for expert-I	Proposed method accuracies for expert-II
Db5	Level 3	24.33	96.93%	96.16%
Coif4	Level 3	31.15	97.01%	96.16%
Sym6	Level 3	24.88	97.10%	96.41%
Fk22	**Level 3**	**30.33**	**97.45%**	**96.42%**
Dmey	Level 1	53.34	96.67%	95.82%
Bior2.4	Level 4	22.83	97.10%	96.16%

From Table 4, it is observed that the result of Ferjer Korovokin wavelet (fk22) with filter order 22 at 3rd decomposition level has obtained a highest result of 97.45% and 96.42% with respect to expert I and II opinions as compared to other wavelets using K-Nearest Neighbor classifier.

Statistical test of Significance-Further, the experimentation was preceded to test the significance level of proposed algorithm using the Chi square test and T-test. These tests are basically performed to check the robusteness and optimality of our proposed algorithm. Initially, Chi-suare test was performed between the results of the proposed algorithm and annotations made by the two experts and it was found that their exists a strong correlation between the results of proposed algorithm and medical expert opinions. Further, T-test was carried out between the annotations made by two experts and it was found that the difference in the annotations of two experts is due to random variations in sample inspections but not due to experts. Therefore, from both the tests it was found that the proposed algorithm is robust and optimal which can be used by medical experts for analysing the disease for appropriate treatment. However, from the experimental results there are some miss-classification observed due to the subjective nature of classification made by medical expert. Precisely, there are no any such appropriate and automatic methods available for exact Osteoarthritis recognition. Thus the algorithm proposed has demonstrated promising results in identifying the ailment at early stage which is validated by medical expert.

5 Conclusion

In this work, the region of interest is automatically identified and extracted based on density. Further feature classification is performed using decision tree and K-NN classifiers utilizing wavelet features. The proposed algorithm is implemented on 1650 knee x-ray images which are manually annotated by two experts as per KL grading system. The experimental results demonstrated an accuracy of 98.54% and 97.93% with respect to expert I and II opinion for Biorthogonal 1.5 wavelet filter at 4th decomposition level using decision tree classifier. For K-nearest neighbour classifier an accuracy of 97.45% and 96.42% with respect to expert I and II opinions is obtained for Ferjer Korovokin wavelet (fk22) at 3rd decomposition level. The emphasis of the proposed algorithm is found to be more significantly correlated with the medical expert opinion which may assist doctors and radiologists for the early assessment of Osteoarthritis treatment. The obtained results are competitive in terms of evaluating Osteoarthritis as per KL grading system. In near future, the analysis of the disease can be done using other radiological parameters like osteophytes and sclerosis which will be helpful for doctors in evaluating the ailment more precisely and provide appropriate treatment to the patients.

Acknowledgement. Authors would like to thank Department of Science and Technology (DST) for financial assistance under Women Scientist-B Scheme. (Ref No: SR/WOS-B/65/2016(G)). Authors are also thankful to Dr. Chetan M. Umarani, Orthopaedic Surgeon, Gokak-Karnataka, and Dr. Kiran S.Marathe, Orthopaedic Surgeon, JSS Hospital, Mysore, Karnataka-India for providing knee X-ray images and validation of the computed results by visual inspection.

Conflict of Interest. The authors declare no conflict of interest regarding the publication of this work.

References

1. Hegadi, R.S., Navale, D.I., Pawar, T.D., Ruikar, D.D.: Multi feature-based classification of osteoarthritis in knee joint X-ray images. In: Medical imaging: Artificial Intelligence, Image Recognition, and Machine Learning Techniques, Chap 5. CRC Press (2019). ISBN 9780367139612
2. Hegadi, R.S., Navale, D.I., Pawar, T.D., Ruikar, D.D.: Osteoarthritis detection and classification from knee x-ray images based on artificial neural network. In: Santosh, K.C., Hegadi, R.S. (eds.) RTIP2R 2018. CCIS, vol. 1036, pp. 97–105. Springer, Singapore (2019). https://doi.org/10.1007/978-981-13-9184-2_8
3. Ruikar, D.D., Santosh, K.C., Hegadi, R.S.: Automated fractured bone segmentation and labeling from CT images. J. Med. Syst. **43**(3), 1–13 (2019)
4. Ruikar, D.D., Santosh, K.C., Hegadi, R.S.: Segmentation and analysis of CT images for bone fracture detection and labeling. In: Medical Imaging: Artificial Intelligence, Image Recognition, and Machine Learning Techniques, p. 131 (2019)
5. Santosh, K.C., Antani, S., Guru, D.S., Dey, N. (eds.) Medical Imaging: Artificial Intelligence, Image Recognition, and Machine Learning Techniques. CRC Press, Boca Raton (2019)
6. Janvier, T., et al.: Subchondral tibial bone texture analysis predicts knee osteoarthritis progression: data from the Osteoarthritis Initiative: tibial bone texture & knee OA progression. Osteoarthritis Cartilage **25**(2), 259–266 (2017)
7. Riad, R., Jennane, R., Brahim, A., Janvier, T., Toumi, H., Lespessailles, E.: Texture analysis using complex wavelet decomposition for knee Osteoarthritis detection: data from the Osteoarthritis Initiative. Comput. Electr. Eng. (2018). https://doi.org/10.1016/j.compeleceng.2018.04.004
8. Kohn, M.D., Sassoon, A.A. and Fernando, N.D.: Classifications in brief Kellgren-Lawrence classification of Osteoarthritis. Clin. Orthop. Relat. Res. **474**(8) (2016). https://doi.org/10.1007/s11999-016-4732-4
9. WeiKoh, J., Tan, T.S., EnChuah, Z., Soh, S.S., Arif, M. and Leong, K.: Genetic algorithm optimized back propagation neural network for knee Osteoarthritis classification. Res. J. Appl. Sci. Eng. Technol. 8(16), 1787–1793 (2014). https://doi.org/10.19026/rjaset.8.1166. ISSN 2040-7459; e-ISSN 2040-7467
10. Gornale, S.S.: Multiresolution Analysis for Image Compression. LAP LAMBERT Academic Publishing, p. 192, 28 July 2016. ISBN-13: 978-3-659-92570-2, ISBN-10: 3659925705, EAN: 9783659925702
11. Ruikar, D.D., Santosh, K.C., Hegadi, R.S.: Contrast stretching-based unwanted artifacts removal from CT images. In: International Conference on Recent Trends in Image Processing and Pattern Recognition, pp. 3–14. Springer, Singapore (2018)
12. Shamir, L., et al.: Knee X-ray image analysis method for automated detection of osteoarthritis. IEEE Trans. Biomed. Eng. **56**(2), 407–415 (2009). https://doi.org/10.1109/TBME.2008.2006025
13. Buckland-Wright, J.C., Macfarlane, D.G., Lynch, J.A., Jasani, M.K., Bradshaw, C.R.: Joint space width measures cartilage thickness in Osteoarthritis of the knee: high resolution plain film and double contrast macro radiographic investigation. Ann. Rheumatic Dis. **54**(4), 263–8 (1995)

14. Sharmir, L., Rahimi, S., Orlov, N., Ferrucci, L., Goldberg, I.G.: Progression analysis and stage discovery in continuous physiological process using image computing. EURASIP J. Bioinform. Syst. Biol. 2010(1) (2010). https://doi.org/10.1155/2010/107036. Article ID 107036. PMCID: PMC3171360
15. Kruti, R., Patil, A., Gornale, S.: Fusion of features and synthesis classifiers for gender classification using fingerprints. Int. J. Comput. Sci. Eng. 7(5), 526–533. https://doi.org/10.26438/ijcse/v7i5.526533
16. Gornale, S.S., Patravali, P.U., Manza, R.R.: A survey on exploration and classification of osteoarthritis using image processing techniques. Int. J. Sci. Eng. Res. 7(6), 334–355 (2016). ISSN 2229-5518
17. Gornale, S.S., Patravali, P.U., Manza, R.R.: Detection of osteoarthritis using knee X-ray image analyses: a machine vision based approach. Int. J. Comput. Appl. (IJCA) 145(1), 20–26 (2016). ISSN-0975-8887
18. Gornale, S.S., Patravali, P.U., Manza, R.R.: Computer assisted analysis and systemization of knee Osteoarthritis using digital X-ray images. In: Proceedings of 2nd International Conference on Cognitive Knowledge Engineering (ICKE), Chapter 42, pp:207–212. Excel Academy Publishers, Aurangabad, December 2016. ISBN 978-93-86751-04-1
19. Gornale, S.S., Patravali, P.U., Marathe, K.S., Hiremath, P.S.: Determination of Osteoarthritis using histogram of oriented gradients and multiclass SVM. Int. J. Image Graph. Signal Process. (IJIGSP) 9(12), 41–49 (2017). https://doi.org/10.5815/ijigsp.2017.12.05
20. Gornale, S.S., Patravali, P.U., Uppin, A.M., Hiremath, P.S.: Study of segmentation techniques for assessment of Osteoarthritis in knee X-ray images. Int. J. Image Graph. Signal Process. (IJIGSP) 11(2), 48–57 (2019). https://doi.org/10.5815/ijigsp.2019.02.06
21. Gornale, S.S., Patravali, P.U., Hiremath, P.S.: Identification of region of interest for assessment of knee Osteoarthritis in radiographic images. Int. J. Med. Eng. Inform. (Accepted)
22. Coppersmith, D., Hong, S.J., Hosking, J.R.M.: Partitioning nominal attributes in decision trees. Data Min. Knowl. Disc. 3(2), 197–217 (1999). https://doi.org/10.1023/A:100986980 4967
23. Nithya, R., Santhi, B.: Computer aided diagnostic system for mammogram density measure and classification. Biomed. Res. (0970-938X) 28(6), 2427–2431 (2017)
24. Cristina Stolojescu, C., Holban, S.: A comparison of X-ray image segmentation techniques. Adv. Electr. Comput. Eng. 13(3), 85–90 (2013). https://doi.org/10.4316/AECE.2013.03014
25. Gornale, S.S., Patravali, P.U.: medical imaging in clinical applications: algorithmic and computer based approaches. In: Engineering and Technology: Latest Progress, Basic Chapter, pp. 65–104 (2017). ISBN 978-81-32850-2-2
26. Frosio, I., Borghese, N.A.: Statistical based impulsive noise removal in digital radiography. IEEE Trans. Med. Imaging 28(1), 3–16 (2009). https://doi.org/10.1109/TMI.2008.922698
27. Semmlow, J.L.: Bio Signal and Biomedical Image Processing: MATLAB-Based Applications (Signal Processing). Taylor & Francis Inc. Annotated edition, 14 January 2004. ISBN: 0–8247-4803-4.
28. Gornale, S.S., Patil, A., Kruthi, R.: Multimodal biometrics data based gender classification using machine vision. Int. J. Innov. Technol. Explor. Eng. (IJITEE) 8(11) (2019). ISSN 2278-3075
29. Norouzi, A., et al.: Medical image segmentation methods, algorithms, and applications. IETE Tech. Rev. 31(3), 199–213 (2014). https://doi.org/10.1080/02564602.2014.906861
30. Gornale, S.S., Patil, A., Hangarge, M., Pardesi, R.: Automatic human gender identification using palmprint. In: Luhach, A.K., Hawari, K.B.G., Mihai, I.C., Hsiung, P.-A., Mishra, R.B. (eds.) Smart Computational Strategies: Theoretical and Practical Aspects, pp. 49–58. Springer, Singapore (2019). https://doi.org/10.1007/978-981-13-6295-8_5

31. Tiulpin, A., Thevenot, J., Rahtu, E., Saarakkala, S.: A novel method for automatic localization of joint area on knee plain radiographs. In: Sharma, P., Bianchi, F.M. (eds.) SCIA 2017. LNCS, vol. 10270, pp. 290–301. Springer, Cham (2017). https://doi.org/10.1007/978-3-319-59129-2_25

32. Paul, S., Sarkar, P.K., Mishra, D.G., Joshi, V.M., Pulhani, V.: Wavelet based spectrum processing for reduction of counting duration for quantitative estimation of ultra-trace activity in environmental matrices. J. Radioanal. Nucl. Chem. **299**(1), 415–426 (2013). https://doi.org/10.1007/s10967-013-2761-y

33. Lim, J., Kim, J., Cheon, S.: A deep neural network-based method for early detection of Osteoarthritis using statistical data. Int. J. Environ. Res. Public Health **16**(7), E1281 (2019). https://doi.org/10.3390/ijerph16071281

34. Gornale, S.S., Patravali, P.U., Hiremath, P.S.: Early detection of Osteoarthritis based on cartilage thickness in knee X-ray images. Int. J. Image Graph. Signal Process. (IJIGSP) **11**(9), 56–63 (2019). https://doi.org/10.5815/ijigsp.2019.09.06

35. Gornale, S.S., Patravali, P.U., Hiremath, P.S.: Detection of Osteoarthritis in knee radiographic images using artificial neural network. Int. J. Innov. Technol. Explor. Eng. (IJITEE) **8**(12), 2429–2434 (2019). https://doi.org/10.35940/ijitee.L3011.1081219. ISSN 2278-3075

DWT Textural Feature-Based Classification of Osteoarthritis Using Knee X-Ray Images

Dattatray I. Navale[1], Darshan D. Ruikar[1($\boxtimes$)], Kavita V. Houde[1],
and Ravindra S. Hegadi[2]

[1] Department of Computer Science, P.A.H. Solapur University, Solapur 413255,
Maharashtra, India
navaledatta@gmail.com, darshanruikar1986@gmail.com, kavitahoude@gmail.com
[2] Department of Computer Science, Central University of Karnataka,
Kadaganchi, Kalaburagi 585367, India
rshegadi@gmail.com

Abstract. Nowadays not only older people but also younger one (having age in between 30 to 45) are also suffering from knee Osteoarthritis (OA). Diagnosis and proper treatment of OA can delay the onset of severe disability. The primary pathological features (reduction in joint space distance and synovial cavity, inflammation and change in bone morphology for instance) to predict knee OA severity is clearly visible in X-ray images. However the X-ray images findings in this disease are often non-productive and may mislead due to the presence of noise in the image and create difficulties for analysis and classification. So to overcome the aforementioned problems an effective OA classification is system is presented.

The proposed classification system is to work in stages. Initially X-ray image is enhanced by using contrast stretching-based intensity transformation function. Next, the histogram modeling-based segmentation method is adapted to localize the desired region that is the knee joint region. Extract ROI is converted to discrete wavelet transform to do multi-resolution analysis which helps to detect interesting patterns those are not visible in an input image. As the inflammation cause the change in texture and morphology in affected knee joint region, DWT based statistical features which represent textural changes are extracted and passed to the SVM model for exact severity grade detection. Experiments are conducted on patient-specific X-ray images procured from local sources (Chidgupkar hospital Pvt. Ltd. India). In total hundred images are collected of various severity grades. By confirming results our model yields 99% accuracy for five-class classification and it will be a beneficial tool for rheumatologists to predict OA severity grade. In future we aim to conduct experiments on Osteoarthritis database (OAI) to confirm the robustness of the proposed system.

Keywords: Osteoarthritis (OA) · X-ray · Discrete wavelet transform (DWT) · Support vector machine (SVM) · Texture features · Classification

© Springer Nature Singapore Pte Ltd. 2021
K. C. Santosh and B. Gawali (Eds.): RTIP2R 2020, CCIS 1381, pp. 50–59, 2021.
https://doi.org/10.1007/978-981-16-0493-5_5

1   Introduction

Knee Osteoarthritis (OA) is most usual knee joint disorder. Majorly it causes pain in knee joints, stiffness, swelling, reduction in synovial fluid and loss of proper functionality in joints. The possibility of people suffering from knee Osteoarthritis (OA) is growing slowing from last two decades. Near about 0.7% to 1.2% of the world, the population is suffering from OA. More than 55% of patients suffering from chronic disability and due to that over all life span of older people reduced to 3 to 18 years [18]. Timely and exact diagnosis of OA in early stage will help in rapid recovery and may lead to drug-free life. Nowadays healthcare field showing interest to adopt technological advancements to diagnose various disease and find out optimal recovery plan/process [13]. Several computer-aided diagnosis (CAD) models are already available for most of the diseases (Fracture detection and analysis for instance) [14,15,17]. Still, the CAD development for OA diagnosis is in its initial stage due to several reasons such as lack of prevalence and availability of public medical imaging dataset with proper annotations to conduct experiments.

Common symptoms of OA are a pain, reduction in joint space distance (JSW) and synovial fluid, swelling of joints, and change in bone morphology. All symptoms (i.e. knee OA severity) increases over time. Swollen joints are one of the criteria used to decide OA severity grade and to initiate suitable recovery therapy. In the initial examination rheumatologist tries to detect joint swelling. Most of the times in the initial stage there could be no specific OA symptoms are observed externally. So experts recommends X-ray or computed tomography (CT) scanning of knee joints. Various medical imaging modalities (X-ray, CT, or Magnetic Resonance (MR) for instance) play an vital role in the analysis, diagnosis and will help experts to decide suitable recovery plan [16]. In addition to joint swelling, it may show other indications such as the morphological changes made in bones or reduction in joint space due to loss in the synovial cavity. Besides this CT, and MRI modalities are quite expensive and availability is also less (in rural regions of India). Moreover, these techniques require a lot of inspection time and patients suffer from heavy ionizing dose or magnetic radiation respectively [13]. So the X-ray imaging will be the more promising and widely used technique for OA analysis. The pathological features such as joint swelling, reduction in JSW and morphological changes in bones (bone spur) are well represented in X-ray images. In addition to medical experts, pathologist and expert orthopedic doctors, nowadays X-ray images are widely used for patient-specific CAD development.

In this paper, the X-ray image analysis-based non-invasive method proposed to provide a quantitative assessment and classification of knee OA. The proposed method will help in accurate diagnosis of knee OA and to decide suitable (patient-specific) recovery plan.

The reminder of the paper is structured as follows. The related research attempts in the literature are explored in Sect. 2. Section 3 describes the detailed explanation on the proposed methodology. It includes the data collection process, and the description of applied images processing techniques to achieve desired results. The experimental results are presented in Sect. 4 and concluding remarks and future scope is discussed in Sect. 5.

2   Related Work

Osteoarthritis (OA) is recurring knee joint disorder. Patients suffering from knee OA suffering from swelling, stiffness, and difficulty in useful knee joint functioning. Till the date no prominent remedy is know. At present prescribing mussel relaxant, pain relief, and anti-inflammatory doses are primary part treatment. Responses, (side) effects of aforementioned drugs are discussed in [22]. However such action plan will not lead to promising recovery. To recover or to stop progress of joint degradation early analysis and diagnosis is important. The X-ray images are widely used to analyze and diagnose the knee OA. A lot of X-ray image-based knee OA identification and classification methods were proposed in the literature. Some of the important and current state-of-art methods are briefly discussed in this section.

Most of the computer vision-based knee OA severity classification system work in states and most of them are based in K&L (Kellgren & Lawrence) severity grading scheme. ROI localization and accurate classification are the primary the main steps [2]. As the severity of the OA is based on features of the knee joint region, ROI localization step plays an important role in accurate severity level detection. Input X-ray image may contain some noise and unwanted artifacts such as patient cloths and wire. Before the application of a suitable ROI localization method, an input X-ray image needs to enhance to the desired level by removing noise and unwanted artifacts. In the literature wiener filter is used to remove noise and the histogram modeling-based contrast limited adaptive histogram equalization (CLAHE) method is adapted to enhance an image to desired level [7].

Researchers have used several image segmentation techniques to accurately extract the knee from radio-graphic image [3]. Initial few experiments adapt manual methods that are with the help of user ROI is extracted [9,11]. These methods are used to create a template. However, these methods are very much time consuming and laborious. The results are also user-dependent. The research attempts discussed in [11] used semi-automatic methods to extract ROI. Method discussed in [19] accepts initial region to segment knee joint region. Though semi-automatic methods show good accuracy but some sort of user intervention is required [21].

From few years researchers are adapting an automatic segmentation method to localize the ROI. Hegadi et al. [12] developed a simple block-based X-ray image segmentation method. The given radio-graphic image is divided into nine equal blocks having thee rows and three columns. By observing these blocks it is evident that only the middle three blocks are required for further processing as those contain knee joint region. In [5] developed a thresholding-based segmentation method to localize ROI from a knee X-ray image Otsu method [20] is adapted to decide image specific threshold value. Further, in [6,8] they used deformable model-based active contour segmentation method extracts the knee joint region.

After desired portion extraction (i.e. ROI localization) next step is an identification of severity grades based on extracted features. Previously most of the research attempts were based on handcrafted features extraction techniques. In [5,8,12] authors respectively used statistical, edge curvature, and gray level co-occurrence matrix (GLCM) texture fractures to decide severity grades. Artificial neural networks (ANN) [7], and support vector machine (SVM) [12] classifiers are used to classify input image to correct class.

By confirming literature, most of the research attempts performed two-class classification (i.e. normal or OA affected) whereas OA severity classification is five-class. For the right diagnosis and better recovery planning, more research attempts must be made to develop a five-class classification system. To improve classification accuracy deep learning-based models would be the more better option as it extracts deep feature those are much similar to pathological features to decide severity grades.

3 Methodology

The proposed approach of a Knee OA severity classification based on knee X-ray image is discussed in this section.

3.1 Data Acquisition

Patient-specific X-ray images are collected form local hospital named "Dr. Chidgupkar Hospital Pvt. Ltd, Solapur, India". In total hundred X-ray images are collected having various severity grades. With the help of expert pathologist, these images are categorized into respective grades. Table 1 provides the details of numbers patient-specific X-ray images are exists per knee OA severity level. In the formed database there are ten images contain normal knee (i.e. grade 0), ten images of grade 1 and 2, thirty images of grade 3 and forty images are grade 4.

Table 1. Summary of formed Knee OA database

OA severity grades	No. of images
0 (Normal)	10
1 (Doubtful)	10
2 (Mild)	10
3 (Moderate)	30
4 (Severe)	40
Total	100

The sample X-ray images of each grade are shown in Fig. 1. By observing database we can conclude that due to lack prevalence patients start examination at the later stage when the severity of pain is intolerable. So X-ray images having severed grade are more than moderate one.

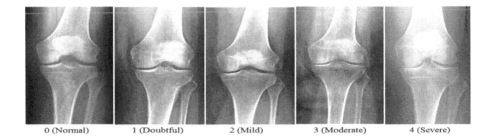

| 0 (Normal) | 1 (Doubtful) | 2 (Mild) | 3 (Moderate) | 4 (Severe) |

Fig. 1. Knee OA severity Grades

3.2 Preprocessing

The collected X-ray images contain some noise. The noise may be added due to several reasons like the technique of capturing the X-Ray, incorrect usage of the X-Ray machine, too much movement of the X-Ray machine due to some defect. Defect in the films and film pocket or inappropriate film processing method may add noise in the input image. Along with this, the patient's movement or the patient's clothes may cause noise in the X-ray images.

This situation leads to the application of promising image enhancement technique which will denoise the X-ray image and enhance the desired region (knee joint) nicely. To achieve this simple contrast stretching-based image enhancement technique is used. In addition to noise removal, the contrast stretching technique also increases the contrast of input image which results in high contrast image [16]. The input X-ray image and enhanced image is shown in Fig. 2 (a) and (b) respectively.

The intensity transformation function shown 1 in Equation is used to improve the dynamic range and to remove noise form input X-ray image [4].

$$S = T(R) = (S_{max} - S_{min})/(R_{max} - R_{min}) \times (R - R_{min}) + S_{min} \qquad (1)$$

where R and S is input and resultant image respectively. In the function T(R) $(S_{max} - S_{min})/(R_{max} - R_{min})$ is the slope of straight line. For 8 bit input X-ray image the values of S_{max} and S_{min} is set 255 and o respectively to cover entire dynamic range. The values of R_{max} and R_{min} are set 100 and 200 by conducting experiments.

3.3 ROI Localization

The OA majorly shows the morphological changes in lower end of femur and upper end tibia and reduction in synovial fluid (i.e. reduction in joint space

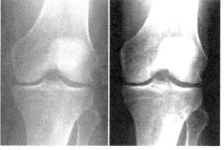

(a) input image (b) enhanced image

Fig. 2. Result of image enhancement method

distance). So instead of considering entire X-ray image for further processing only knee joint is considered. To extract knee joint area histogram modeling-based segmentation method is used.

Algorithm 1. RowSum(I)

Require: I // I must be in binary form.
Ensure: Histogram of row sum
1: [x,y] := size (I); Incr :=1; c := zeros(y,1); // array of row sum of each row
2: for i : = 1 to x do
3: for = j : = 1 to y do
4: if (I(x,y)==255) then
5: c(Incr, 1) := c(Incr, 1) +1;
6: end if
7: end for
8: Incr = Incr +1;
9: end for
10: plot (c);

At first Otsu threshoding method is used to find out image-specific global threshold value. That value is used convert input X-ray image to a binary image. After that row sum of each row is calculated by using Algorithm 1 and the histogram is plotted. The Fig. 3 (a) and (b) respectively shows the input image and the histogram of the calculated row sum. By analyzing histogram plots we can conclude that at the middle part of the histogram row sum is less because at that location a joint gap is present. At the middle part femur bone, ends and tibia bone will start by kipping some gap. That gap is nothing but the joint space between two bones having the presence of synovial fluid which has the least intensity value (zero) in the binary image. After getting a row with minimum row sum a bounding box is formed by considering fifteen rows in both the directions (Up and down) to extract the knee joint region. At last extracted

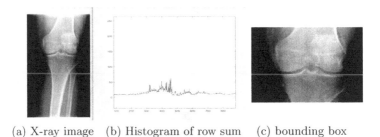

(a) X-ray image   (b) Histogram of row sum   (c) bounding box

Fig. 3. Histogram modeling-based segmentation

region is resized to 300×300 pixels. Figure 3 (c) shows the exacted ROI using Histogram modeling-based method.

3.4 Texture Feature Extraction Through DWT

Morphological changes of bones in the knee joint region, reduction in joint space distance, lack of synovial fluid and swelling are the primary pathological features used to diagnose and decide the severity of knee OA. These pathological features are well represented by textural descriptors. To solve the variability problem and to decide exact severity grade, we proposed the Discrete wavelet transform (DWT) based texture feature analysis system. DWT features give the best feature discrimination between five grades and results in high accuracy with SVM classifier.

Mostly DWT is used to perform multi-resolution analysis which is beneficial to detect patterns that are not visible in the raw data [1]. To extract DWT features at first image is converted to the frequency domain using fast Fourier transform. DWT results in an approximate image having desired features only moreover it helps to compare the distribution of energy in signals across frequency bands. In the experiment DWT parameters wavelet, order, and level are set to values Daubechies, 2 and 2 respectively and its horizontal detail which gives enough information is taken. Four DWT texture features: Entropy, mean, standard deviation, and variance are considered for the classification.

3.5 Classification

Knee OA severity grading based on X-ray images is multi-class (five fold) classification problem. Support vector machine (SVM) can be effectively used to solve multi-class classification problem. In our experiment we used SVM with quadratic kernel. The reason behind to choose quadratic kernel is knee OA severity classification problem is five fold so liner classifier is not enough to classify image to correct class.

SVM non-linearly maps given input feature space to hyper-dimensional feature space and finds a separating hyperplane which separates feature space two classes with maximum margin. This can achieved by applying suitable kernel function (Gaussian, Quadratic, and Cosine for instance). By this way SVM achieve better generalization performance without any intimate information about data [10]. SVM is binary classification, however it can be extended to multi-class classification using one-against rest approach.

4   Experimental Results

In addition to DWT features, gray level co-occurrence matrix (GLCM) features and some statistical features are considered for the comparison. Contrast, correlation, energy, entropy and homogeneity GLCM features and statistical features like skewness and kurtosis are extracted. To maintain uniformity same ROI localized image is considered to extract all types features (DWT, GLCM and statistical feature)s and five-class SVM classifier is applied to decide severity grade. By conforming results DWT-based texture features give promising results. The comparative results of both the methods are tabulated in the Table 2.

Table 2. Comparative classification results of DWT, GLCM and statistical features

Sr. No	Method	Result
1	DWT features	99%
2	GLCM features	94%
3	Statistical features	92%

5   Conclusion and Future Work

Knee Osteoarthritis is a common knee disease not only in older people but also in younger ones too due to overuse or adaption wrong lifestyle and food habits. Early detection severity grade will be always helpful to decide the optimal recovery plan. So there is a drastic need of automated classification system based on severity level. Here a DWT textural feature-based automated classification system is proposed. Initially, the system used contrast stretching based image enhancement technique to denoise the image and to remove unwanted artifacts. Later desired ROI (knee joint region) is extracted by developing a histogram modeling based segmentation technique. DWT textural features based classification system yield 99% accuracy on locally collected knee X-ray images. We compared the results of the classification obtained by DWT textural features with the GLCM and statistical features. The result of comparison says DWT features provide promising result.

National Institute of Health (NIH) made knee image database named Osteoarthritis Initiative (OAI) publically available to accelerate research. However, that database is not classified according to severity levels. In the future, we aim to classify the OAI dataset with the help of expert orthopedic surgeons.

In addition to this, we aim to design a deep neural network (DNN) based supervised classification system to produce robust results. We aim to conduct construct validity and adaptability of the developed classifier model in the healthcare community.

Acknowledgments. Authors thank the Ministry of Electronics and Information Technology (MeitY), New Delhi for granting Visvesvaraya Ph.D. fellowship through file no. PhD-MLA \ 4(34) \ 201-1 Dated: 05/11/2015.

The authors would like to thank, Chidgupkar hospital Pvt. Ltd., Solapur, India, for providing patient-specific X-ray images.

References

1. Akansu, A.N., Haddad, R.A., Caglar, H.: Perfect reconstruction binomial QMF-wavelet transform. In: Visual Communications and Image Processing 1990: Fifth in a Series, vol. 1360, pp. 609–618. International Society for Optics and Photonics (1990)
2. Antony, A.J.: Automatic quantification of radiographic knee osteoarthritis severity and associated diagnostic features using deep convolutional neural networks. Ph.D. thesis, Dublin City University (2018)
3. Bankman, I.: Handbook of Medical Image Processing and Analysis. Elsevier, Amsterdam (2008)
4. Gonzalez, R.C., Woods, R.E.: Digital image processing (2012)
5. Hegadi, R.S., Chavan, U.P., Navale, D.I.: Identification of knee osteoarthritis using texture analysis. In: Nagabhushan, P., Guru, D.S., Shekar, B.H., Kumar, Y.H.S. (eds.) Data Analytics and Learning. LNNS, vol. 43, pp. 121–129. Springer, Singapore (2019). https://doi.org/10.1007/978-981-13-2514-4_11
6. Hegadi, R.S., Navale, D.I., Pawar, T.D., Ruikar, D.D.: Multi feature-based classification of osteoarthritis in knee joint X-ray images, Chap 5. In: Medical imaging: Artificial Intelligence, Image Recognition, and Machine Learning Techniques, CRC Press. ISBN: 9780367139612 (2019)
7. Hegadi, R.S., Navale, D.I., Pawar, T.D., Ruikar, D.D.: Osteoarthritis detection and classification from knee X-ray images based on artificial neural network. In: Santosh, K.C., Hegadi, R.S. (eds.) RTIP2R 2018. CCIS, vol. 1036, pp. 97–105. Springer, Singapore (2019). https://doi.org/10.1007/978-981-13-9184-2_8
8. Hegadi, R.S., Pawar, T.D., Navale, D.I.: Classification of osteoarthritis-affected images based on edge curvature analysis. In: Nagabhushan, P., Guru, D.S., Shekar, B.H., Kumar, Y.H.S. (eds.) Data Analytics and Learning. LNNS, vol. 43, pp. 111–119. Springer, Singapore (2019). https://doi.org/10.1007/978-981-13-2514-4_10
9. Hirvasniemi, J., et al.: Quantification of differences in bone texture from plain radiographs in knees with and without osteoarthritis. Osteoarthritis Cartilage **22**(10), 1724–1731 (2014)
10. Kuo, B.C., Ho, H.H., Li, C.H., Hung, C.C., Taur, J.S.: A kernel-based feature selection method for SVM with RBF kernel for hyperspectral image classification. IEEE J. Sel. Top. Appl. Earth Observations Remote Sens. **7**(1), 317–326 (2013)
11. Marijnissen, A.C., et al.: Knee images digital analysis (KIDA): a novel method to quantify individual radiographic features of knee osteoarthritis in detail. Osteoarthritis Cartilage **16**(2), 234–243 (2008)

12. Navale, D.I., Hegadi, R.S., Mendgudli, N.: Block based texture analysis approach for knee osteoarthritis identification using SVM. In: 2015 IEEE International WIE Conference on Electrical and Computer Engineering (WIECON-ECE), pp. 338–341. IEEE (2015)
13. Ruikar, D.D., Hegadi, R.S., Santosh, K.C.: A systematic review on orthopedic simulators for psycho-motor skill and surgical procedure training. J. Med. Syst. **42**(9), 168 (2018)
14. Ruikar, D.D., Santosh, K.C., Hegadi, R.S.: Automated fractured bone segmentation and labeling from CT images. J. Med. Syst. (2019). https://doi.org/10.1007/s10916-019-1176-x
15. Ruikar, D.D., Santosh, K.C., Hegadi, R.S.: Segmentation and analysis of CT images for bone fracture detection and labeling, Chap 7. In: Medical imaging: Artificial Intelligence, Image Recognition, and Machine Learning Techniques, CRC Press. ISBN: 9780367139612 (2019)
16. Ruikar, D.D., Santosh, K.C., Hegadi, R.S.: Contrast stretching-based unwanted artifacts removal from CT images. In: Santosh, K.C., Hegadi, R.S. (eds.) RTIP2R 2018. CCIS, vol. 1036, pp. 3–14. Springer, Singapore (2019). https://doi.org/10.1007/978-981-13-9184-2_1
17. Ruikar, D.D., Sawat, D.D., Santosh, K.C., Hegadi, R.S.: 3D imaging in biomedical applications: a systematic review, Chap 8. In: Medical imaging: Artificial Intelligence, Image Recognition, and Machine Learning Techniques, CRC Press. ISBN: 9780367139612 (2019)
18. Scott, D., et al.: The links between joint damage and disability in rheumatoid arthritis. Rheumatology **39**(2), 122–132 (2000)
19. Seise, M., McKenna, S.J., Ricketts, I.W., Wigderowitz, C.A.: Segmenting tibia and femur from knee x-ray images. In: Proceedings of Medical Image Understanding and Analysis, pp. 103–106 (2005)
20. Vala, H.J., Baxi, A.: A review on OTSU image segmentation algorithm. Int. J. Adv. Res. Comput. Eng. Technol. (IJARCET) **2**(2), 387–389 (2013)
21. Zhao, F., Xie, X.: An overview of interactive medical image segmentation. Ann. BMVA **2013**(7), 1–22 (2013)
22. Zhu, Z., Li, J., Ruan, G., Wang, G., Huang, C., Ding, C.: Investigational drugs for the treatment of osteoarthritis, an update on recent developments. Expert Opin. Investig. Drugs **27**(11), 881–900 (2018)

A Deep Learning Based Visible Knife Detection System to Aid in Women Security

Himadri Mukherjee[1], Sahana Das[1], Ankita Dhar[1(✉)], Sk Md Obaidullah[2],
K.C. Santosh[3], Santanu Phadikar[4], and Kaushik Roy[1]

[1] Department of Computer Science, West Bengal State University, Kolkata, India
himadrim027@gmail.com, sahana.das73@gmail.com, ankita.ankie@gmail.com,
kaushik.mrg@gmail.com
[2] Department of Computer Science and Engineering, Aliah University, Kolkata, India
sk.obaidullah@gmail.com
[3] Department of Computer Science, The University of South Dakota,
Vermillion, SD, USA
santosh.kc@ieee.org
[4] Department of Computer Science and Engineering,
Maulana Abul Kalam Azad University of Technology, Kolkata, India
sphadikar@yahoo.com

Abstract. Criminal activities have increased largely over the past couple of years and the security of the commoners especially women have been hugely jeopardized. There has been multifarious cases of threats and assaults with weapons in the present days especially with knives which are one of the most common and readily available household items. Such cases have made CCTV cameras a common sighting in the neighbourhood. The prime idea behind their installation is surveillance. The footage from such cameras can serve as an extremely important source of evidence during investigation. However, such systems only make themselves useful as evidences of a crime and do not aid in prevention of a crime in progress. Standing in such times, making CCTV cameras intelligent can be a solution which can detect weapons and thereafter alert authorities. Here, a deep learning based system is presented which can automatically detect visible knives to alert authorities of a prospective crime and thereby aid in women security. The system has been tested on a freely available dataset [5] consisting of over 12000 frames and a highest accuracy of 96.11% has been obtained. We have also tested the performance of handcrafted feature-based framework with grey level co-occurrence matrix (GLCM) and our system produced better result.

Keywords: CCTV footage · Women security · Visible knife detection · Surveillance · CNN

K. C. Santosh—IEEE Senior Member.

K. C. Santosh and B. Gawali (Eds.): RTIP2R 2020, CCIS 1381, pp. 60–69, 2021.
https://doi.org/10.1007/978-981-16-0493-5_6

1   Introduction

Social security is a major concern in the present days, especially the security of women. The security of women and children has been highly jeopardized in the present times. They have been subject to several incidents of violence. One of such common incidents is knife stabbing. Knives are one of the most readily and easily available household items thereby making it an ideal choice of weapon for threats and assault. To tackle such situations, CCTV cameras have now become a common sighting. Many buildings now have CCTV surveillance systems and so do roads and avenues. However, such cameras can only provide feeds which can be used as a source of evidence during investigation of a crime. The CCTV systems can be made more functional and useful if they are equipped with automatic weapon and unusual activity detection mechanisms. In this way, the CCTV cameras can truly be made functional rather than being just dumb devices which can act during a crime or sense a prospective crime instead of being only helpful after a criminal incident.

Dee and Velastin [9] have discussed about different aspects related to automatic visual surveillance. They have talked about different algorithms which are suitable for surveillance in the thick of tracking, occlusion reasoning, scene modelling, behaviour analysis and alarming event detection. They have also provided pointers to several evaluation metrics for surveillance systems whose details can be found in [9]. Kmiec et al. [1] used active appearance model to detect presence of knives in images. They worked with a dataset having 40 positive and negative images and reported a highest accuracy of 92.5%. Glowacz [3] also made use of the same framework for knife detection. They further extended their system to aid in baggage scanning and obtained good results. Their experiments are detailed in [3]. Buckchash and Raman [2] used an object detection based technique for detection of knives in images. They worked on a dataset consisting of over 1500 images. They had experimented with several setups as well as features and reported accuracies in the higher 90s. Zywicki et al. [4] used haar cascade-based object detection for detecting knives. They performed 3 trainings with different sets of data and reported a positive detection ratio of 0.46. Grega et al. [5] attempted to identify presence of firearms or knives in images. They used a sliding window approach for extracting features which were fed to a SVM-based classifier for detection of the knives. They experimented on a collected dataset of CCTV recordings totaling to 12899 images. A highest accuracy of 91% was reported by them for knife detection with specificity and sensitivity values of 94.93% and 81.18% respectively.

Yuenyong et al. [6] presented a system to detect knives from infrared images. They worked on a generated dataset consisting of 8527 images (training). An accuracy of 97.91% was reported using a deep neural network for a test set having 527 images out of which 217 had knives. Maksimova et al. [7] proposed fuzzy classification-based technique for detecting knives from images. They experimented with a dataset consisting of over 12000 images which were characterized using MPEG-7 descriptors. They performed 5 different experiments and their best result had less than 15% miss labels. They also made use of probabilistic shell clustering for detection of knives [8]. They experimented with a dataset of 90 images out of which 30 had knives while the remaining did not. They obtained an accuracy of 97.7% using the proposed probabilistic technique. They had also experimented with a fuzzy-based technique and obtained an accuracy of 94%.

Here, a system is proposed towards detection of visible knives using deep learning to improve the functionality of the CCTV cameras and thereby aid in women security. The system has been tested on over 12000 frames whose details are presented in the subsequent paragraphs. The performance of handcrafted feature-based framework has also been analyzed using GLCM [13] which is a very popular textural feature. The experimental framework is illustrated in Figure 1.

Hereafter, the dataset is discussed in Sect. 2 followed by the proposed method in Sect. 3. The results are discussed in Sect. 4 and the conclusion is finally drawn in Sect. 5.

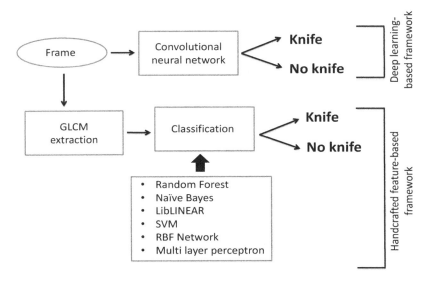

Fig. 1. Experimental framework.

2   Dataset

Data is an extremely important aspect of an experiment. Its quality is very important for the development of any system. It needs to be ensured that the data is not biased and it portrays real-world characteristics. This is an essential factor which aids in the development of any robust system. Experiments were performed in a dataset [5] consisting of 12899 images. It had 2 types of images, the former having knife (3599) and the later without knife (9340). The images had a dimension of 100 × 100 pixels. The dataset was put together with the aid of CCTV recordings. The images were cropped from the original frames. Images of such small dimension ensured real world condition where quality of recordings is often not very good and blurred at times as well. It is also often observed that the object of interest within a CCTV frame is very small. This was also upheld by the present dataset due to the small image dimensions. Some instances from the knife and non-knife classes are presented in Fig. 2.

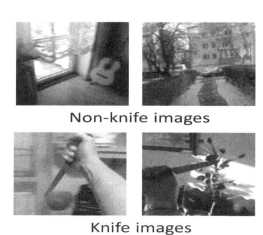

Fig. 2. Samples of knife and non-knife images.

3   Proposed Method

3.1   Deep Learning-Based Framework

Deep learning [10] has been one of the most widely explored avenues of machine learning in the recent past. It involves analyzing data in various layers. Each of such layers aid in extraction of different features. Here, we have adopted a deep learning-based approach with Convolutional Neural Networks (CNNs) [11]. The images were fed to a 256 dimensional 5×5 convolution layer having a ReLU activation function. The same is presented in Eq. (1). The output of this layer was subjected to a 3×3 max pooling layer followed by another 128 dimensional convolution of 3×3. The output from this layer was pooled and passed on to the last convolution layer of 64 dimension with 2×2 filters. Finally, the output from this layer was pooled and passed through 2 dense layers of 256 and 50 dimensions both having ReLU activation. The last dense layer (also termed as output layer) had a softmax activation (shown in Eq. (2)) and was of dimension 2 (the number of classes). The activation functions and the parameters of the CNN were chosen based on trial runs. The CNN architecture is illustrated in Fig. 3.

$$f(y) = max(0, y), \tag{1}$$

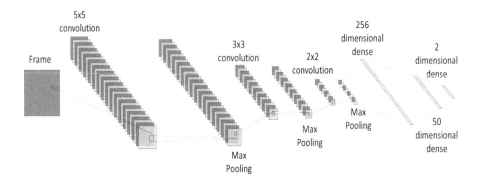

Fig. 3. The used CNN architecture.

where, y is the input to a neuron.

$$\sigma(a)_j = \frac{e^{a_j}}{\sum_{k=1}^{K} e^{a_k}},\tag{2}$$

where, a is denotes a K length input vector.

The CNN ultimately had 507992 parameters which were involved in the training phase. The distribution of parameters across different layers is tabulated in Table 1.

Table 1. Number of trainable parameters in each of the convolution and dense layers.

Layer	# Parameters
Convolution 1	19456
Convolution 2	295040
Convolution 3	32832
Dense 1	147712
Dense 2	12850
Dense 3	102

3.2 Handcrafted Feature-Based Framework

In this method, standard GLCM features [13] were extracted from the frames which were then fed to popular classifiers to distinguish knife and non knife images which are detailed in the subsequent paragraphs.

Gray Level Co-Occurrence Matrix (GLCM) [13]: It is a statistical technique of investigating texture by considering the spatial relationship of pixels. Here, the image texture is modelled by computing the frequency of disparate pixel pairs having a particular value and having a certain spatial relationship. Each entry of the GLCM[i,j] matrix consists of the frequency of occurrence of a particular intensity pair having a certain spatial relationship.

Random Forest [14]: It is an ensemble learning method for classification that works by constructing an assembly of decision trees. During training each tree provides a class prediction and the mode of the predictions is the classification for the given model. Ensemble prediction has higher accuracy than individual predictions. The prerequisite for random forest is that the individual trees should have a low correlation.

Naïve Bayes [15]: It is a probabilistic classifier that is based on Bayes theorem. It works on the assumption that the features are independent of each other. It is fast and easy to implement. It, however, is not suitable for real life problems as features are usually inter-dependent.

LibLINEAR [16]: It is an open source machine learning library that implements linear support vector machine (SVM) and logistic regression models trained using coordinate descent algorithm.

SVM [18]: It is a discriminative classifier that is defined by a separating hyperplane. When a set of training data is provided, SVM outputs an optimal hyperplane which is a line in two-dimensional plane that divides it in two parts where each class lie on the either side of it.

RBF Network [17]: It is a neural network based on radial basis function activation function. It classifies by computing the similarity of the input to examples from the training data. Each neuron of the RBF stores a prototype which is one of the examples from the training set. When a new input is given, each neuron calculates Euclidean distance between the input and the prototype. The input with the shortest Euclidean distance is classified accordingly.

Multi Layer Perceptron [19]: It is a feedforward artificial neural network. MLP consists of an input layer that receives the signal and an output layer that provides a decision. In between these two there may be numerous hidden layers that perform the computations. MLP is used in supervised learning problems. They are trained using a set of input-output pair and learn to model the correlation between them. During the training process the weights are adjusted to reduce the error in classification.

4 Result and Discussion

4.1 Our Result

In the current experiment, we had used n-fold cross validation for evaluation. This was used in order to ensure that every instance of the dataset was subjected to testing and training phases at least once. Initially, the value of n was fixed at 5 and the batch size was set to 100 instances. The value of training iteration was varied from 50 to 250 iterations (results in Table 2). It is noted that 200 iterations produced the best result.

Table 2. Performance of the proposed system for different training iterations.

Iteration	Accuracy(%)
50	95.09
100	95.49
150	94.90
200	**95.64**
250	95.54

Next, the batch size was varied for 200 training epochs and 5 fold cross validation. The batch size was varied from 25 instances to 200 instances whose obtained accuracies are shown in Table 3. It is noted that 50 instance batch size produced the best performance.

Table 3. Performance of the proposed system for different batch sizes with 200 training iterations.

Batch size	Accuracy (%)
25	89.19
50	**96.11**
100	95.64
150	95.68
200	96.03

Finally, the cross fold validation was also varied whose results are presented in Table 4 where it is observed that the overall accuracy did not improve on varying the number of folds.

The confusion matrix for the best result (5 fold) is presented in Table 5. It is observed from Table 5 that the misclassification for knife class was higher than that of the non-knife class. One of the most probable reasons is the image size and quality. The images were only of 100 * 100 pixels dimension which made it difficult to distinguish knives. Moreover, in some images there were shadows along with knives which also confused the system. It was also observed in some images that the knife was not very prominent mostly due to the angle of the image. In some instances, the knife was somewhat similar in colour to that of the background.

Table 4. Performance of the proposed system for different folds of cross validation with 50 instance batch size and 200 training iterations.

Folds	Accuracy (%)
2	77.64
3	89.00
4	93.80
5	**96.11**
10	95.16
15	88.06
20	80.33

Table 5. Confusion matrix for the best result.

	Non-knife	Knife
Non-knife	9173	167
Knife	335	3224

4.2 Comparative Study

The performance of our system was compared with an available work [5] which reported an accuracy of 91% using edge histogram-based features. We obtained an accuracy of 96.11%. In the case of the handcrafted feature-based framework, the GLCM feature set was fed to the different classifiers [12]. The obtained results for these classifiers is tabulated in Table 6 where it is noted that the performance of the proposed deep learning-based system was better.

Table 6. Performance of different classifiers with GLCM (Handcrafted feature-based framework).

Classifier	Accuracy (%)
Random forest	75.38
Naïve bayes	44.60
LibLINEAR	72.38
RBF Network	72.41
SVM	72.41
Multi layer perceptron	73.67
Our approach	**96.11**

5   Conclusion

Here, a system is presented towards automatic detection of visible knives. Such a system can empower CCTV cameras to be more functional and aid in social security, especially for those of women and children. The system works with a deep learning-based framework and was tested on a freely available dataset [5] of over 12000 frames and encouraging results have been obtained. In future, the data will be further processed to remove background details for better identification of knives. Further, real time data processing will also be incorporated in the system for integration with CCTVs and deployment in real-World environments. Other deep learning architectures and handcrafted features will also be experimented with in the future to obtain further performance improvements. We also plan to experiment with a larger dataset and work with partially visible weapons.

References

1. Kmieć, M., Głowacz, A., Dziech, A.: Towards robust visual knife detection in images: active appearance models initialised with shape-specific interest points. In: Dziech, A., Czyżewski, A. (eds.) MCSS 2012. CCIS, vol. 287, pp. 148–158. Springer, Heidelberg (2012). https://doi.org/10.1007/978-3-642-30721-8_15
2. Buckchash, H., Raman, B.: A robust object detector: application to detection of visual knives. In: 2017 IEEE International Conference on Multimedia & Expo Workshops (ICMEW), pp. 633–638. IEEE (2017)
3. Glowacz, A., Kmieć, M., Dziech, A.: Visual detection of knives in security applications using Active Appearance Models. Multimedia Tools Appl. **74**(12), 4253–4267 (2015). https://doi.org/10.1007/s11042-013-1537-2
4. Żywicki, M., Matiolański, A., Orzechowski, T.M., Dziech, A.: Knife detection as a subset of object detection approach based on Haar cascades. In: Proceedings of 11th International Conference on Pattern Recognition and Information Processing, pp. 139–142 (2011)
5. Grega, M., Matiolański, A., Guzik, P., Leszczuk, M.: Automated detection of firearms and knives in a CCTV image. Sensors **16**(1), 47 (2016)
6. Yuenyong, S., Hnoohom, N., Wongpatikaseree, K.: Automatic detection of knives in infrared images. In: 2018 International ECTI Northern Section Conference on Electrical, Electronics, Computer and Telecommunications Engineering (ECTI-NCON), pp. 65–68. IEEE (2018)
7. Maksimova, A., Matiolański, A., Wassermann, J.: Fuzzy classification method for knife detection problem. In: Dziech, A., Czyżewski, A. (eds.) MCSS 2014. CCIS, vol. 429, pp. 159–169. Springer, Cham (2014). https://doi.org/10.1007/978-3-319-07569-3_13
8. Maksimova, A.: Knife detection scheme based on possibilistic shell clustering. In: Dziech, A., Czyżewski, A. (eds.) MCSS 2013. CCIS, vol. 368, pp. 144–152. Springer, Heidelberg (2013). https://doi.org/10.1007/978-3-642-38559-9_13
9. Dee, H.M., Velastin, S.A.: How close are we to solving the problem of automated visual surveillance? Mach. Vis. Appl. **19**(5–6), 329–343 (2008). https://doi.org/10.1007/s00138-007-0077-z
10. LeCun, Y., Bengio, Y., Hinton, G.: Deep learning. Nature **521**(7553), 436 (2015)

11. Krizhevsky, A., Sutskever, I., Hinton, G.E.: ImageNet classification with deep convolutional neural networks. In: Advances in Neural Information Processing Systems, pp. 1097–1105 (2012)
12. Hall, M., Frank, E., Holmes, G., Pfahringer, B., Reutemann, P., Witten, I.H.: The WEKA data mining software: an update. ACM SIGKDD Explor. Newslett. **11**(1), 10–18 (2009)
13. Mohanaiah, P., Sathyanarayana, P., GuruKumar, L.: Image texture feature extraction using GLCM approach. Int. J. Sci. Res. Publ. **3**(5), 1 (2013). Neural networks. In: Advances in Neural Information Processing Systems, pp. 1097–1105
14. Breiman, L.: Random forests. Mach. Learn. **45**(1), 5–32 (2001). https://doi.org/10.1023/A:1010933404324
15. Rish, I.: An empirical study of the naive Bayes classifier. In: IJCAI 2001 Workshop on Empirical Methods in Artificial Intelligence, vol. 3, no. 22, pp. 41–46 (2001)
16. Fan, R.E., Chang, K.W., Hsieh, C.J., Wang, X.R., Lin, C.J.: LIBLINEAR: a library for large linear classification. J. Mach. Learn. Res. **9**(Aug), 1871–1874 (2008)
17. Yun, Z., Quan, Z., Caixin, S., Shaolan, L., Yuming, L., Yang, S.: RBF neural network and ANFIS-based short-term load forecasting approach in real-time price environment. IEEE Trans. Power Syst. **23**(3), 853–858 (2008)
18. Chang, C.C., Lin, C.J.: LIBSVM: a library for support vector machines. ACM Trans. Intell. Syst. Technol. (TIST) **2**(3), 27 (2011)
19. Gardner, M.W., Dorling, S.R.: Artificial neural networks (the multilayer perceptron)-a review of applications in the atmospheric sciences. Atmos. Environ. **32**(14–15), 2627–2636 (1998)

Computerized Medical Disease Identification Using Respiratory Sound Based on MFCC and Neural Network

Santosh Gaikwad[1]([✉]) [ID], Mohammad Basil[2] [ID], and Bharti Gawali[3] [ID]

[1] Department of Computer Science, Model Degree College, Ghansawangi, Jalna, M.S., India
santosh.gaikwadcsit@gmail.com
[2] Department of Computer Science and Engineering, AIMaari University College, Mosul, Iraq
dr.mohd.alnaqeeb@gmail.com
[3] Department of Computer Science and Information Technology, Dr. Babasaheb Ambedkar Marathwada University, Aurangabad, M.S., India
bharti_rokade@yahoo.co.in

Abstract. For the medical domain, computer assumes a significant job in computerization and determination of the disorder. The stethoscope is an eminent and widely available traditional diagnostic instrument for the medical professionals. The computer system is used in medical science for collection and analysis of large amounts of massive data and concern accurate decision making. The respiratory sound database has been available from research community. However, full utilization of available recording device or database, there is a need to design and development of the respiratory disease identification. This paper explained the respiratory data creation and application of this data over the respiratory disorder identification. The database is collects with the help of local government hospital. The data is recorded with directional stethoscope with 3.5 jack based microphone connected with laptop or computer. The database includes 1000 recording of 7.5 h. The data is collected from 50 patients. The Mel Frequency Cepstral Coefficient technique is applied over the database for feature extraction. The pitch, energy and time are the dominant features for the disorder identification. The neural network has been used for the classification of the disorder identification. The experiment has been achieved accuracy of 91% over the two class classification. The precision of the experiment is 88% whereas sensitivity is 87%. The 9% error rate has been shows the experimental system. From the experimental analysis the author recommended the MFCC and neural network are the strong and dynamic approach in respiratory dieses determination.

Keywords: Respiratory · Stethoscope · MFCC · Neural network · Precision · Recall · Sensitivity · Specificity

1 Introduction

In the medical science the most of the respiratory disorder impacted as 15% for overall world population. It is identified that the approximately 230 million patients undergo

© Springer Nature Singapore Pte Ltd. 2021
K. C. Santosh and B. Gawali (Eds.): RTIP2R 2020, CCIS 1381, pp. 70–82, 2021.
https://doi.org/10.1007/978-981-16-0493-5_7

from respiratory disease which causes excess of 300000 deaths per year [1]. In the traditional system doctor predicts the diseases on the basis of respiration sound occurred in the stethoscope. The respiratory diseases are classified into the asthma, COPD, and pneumonia amongst others [2]. For the medical database more than 230 million cases reported for respiratory diseases universally [3]. The researcher predicted that the respiratory diseases are third most leading cause for human death over the world in 2030 just nearest with heart diseases [3]. The survey of National Health Service, UK describes the respiratory diseases such as GBP, asthenia and COPD attain more than 3 million patients for their diseases [4]. The respiratory diseases are generally caused by indication of wheezing, breathlessness, and coughing. The symptom of the respiratory diseases affect the patients day to day life and has inferior quality when patients performing physical task [1, 3]. The intimation of this cause is effective at the night and morning environment which causes the sleeping disruptions in patients [5]. It is challenging task to identify this symptoms at a premature stage for improving the patients' health to reduce the ratio of death [6]. The monitoring and identification of symptom for respiratory diseases is the challenging task for the doctor due to available resources and traditional techniques [7, 8]. All respiratory diseases caused due to narrowing the airways in the respiration [9]. This diseases caused in at home, public or workplace for the people so, it is not possible to avail the medical professional everywhere and their resources. The amount of work related to automated respiratory diseases identification has not yet reached critical level to be used in real time use for anywhere respiratory diseases identification.

The excitation respiratory sound is spectrally formed by a verbal region corresponding strain. The effect of this process is become output as a respiratory sound which is listenable using electronic stethoscope. The process of listen the internal sound of human body using a stethoscope is known as auscultation. The auscultation is the valuable tool for the early finding of respiratory disorder and abnormality. Using this techniques the medical professional has listen the normal sound of respiration and absent sound of the cycle and predict the abnormalities [10, 11]. The respiratory sound based qualitative assessment of the people with their diagnosis is used by this non-invasive stethoscope method [12]. The respiration diseases have been developed by the researcher from which data has been collected from the microphone in both inspiration and in expiration condition. The recognition of abnormal sound has been developed [13]. The MFCC techniques are the strong and dynamic feature extraction techniques in speech processing [14]. The MFCC is the combination of set of coefficients which represent the spectrum of the signal and it is calculated using logarithm of discrete cosine transform to the signal spectrum [15]. The MFCC is mostly used in the dominant feature extraction for the normal and respiratory diseases sound identification [15, 16]. In the current era of speech recognition the researcher extract the 13 universal features with combination of 26 and 39 feature derivatives. For the respiratory sound identification the MFCC with 13 feature vector is good choice [15]. For the experiment the researcher stand towards the hybrid combination of MFCC with LPC and Wavelet approach for respiratory diseases identification [17, 18]. The neural network is proved as robust and dynamic classification measure for speech and sound identification [19]. The traditional method for respiratory diseases identification need the computer based automation. This research is explained the respiratory sound database creation and application of this database for automated diseases identification. This article is structured as follows: Sect. 2 depicts to previous research work. Section 3 is explaining respiratory data creation. Section 4 explained the experimental outcome and its observation. Section 5 illustrated the conclusion followed by references.

2   Previous Research Work

This section explained the summarization of various approaches for analysis used by the researcher such as feature and method towards the respiratory diseases identification. The outcome of current studies shows a high level of concurrence in traditional non-automatic identification. Current literature of these studies describes that need of automated promising solution for relevant diseases identification or classification using the current robust techniques. Detail related experiment with number of patients, types of disease, methodology and accuracy is explained in Table 1.

Table 1. Summarized previous related work in respiratory diseases domain.

Year and reference	Type of disease classified	Feature ex-traction method	Classification method	Accuracy
2000 [20]	Asthmatic patients	Spectral analysis	k-NN	60%–90%
2000 [21]	Normal or abnormal	Fourier Transformation	MLP	73%
2016 [22]	Normal and disorder	Spectral features	SVM, LRM	88.9%
2010 [23]	Asthma	MFCC	GMM	52.5%
2009 [24]	Normal and Disorder	MFCC	HMM	93.2%
2007 [25]	Diseases identification	MFCC	GMM	94.9%
2004 [26]	Disorder identification	Energy	Threshold	82.7%
2000 [26]	Normal and disorder	Average power spectrum	BPNN	59%
2007 [28]	Not Mention	Peak selection	Threshold	95.5%
2003 [29]	Wheeze	MFCC, FFT	VQ	75.80%
2008 [30]	Normal crackles	GMM	VQ	95.1%
2008 [31]	Lung sounds	AR Model	KNN	96%
2011 [32]	Crackles WT	DWT	SVM	97.20%
2015 [33]	Pathological	MFCC LFCC	KNN	96%
2015 [34]	Wheeze	MFCC	KNN	93.2%

From the above related work observation, it is understand that there is scope to develop own respiratory sound database and develop an automatic system over it using MFCC and neural network.

3 Database Creation

The respiratory sound brings the important decision for patient disease identification. Sound acts as an indicator for the health and respiratory disorder identification using the sound feature. The respiratory sound is emitted output from lung tissues and position of secretion within the lung when person breathes with the help of air moment [35]. This database is recorder using the digital stethoscopes and recording software system. For the database recording the software interface has been developed which stored the basic information of the patient and its related recording. The collected database is used for the respiratory disorder identification and its use for the research in concern domain. There are two types of electronic stethoscopes available in commercial market named as Littman 2100 and Think labs electronic stethoscope [36, 37]. For the sound recording in Littman 2100 electronic stethoscope required proprietary software developed by it [38]. For this data recording the directional stethoscope is connected with 3.5 jack microphone towards connection for laptop or desktop. The data is collected with the help of local government hospital [39, 40]. This data is collected in office and hospital environment so, data has a noisy effect. The graphical representation of data recording software interface is shown in Fig. 1. This database includes 1000 annotated recording 0f 10 s to 40 s length variation. The database is collected from 50 patients. The database has 7.5 h recording with 4560 respiratory cycles. The database is collected from teenager and elder patients.

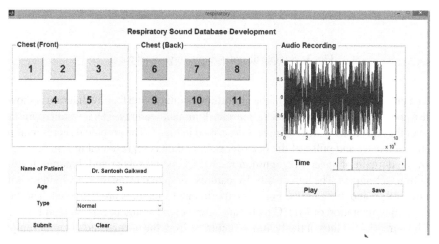

Fig. 1. Graphical representation of the software interface for respiratory data

4   Experiment Analysis

For this experiment the respiratory sound data is collected from 50 patients. The database is collected from normal respiratory sound and disorder respiratory sound. The sample normal respiratory sound representation is shown in Fig. 2. The graphical representation of disorder respiratory sound data sample is described in Fig. 3.

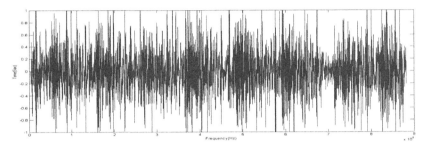

Fig. 2. Sample normal respiratory sound data

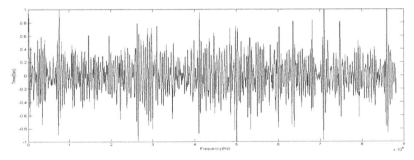

Fig. 3. Sample disorder respiratory sound data

The graphical representation of the multidimensional sound feature diagram is shown in the Fig. 4. This figure represents the various dominant features. The formant representation of the recorded disorder sample is described in Fig. 5. The graphical representation of pitch, intensity and pulse of the disorder sample data is represented in Fig. 6.

In the domain of speech recognition the MFCC is the robust and dynamic feature extraction techniques. The universally 13 features computed using these techniques with variation of 26 and 39 features. For this experiment 13 features are extracted. The step by step implementation of MFCC techniques for sample sound is explained in Fig. 7.

The numerical values of the features calculated from the normal and disorder sample for respiratory sound are shown in the Table 2 and 3 respectively.

The neural network is the strong and dynamic grouping approach in speech processing. In this investigation the neural system is applied over the two classes based classification such as normal and disorder. The classification involves the training the 70%, 15% testing and 15% validation approach. The exhibition of the neural system is

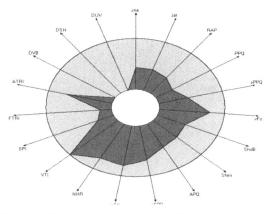

Fig. 4. The multidimensional sound feature diagram for disorder sample

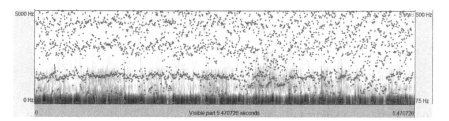

Fig. 5. Formant representation of disorder sample

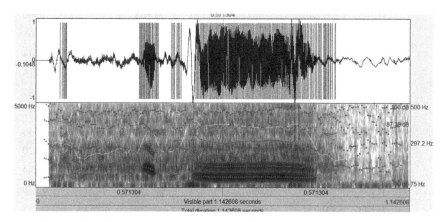

Fig. 6. The pitch, intensity and pulse of the disorder respiratory sound sample

tried over the hidden layer. From the perception the variety of the presentation relies upon the hidden layer. The framework is tried over the 10, 15 and 20 concealed layer approach. The graphical demonstration of the neural system training such preparing state is represented in the Fig. 8.

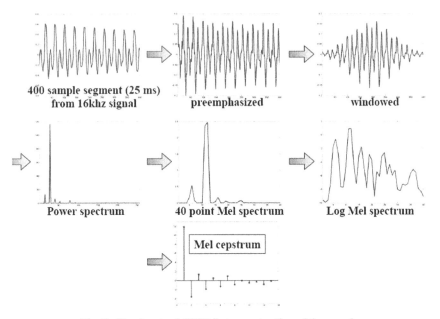

Fig. 7. Step by step MFCC feature extraction of the sample

Table 2. Values of MFCC 13 coefficients for first seven frame of normal sample sound

Feature	F1	F2	F3	F4	F5	F6	F7
1	0.648	0.540	0.554	0.569	0.565	0.532	0.590
2	0.480	0.421	0.434	0.466	0.442	0.417	0.465
3	0.345	0.321	0.351	0.374	0.343	0.328	0.344
4	0.184	0.211	0.245	0.266	0.236	0.219	0.215
5	0.030	0.067	0.111	0.129	0.098	0.074	0.087
6	−0.079	−0.069	−0.019	0.004	−0.034	−0.060	−0.037
7	−0.160	−0.171	−0.108	−0.099	−0.143	−0.167	−0.146
8	−0.222	−0.253	−0.193	−0.185	−0.227	−0.244	−0.220
9	−0.220	−0.284	−0.245	−0.231	−0.258	−0.284	−0.244
10	−0.188	−0.284	−0.263	−0.239	−0.256	−0.283	−0.230
11	−0.143	−0.257	−0.252	−0.215	−0.230	−0.259	−0.198
12	−0.091	−0.199	−0.212	−0.159	−0.182	−0.215	−0.167
13	−0.042	−0.116	−0.151	−0.083	−0.110	−0.130	−0.113

The performance of ROC curve for the neural network experiment is shown in Fig. 9. The graphical representation of confusion matrix and error histogram for the concern experiments has been shown in Fig. 10 and 11 respectively.

Table 3. Values of MFCC 13 coefficients for first seven frame of disorder sample sound

Feature	F1	F2	F3	F4	F5	F6	F7
1	0.610	0.555	0.565	0.527	0.545	0.569	0.584
2	0.538	0.493	0.472	0.461	0.473	0.457	0.477
3	0.424	0.408	0.346	0.359	0.370	0.315	0.274
4	0.299	0.312	0.212	0.225	0.238	0.171	0.048
5	0.174	0.196	0.067	0.081	0.094	0.027	−0.144
6	0.057	0.070	−0.079	−0.056	−0.046	−0.106	−0.261
7	−0.034	−0.031	−0.175	−0.164	−0.154	−0.209	−0.299
8	−0.087	−0.106	−0.230	−0.230	−0.226	−0.269	−0.282
9	−0.105	−0.158	−0.250	−0.260	−0.253	−0.285	−0.234
10	−0.097	−0.182	−0.237	−0.259	−0.244	−0.260	−0.170
11	−0.070	−0.180	−0.202	−0.233	−0.209	−0.205	−0.100
12	−0.035	−0.157	−0.159	−0.187	−0.156	−0.131	−0.027
13	0.000	−0.117	−0.118	−0.131	−0.093	−0.050	0.046

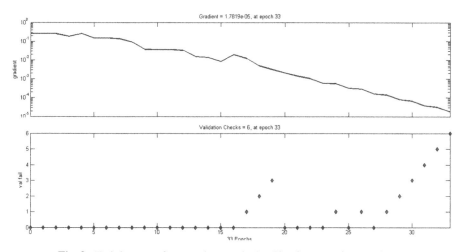

Fig. 8. Training state for neural network classification over the two classes

The performance of the neural network is tested with the help of statistical measure such as precision, recall, sensitivity and specificity [41, 42]. The outcome for this experiment is extracted for 10, 15 and 20 hidden layers. The classification has been done on 100 normal and 100 disorder sound data. The detail outcome for the training experiment is explained at Table 4. Table 5 explained the detailed performance of neural network for testing state or mode.

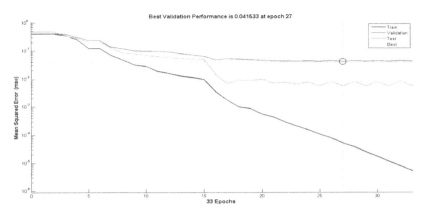

Fig. 9. The performance ROC curve for the neural network

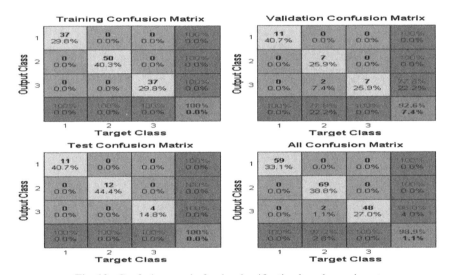

Fig. 10. Confusion matrix for the classification based experiment

From the above experiment the neural network is tested over the training and testing state. In the testing state the neural network achieved the 91% accuracy for normal and disorder two class classifications.

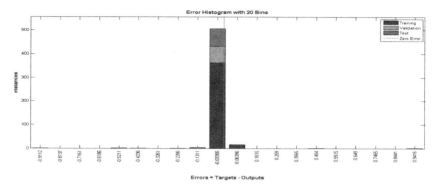

Fig. 11. Error histogram of classification based experiment

Table 4. Performance of neural network training state

Hidden layer	Training accuracy	Training precision	Training recall	Training sensitivity	Training specificity
10	88%	90%	89%	89%	95%
15	92%	94%	87%	87%	87%
20	93%	93%	88%	89%	88%
Average	91%	92%	88%	88%	90%

Table 5. Testing state performance of neural network

Hidden layer	Test accuracy	Test precision	Test recall	Test sensititivity	Test specificity
10	87%	86%	86%	86%	86%
15	93%	89%	87%	87%	84%
20	92%	89%	87%	87%	85%
Average	91%	88%	87%	87%	85%

5 Conclusion

Respiratory sound database creation is the challenging task over the medical science. This experiment has been done for designing and development of e respiratory sound . database for the 1000 annotated recording with 50 patients. The database is of 7.5 h. The database is recorded using the direction stethoscope with 3.5 jack microphone in hospital environment. The feature extraction of the collected sound sample has been done using the MFCC techniques. For Each sound 13 universal feature is extracted. For normal and disorder two class classification neural network techniques has been used. The neural network techniques tested for 10, 15 and 20 hidden layer approach. The outcome of the experimental analysis is deliberated using statistical measure such as precision, recall, sensitivity and specificity. The system proves the 90% for training and 91% for testing accuracy. The author suggested that the MFCC and neural system has strong and dynamic methods for the respiratory disorder identification.

References

1. World Health Organization. Asthma Fact Sheet (2019). https://www.who.int/mediacentre/fac tsheets/fs307/en/
2. Pramono, R.X.A., Bowyer, S., Rodriguez-Villegas, E.: Automatic adventitious respiratory sound analysis: a systematic review. PLoS ONE **12**(5), e0177926 (2017). https://doi.org/10. 1371/journal.pone.0177926
3. World Health Organization. Chronic obstructive pulmonary disease (COPD) Fact Sheet (2017). https://www.who.int/mediacentre/factsheets/fs315/en
4. World Health Organization. World Health Statistics (2008). https://www.who.int/gho/public ations/world_health_statistics/
5. Trueman, D., Woodcock, V., Hancock, E.: Estimating the economic burden of respiratory illness in the UK (2017). https://www.blf.org.uk/what-we-do/our-research/economic-burden
6. Balachandran, J., Teodorescu, M.: Sleep Problems in Asthma and COPD. Am. J. Respire. Crit. Care Med. (2013)
7. National Health Service England. Overview of potential to reduce lives lost from Chronic Obstructive Pulmonary Disease (COPD) (2014). https://www.england.nhs.uk/wp-content/upl oads/2014/02/rm-fs-6.pdf
8. British Thoracic Society and Scottish Intercollegiate Guidelines Network. British guideline for the management of asthma; a national clinical guideline (SIGN 153) (2016). https://www. brit-thoracic.org.uk/guidelines-and-quality-standards/asthma-guideline/
9. British Thoracic Society and Scottish Intercollegiate Guidelines Network. Chronic obstructive pulmonary disease in over 16s: diagnosis and management (2010). https://www.nice.org.uk/ guidance/CG101
10. Pasterkamp, H., Kraman, S.S., Wodicka, G.R.: Respiratory sounds: advances beyond the stethoscope. Am. J. Respir. Crit. Care Med. **156**(3), 974–987 1997. https://doi.org/10.1164/ ajrccm.156.3.9701115
11. Loudon, R.G.: The lung exam. Clin. Chest Med. **8**(2), 265–272 (1987)
12. Reichert, S., Gass, R., Brandt, C., Andres, E.: Analysis of respiratory sounds state of the art. Clin. Med. **2**, 45–58 (2008)
13. Deng, L.: Three classes of deep learning architectures and their applications: a tutorial survey. APSIPA Trans. Signal Inf. Process. (2012)
14. Murphy, R.L.H.: U.S. Patent 6, 139, 505, 31 October 2000
15. Gawali, B.W., Gaikwad, S., Yannawar, P., Mehrotra, S.C.: Marathi isolated word recognition system using MFCC and DTW features. In: Proceedings of International Conference on Advances in Computer Science (2010)
16. Orjuela-Cañón, A.D., Gómez-Cajas, D.F., Jiménez-Moreno, R.: Artificial neural networks for acoustic lung signals classification. In: Bayro-Corrochano, E., Hancock, E. (eds.) Progress in Pattern Recognition, Image Analysis, Computer Vision, and Applications. Iberoamerican Congress on Pattern Recognition. LNCS, pp. 214–221. Springer, Cham (2014). https://doi. org/10.1007/978-3-319-12568-8_27
17. Bahoura, M.: Pattern recognition methods applied to respiratory sounds classification into normal and wheeze classes. Comput. Biol. Med. **39**(9), 824–843 (2009). https://doi.org/10. 1016/j.compbiomed.2009.06

18. Sankur, B., Kahya, Y.P., Güler, E.Ç., Engin, T.: Comparison of AR-based algorithms for respiratory sounds classification. Comput. Biol. Med. **24**(1), 67–76 (1994). https://doi.org/10.1016/0010-4825(94)90038-8

19. Kandaswamy, A., Kumar, C.S., Ramanathan, R.P., Jayaraman, S., Malmurugan, N.: Neural classification of lung sounds using wavelet coefficients. Comput. Biol. Med. **34**(6), 523–537 (2004). https://doi.org/10.1016/S0010-4825(03)00092-1

20. Kandaswamy, A., Sathish Kumar, C., Ramanathan, R.P., Jayaraman, S., Malmurugan, N.: Neural classification of lung sounds using wavelet coefficients. Comput. Biol. Med. **34**(6), 523–537 (2004). https://doi.org/10.1016/S0010-4825(03)00092-1(2004)

21. Oud, M., Dooijes, E.H., van der Zee, J.S.: Asthmatic airways obstruction assessment based on detailed analysis of respiratory sound spectra. IEEE Trans. Biomed. Eng. **47**, 1450–1455 (2000)

22. Waitman, L.R., Clarkson, K.P., Barwise, J.A., King, P.H.: Representation and classification of breath sounds recorded in an intensive care setting using neural networks. J. Clin. Monit. Comput. **16**(2), 95–105 (2000)

23. Bokov, P., Mahut, B., Flaud, P., Delclaux, C.: Wheezing recognition algorithm using recordings of respiratory sounds at the mouth in a pediatric population. Comput. Biol. Med. **70**, 50 (2016)

24. Mayorga, P., Druzgalski, C., Morelos, R., Gonzalez, O., Vidales, J.: Acoustics based assessment of respiratory diseases using GMM classification. In: 32nd Annual International Conference of the IEEE EMBS. IEEE 2010. p. 6312–6316 (2010)

25. Matsunaga, S., Yamauchi, K., Yamashita, M., Miyahara, S.: Classification between normal and abnormal respiratory sounds based on maximum likelihood approach. In: IEEE International Conference on Acoustics, Speech and Signal Processing. IEEE 2009. pp. 517–520 (2009)

26. Chien, J.C., Wu, H.D., Chong, F.C., Li, C.I.: Wheeze detection using cepstral analysis in gaussian mixture models. In: 29th Annual International Conference of the IEEE EMBS. IEEE 2007, pp. 3168–3171 (2007)

27. Homs-Corbera, A., Fiz, J.A., Morera, J., Jané, R.: Time-frequency detection and analysis of wheezes during forced exhalation. IEEE Trans. Biomed. Eng. **51**, 182–186 (2004)

28. Waitman, L.R., Clarkson, K.P., Barwise, J.A., King, P.H.: Representation and classification of breath sounds recorded in an intensive care setting using neural networks. J. Clin. Monitor. Comp. **16**(2), 95–105 (2000)

29. Taplidou, S.A., Hadjileontiadis, L.J.: Wheeze detection based on time-frequency analysis of breath sounds. Comput. Biol. Med. **37**(8), 1073–1083 (2007)

30. Bahoura, M., Pelletier, C.: New parameters for respiratory sound classification. In: Canadian Conference on Electrical and Computer Engineering, IEEE CCECE, IEEE, vol. 3, pp. 1457–1460 (2003)

31. Lu, X., Bahoura, M.: An integrated automated system for crackles extraction and classification. Biomed. Signal. Process. Contr. **3**, 244–254 (2008)

32. Alsmadi, S., Kahya, Y.P.: Design of a DSP-based instrument for real-time classification of pulmonary sounds. Comput. Biol. Med. **38**, 53–61 (2008)

33. Serbes, G., Sakar, C.O., Kahya, Y.P., Aydin, N.: Feature extraction using time–frequency/scale analysis and ensemble of feature sets for crackle detection. In: 33rd Annual International Conference of the IEEE EMBS, Boston, Massachusetts, USA, pp. 3314–3317 (2011)

34. Aras, S., Gangal, A., Bülbül, Y.: Lung sounds classification of healthy and pathologic lung sounds recorded with electronic auscultation. In: 2015 23th Signal Processing and Communications Applications Conference (SIU), pp. 252–255. IEEE (2015)

35. Chen, C.H., Huang, W.T., Tan, T.H., Chang, C.C., Chang, Y.J.: Using K-nearest neighbor classification to diagnose abnormal lung sounds. Sensors **15**, 13132–13158 (2015)

36. Rocha, B.M., et al.: A respiratory sound database for the development of automated classification. In: Maglaveras, N., Chouvarda, I., de Carvalho, P. (eds.) Precision Medicine Powered by pHealth and Connected Health, vol. 66, pp. 33–37. Springer, Singapore. https://doi.org/10.1007/978-981-10-7419-6_6

37. Littmann, Digital stethoscope. https://www.littmann.com/wps/portal/3M/enUS/3M-Littmann/stethoscope/stethoscope-catalog/catalog/~/3M-Littmann-Electronic-Stethoscope-Model-3200-Black-Tube-27-inch-3200BK27?N=5932256+4294958300&rt=d. Accessed 26 May 2019

38. Song, H.A., Lee, S.Y.: Hierarchical representation using NMF. In: Lee, M., Hirose, A., Hou, Z.G., Kil. R.M. (eds.) Neural Information Processing, pp. 466–473 Springer, Heidelberg (2013). https://doi.org/10.1007/978-3-642-42054-2_58

39. Ruikar, D.D., Santosh, K.C., Hegadi, R.S.: Automated fractured bone segmentation and labeling from CT images. J. Med. Syst. **43**(3), 1–13 (2019)

40. Ruikar, D.D., Santosh, K.C., Hegadi, R.S.: Segmentation and analysis of CT images for bone fracture detection and labeling. In: Medical Imaging: Artificial Intelligence, Image Recognition, and Machine Learning Techniques, p. 131 (2019)

41. Hegadi, R.S., Navale, D.I., Pawar, T.D., Ruikar, D.D.: Osteoarthritis detection and classification from knee X-ray images based on artificial neural network. In: Santosh, K.C., Hegadi, R.S. (eds.) RTIP2R 2018. CCIS, vol. 1036, pp. 97–105. Springer, Singapore (2019). https://doi.org/10.1007/978-981-13-9184-2_8

42. Santosh, K.C., Antani, S., Guru, D.S., Dey, N. (eds.): Medical Imaging: Artificial Intelligence, Image Recognition, and Machine Learning Techniques. CRC Press, Boca Raton (2019)

Keywords Recognition from EEG Signals on Smart Devices a Novel Approach

Sushil Pandharinath Bedre[1(✉)], Subodh Kumar Jha[1], Prashant Borde[1], Chandrakant Patil[1], Bharati Gawali[2], and Pravin Yannawar[1]

[1] Vision and Intelligent System Laboratory, Dr. Babasaheb Ambedkar University, Aurangabad 431004, Maharashtra, India
sushil.bedre@gmail.com, borde.prashantkumar@gmail.com, plyannawar.csit@bamu.ac.in
[2] SCMRL Lab, Dr. Babasaheb Ambedkar University, Aurangabad 431004, Maharashtra, India
drbhartirokade@gmail.com

Abstract. The Advancement of communication system has given us the freedom to think beyond traditional communication system and stage is set for thought oriented communication system. There are thousands of thoughts generated and vanished in a timeframe but out of these some prominent thoughts persist and we proceed with the same in our day to day activities. The advancement in Electroencephalogram has provided a chance to see the activity in the human brain in non-invasive manner. The proposed research work presents the method for Digit recognition using the EEG signals acquired and processed on smart devices. The results show the implementation of Computation neural network for the recognition of digits from EEG signals. It was seen that, the 90.64% correct classification was achieved.

Keywords: EEG · FAST independent component analysis · Computational neural network

1 Introduction

We are on the verge of next communication evaluation where communication channel as well as mechanism is changing with a great extent. When it comes to human to machine communication, it has already come up a long way from punch card through keyboard/Mouse then touch screens and Gesture Recognitions and now BCI Devices. With the advancements in computing and network powers, we can now process large real time data in many complex algorithms to derive required information. This was demonstrated in previous work presented in [1 previous paper].

Central nervous system signals can be decoded with the help of neuro-technology for instance a brain-computer interface (BCI). Since, BCI is a non-invasive technology, it provide us direct thought based communication to control various electric appliances (i.e. direct brain robot interface) or it can be used to communicate with other users directly without any collaboration of efferent peripheral nervous system muscles or fibers

© Springer Nature Singapore Pte Ltd. 2021
K. C. Santosh and B. Gawali (Eds.): RTIP2R 2020, CCIS 1381, pp. 83–94, 2021.
https://doi.org/10.1007/978-981-16-0493-5_8

[2]. BCIs present us effective and feasible alternatives and in some cases they can be more appropriate communication augmentation choices in case of pseudocoma, in which patient can not focus or control their eye movements [3]. EEG possesses great potential as it contains lot of un-explored information which is directly linked with the brain activity at real time. This information is being analyzed with sophisticated methodologiesleading to new usages of EEG interpreted information. Few such examples are - control of smart devices using brain signals, performance benchmarking of individuals using EEG analysis on induced cognitive work load, control of prostate organs using EEG, etc. [4, 5].The Brain Computer Interface (BCIs) is upcoming potential area of research. The researchers working in this domain are trying to make system more robust and scalable. The major challenges faced by the researchers are like computational inaccuracies, delays, high cost, inter people variance, false positive detection of signals, and constraints on invasive technology that needs more research in this area. The fundamental research which exploit EEG technology was based on the fact that mental state is responsible for electric impulse rhythmic activity and can be regulated by level of alertness or different mental diseases. The major cause of occurring artifacts is eye blinking and eye movements, although other sources can include poor contact between the electrodes and scalp, use of neck or other muscles [6]. Gotman et al. [7] proposed a system for automatic detection of the seizures using EEG signals. Described methods can automatically detect the spikes, depend on the decomposition of the EEG signals into elementary waves and detection of the paroxysmal bursts of rhythmic activity of the brain. EEG signals were collected from body and brain sensors placed on the users scalp. This scenario requires continuous attention and cooperation of humans, data acquisition is depend upon individuals and the public's willingness which is not addressed here.

Similarly, in numerous automated EEG signal classification and seizure alert or detection systems were also in-place and designed by using different approaches. In-line with reported studies, Gotman et al. [8] proposed a computer based automated system for identifying a variety of seizures, while nearest neighbor classifier was proposed by Qu and Gotman [9], on both time and frequency domains of the EEG features extracts to recognize the onset of epileptic seizures. Gigola et al. [10] employed wavelet analysis method on accumulated energy of EEG signals to predict the epileptic seizure onset from intracranial epliptic EEG signals. While, the potential of nonlinear time series analysis in seizure detection was discussed by Adeli et al. [11], Guler et al. [12] and Ubeyli et al. [13] in their research. Several researchers [14, 15] recommended Artificial Neural Network (ANN) based recognition systems for diagnosis of epilepsy. The method introduced by Weng and Khorasani [16] employs the features proposed by Gotman et al. [17], viz, dominant frequency, average EEG amplitude, variation coefficient, average EEG duration, and average power spectrum, passed to an Adaptive Structured Neural Network (ASNN). However, raw EEG signal input was adopted by Pradhan et al. [18] to utilize Learning Vector Quantization Network. Nigam et al. [19] introduced a novel neural network model, LAMSTAR (Large Memory Storage and Retrieval) network and two time-domain attributes of EEG signals; viz, spike rhythmicity and relative spike amplitude to detect seizures.

In-line with the methods presented by the researches in the existed literature is specifically in the areas of study of EEG recordings in context of cognitive science.

This literature presents an entirely new method of processing EEG signals using Smart Computing/Communication Devices such as smart phone/tablets/notebook based on data extraction, preprocessing, feature extraction methods and classifiers. Using mentioned methodologies team has carried out the research work to identify keywords from the real time EEG signals which can further be used in controlling smart devices itself or can further be used to provide as input to external systems which can further use it to carry out user actions in context of the applications. The content of this paper is organized in following section. Section I has provided the background, Section II addresses database specification, Section III presents detailed methodology adopted and performance analysis of the method, Section IV presents conclusion of the work followed acknowledgement and references.

1.1 Database

In order to design robust EEG recognition system, the researchers have designed the database as per requirement that meets to their research problem in general. EmoEngine device was used to collect the EEG signals of each mode, Fig. 1(a) and (b) depicts the brain lobes and emotive EPOC device, respectively.

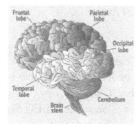

Fig 1. (a) Brain lobes   **Fig (b)** Emotive EPOC device for
brain wave data acquisition

Fig. 1. (a) Brain lobes. (b) Emotive EPOC device for brain wave data acquisition

The data was acquired from Emotive EPOC headset and saved in an output file for analysis. The subject is asked to wear the Emotive head set which sends the data about the activity performed by the subject to the remote smart device through the available communication mechanism. The data is stored on the smart device and further be used for training and testing of the samples over Smart Devices. Traditionally, the data received from the subject is seen for five broad spectral sub-bands, namely, *delta* (0–4 Hz), *theta* (4–8 Hz), *alpha* (8–16 Hz), *beta* (16–32 Hz) and *gamma waves* (32–64 Hz) and these sub-bands have been frequently used EEG signals which are generally adopted in clinical interest. The accurate information regarding neuronal activities can be extracted from these five sub-bands and, consequently, there could be some changes in the EEG signal, which are not so obvious in the original full-spectrum signal, can be amplified when each sub band is considered independently. Each EEG segment was considered as a separate EEG signal resulting in a total of 125 EEG data segments.

The brain is formed using five lobes, these lobes perform all the critical neurological activities such as *frontal lobe* controls the activity of Speech, Thought, Emotions, Problem solving and skilled movements, *Parietal lobe* identifies and interprets sensations such as touch, pain etc. *Occipital lobe* collects and interprets visual images that is sight, *Temporallobe* controls the activities related to hearing and storing memory and *Cerebellum* controls the coordinate's familiar movements. Similarly, the relationship between brain lobes that the excreted (energy) frequency of signal are as below (Table 1).

Table 1. Signal type, frequency and its origin

Type	Frequency range	Origin
Delta	0 Hz–4 Hz	Cortex
Theta	4 Hz–8 Hz	Parietal and temporal
Alpha	8 Hz–13 Hz	Occipital
Beta	13 Hz–20 Hz	Parietal and Frontal
Gamma	20 Hz–40 Hz	Parietal and Frontal

The data set designed in this research work is basically developed on two modalities that is *KEYWORD* and *DIGITS*. The dataset of *KEYWORDS* have been utilized for all experiments. This set contains 12 keywords *{'Close', 'Copy', 'Cut', 'Delete', 'New', 'Ok', 'Open', 'Paste', 'Pause', 'Play', 'Start', 'Stop'}* and EEG Signal recordings of 10 subjects (i.e. 7 Male and 3 Female) from the range of age group (20–25) were considered. The cumulative size of the database is $12 \times 10 \times 10 = 1200$ samples. All participating subjects in the process of data collection/acquisition were normal and away from any physical and mental disorder. The data acquisition set up was developed at Vision and Intelligent System Laboratory of Department of Computer Science and Information Technology, Dr. Babasaheb Ambedkar Marathwada University, Aurangabad. All subjects were asked to sit comfortably on an arm chair and screen was placed next to it in electromagnetically protected room. The written consent was taken from each subject for recording of EEG signals before collecting EEG data. We have selected only subject which are familiar with keywords and have good knowledge of it. All subjects were instructed that this experiment has been designed to be used for Brain Computer Interface applications. A simple display screen in power point was arranged for the data collection under proposed research work. This system generates keyword signal with interval of 2 s. After every 2 s a next number was showed on the screen. In order to collect proper signals from EEG, all subjects were demonstrated the display keyword screen, so that they become more familiar to the task. Same process was repeated five times. So the total volume of keyword dataset is 1000 samples. After extracting the EEG Signal from all subject it will be processed as per methodology.

1.2 Methodology

To implement the above keyword recognition system on smart devices, the critical aspect of consideration is the accuracy of the EEG based thought recognition algorithm. This paper presents a method of acquiring, preprocessing, feature extraction, normalization and classification model from raw EEG signal. In this section, we provides an overview of proposed approach and ANTARANG framework for interpretation of EEG data.

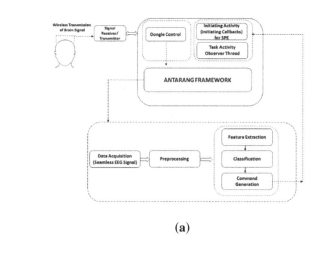

(a)

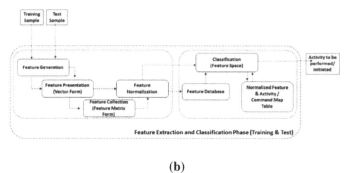

(b)

Fig. 2. (a) and (b) Block diagram of thought processing system 'ANTARANG'

a. **Overview**

The above Fig. 2(a) and 2(b) illustrate the organization of various steps involved in the recognition of keywords. The functionalities of the systems are as discussed below,

1. ***Data acquisition***: the method of data acquisition is as discussed in Sect. 2 is followed. The subject is required to place EMOTIVE EPOC head gear in order to acquire EEG signals corresponding to the activity. The data is acquired by

EMOTIVE head set is transferred to the wireless dongle connected to the device seamlessly via Bluetooth mechanism. The dongle control mechanism connected to the smart device acts as receiver. This will store the test sample primarily in the storage of Smart device and handed over to ANTARANG framework.

2. *Preprocessing*: the captured test sample was preprocessed using, FAST *Independent component analysis* (ICA) was carry out on EEG sample data to removing artifact and resulting ICs were passed for feature extraction.

3. *Feature Extraction*: the objective of this phase is to generate unique set of features such that the overall performance of classification is improved. In this research work stack of feature extraction methods were used which contains methods like *Discrete Cosine Transform (DCT) discrete wavelet transform (DWT)and Short Time Fourier Transform (STFT)* were utilized towards computation of features.

4. *Feature Normalization*: the computed features are normalized. This is required to reduce the size of feature space and speedup the classification of the system. The Linear Discriminant analysis were utilized towards reducing the feature space. This is feature normalization is performed with all training vectors as well the test sample is also normalized before classification.

5. *Classification:* the classification phase has immense potential in the design of any automated system. The proposed system is developed with the stack of classifiers namely, *-Layer perceptron, Random forest, Support vector machine, k-Nearest Neighbor, Naïve Bias classifier, Multi, and Convolution neural network.* The result of classifier will be handed over to the native command translation mechanism which initiate the activity in the smart processing elements (Smart Devices).

6. *Command Map Table and Task observer thread:* The command map table contains information about the mapped callback corresponding the thought. The task observer thread observers the activity and invoke/dispatch the task for execution on the smart devices.

7. *Tools and Software:* As part of this work, the preprocessing and feature extraction was implemented in the *SciPy* and *Numpy* library of Python language. Convolutional neural network models were designed using the *Keras library* and run using *Tensorflow* in an attempt to classify the time-frequency representations. The *matplotlib* library was used to create plots the figures and data visualization.

b. **Working**

The subject is required to gear with Emotive EEG set at the time of data acquisition as well as during testing samples. The electrode placed on the scalp or subset of electrodes in an EEG device may dislocate during data acquisition this may leads into bad contact with the scalp and therefore a artifacts or poor quality signal may be obtained. However, there might be a chance, electrodes placed on the scalp may also have mechanical faults, for instance frayed wiring, which can completely or partially reduce the signal obtained. Such electrodes can induce artifacts into the signals. So in a preprocessing step, FAST Independent Component Analysis (ICA) was applied on EEG sample data to removing artifact and resulting ICs were pass for feature extraction. Basically, in biomedicine, extraction and separation of statistically independent sources underlying multiple measurements of biomedical signals extracted and analyzed with the help of ICA.

1. **Feature Extraction Using DCT**
   Time series signals converted into basic frequency components using the Discrete Cosine Transform (DCT) algorithm. DCT transforms low frequency components into first coefficients and high frequency signals transformed into last ones. The one-dimensional DCT for a list of N real numbers is expressed by Eq. (1) as,

$$Y(u) = \sqrt{\tfrac{2}{N}}\alpha(u) \sum_{x=0}^{N-1} f(x)\,cos\left(\frac{\pi(2x+1)}{2N}\right) \tag{1}$$

   Where u = 0, 1, 2, 3... N − 1;

$$\alpha(0) = \frac{1}{\sqrt{2}}$$

$$a(j) = 1, j \neq 0;$$

   An acquired input EEG sample from training set is a set of 'N' data values and the output is a set of N-DCT transform coefficients Y(u). The first coefficient Y(0) is called the DC coefficient and holds average signal value. The rest coefficients are referred to as the AC coefficients [19]. DCT exhibits good energy compaction for highly correlated signals. If the input data consists of correlated quantities, then most of the N transform coefficients produced by the DCT are zeros or small numbers, and only a few are large. Compressing data with the DCT is therefore accomplished by quantizing the coefficients. The small ones are quantized coarsely and the large ones can be quantized finely to the nearest integer. Applying this feature for EEG signals allow compressing useful data to the first few coefficients. Therefore, only these coefficients can be used for classification using machine learning algorithms. This kind of data compression may dramatically reduce input vector size and decrease time required for training and classification. These feature were calculated for all the samples of 'Keyword set'. The 'DCT Feature Matrix' for the samples of 'Keyword set', are as shown in Table 2.

2. **Feature Selection Using LDA**
   After feature extraction and signal analysis using DCT, the feature set, Y = [y1, y2, y3, ..., yn] is constructed. Feature set with high dimensions (N) need to be reduced, since it contains noise and design of the appropriate classifiers for higher dimensions may find complex difficulties. Those are mostly numerical problems that involve operation with high-order matrices. Such classifiers in N-dimensional feature space are very tedious to analyze and time consuming in terms of computations. Thus Linear Discernment Analysis (LDA) was applied on feature vector to deduce the feature and selecting most prominent features for classification. The data from the different classes are separated with help of hyper

Table 2. The 'DCT Feature Matrix'

Close (0)	Copy (1)	Cut (2)	Delete (3)	New (4)	OK (5)	..	Stop (12)
0.366515	1.081741	1.355358	4.177025	8.047682	1.687529	..	2.732452
0.639972	0.688207	1.752748	9.701809	3.125768	21.37542	..	31.93106
0.061715	0.542834	0.970531	2.899618	2.956171	7.894226	..	6.532051
0.031866	1.167931	0.613204	4.922474	6.982062	21.8817	..	1.435193
0.567666	0.526758	0.790951	7.802225	2.726657	22.01105	..	51.35523
1.073398	0.276145	0.029724	0.07606	0.027696	0.03968	..	0.680563
3.476802	2.356264	0.774597	0.754562	0.40443	0.505899	..	35.93576
0.194747	0.033649	0.033631	0.017214	0.009243	0.041233	..	0.107949
0.056582	0.016408	0.020714	0.024856	0.006497	0.016135	..	0.082541
..	..	..	..	..	..	..	..
0.402641	0.091293	0.008991	0.04552	0.04046	0.051385	..	13.32203

planes using LDA [20, 21] The separating hyper plane is achieved by seeking the projection that maximize the distance between the two classes means and minimize the inter classes variance [21]. To solve an N-class problem (N > 2) several hyper planes are used. This technique has a very low computational requirement which makes it suitable for BCI system. So all the sample of 'Digit database' and 'Keyword dataset' normalized using LDA and selected 100 features of each sample for classification.

2 Results and Discussion

The recognition of EEG Signal sample was carried out by DCT and LDA. The feature set was constructed from all samples of training set and saved for recognition purpose. The entire preprocessed features data set of EEG Keywords were divided into 70:30 ratio that is 70% (Training samples), 30% (Test samples) and evaluated using Convolution Neural Network (CNN). This artificial neural network is improved in both parameters that is shift and translational invariance [22]. In the recent years CNN has attracted a significant attention of the researchers since it is a subset of Deep Learning and employed in several experiments of image recognition *viz*, x-ray analysis in medical imaging [23], Magnetic Resonance Images (MRI) [24], histopathological images [25], retina images [26], and computed tomography images [26]. Although, very few amount of research have been carried out using CNN for physiological signals. The architecture of the CNN is composed of three layers such as convolutional layer, pooling layer, and a fully connected layer. CNNs are very effective models for Image classification tasks.

For the proposed work CNN model was designed, where EEG keyword dataset data first Convolution layer takes this 1-dimensional array as input and the Convolution operation uses 10 initial convolution filters and a convolutional kernel of size 11. Where, as the Activation function the first convolution layer uses *'relu'* (*'Relu'* or *Rectilinear units*

Table 3. Confusion matrix for keyword classification using CNN

Keywords	Total test sample	Training samples												Correct classified	Miss-classified	Accuracy
		Close	Copy	Cut	Delete	New	Ok	Open	Paste	Pause	Play	Start	Stop			
Close	17	15	0	0	0	0	1	0	0	0	0	0	1	15	02	88.24
Copy	13	0	13	0	0	0	0	0	0	0	0	0	0	13	00	100.00
Cut	13	0	0	13	0	0	0	0	0	0	0	0	0	13	00	100.00
Delete	10	0	1	0	9	0	0	0	0	0	0	0	0	09	01	90.00
New	12	1	0	0	0	10	0	0	0	0	1	0	0	10	02	83.33
Ok	14	0	0	0	0	0	12	0	0	1	0	1	0	12	02	85.71
Open	14	0	0	1	0	0	0	13	0	0	0	0	0	13	01	92.86
Paste	12	0	0	0	0	0	0	1	11	0	0	0	0	11	01	91.67
Pause	07	0	0	0	0	0	0	0	0	7	0	0	0	07	00	100.00
Play	12	0	0	0	0	0	0	0	0	0	10	1	1	10	02	83.33
Start	07	0	0	0	0	0	0	0	0	0	0	7	0	07	00	100.00
Stop	13	0	0	0	0	0	0	0	0	0	1	1	11	11	02	84.62
	144	Classification result												131	13	90.97

as Activation for Arousal model). In these proposed experiments, the choice of activation functions for this first layer are of cardinal importance, as some functions like *sigmoid or softmax* might be failed to activate neurons of later layers consistently this improper activation function contributed towards making model defective. The successive layer is another Convolutional Neural Layer, which again with 100 filters and 3*3 size kernel. This layer adopts '*relu*' as the Activation function for both Valence and Arousal classification. Thus, Max pooling Layer produce a flat 1 dimensional layer with dropout on outputs with probability of 0 and 5. The final dense layer utilizes '*softmax*' layer as its activation function. Categorical Cross Entropy used as loss function in this model and '*rmsprop*' used as the optimizer. The experiment were carried out up to 500 epochs and train the model using batches of 32 experiments each. Following is confusion matrix of CNN classification of EEG keyword signal.

The confusion matrix as shown in Table 3, the total 144 test samples of tested on the training data set. It was see that out of 17 samples of 'Close, 15 were classified correctly and 2 samples were misclassified so that the found in class of two and four. Similarly, all test samples of 'copy, 'cut', 'pause' and 'start' were completely classified. Out of these all 144 test samples, the 10 test samples of number 'delete', 09 were classified correctly and 01 were misclassified into 'copy'. The average classification resulting into 90.97% classification accuracy that is out of 144 samples, 131 samples were classified into correct classes and only 13 samples were misclassified.

3   Conclusion

The proposed research work presents the system for automatic classification of EEG signal of Keywords for smart devices. The proposed work evaluates the performance of CNN classifier evaluated over normalized features of Discrete Cosine Transform. The work also signifies method of feature minimization using linear discriminant analysis. The overall accuracy was observed to be 90.97% and the work will be also extended towards automatic classification of 'keywords'. The proposed work also is extended towards design of EEG operated smart devices.

Acknowledgment. This piece of work is registered under Indian **Patent application number** 201621005217 **Titled** SYSTEM FOR THOUGHT BASED COMMUNICATION **Year** - 2016.

The Authors would like to thanks Dept. of Computer Science and IT, Dr. Babasaheb Ambedkar Marathwada University, Aurangabad for providing necessary infrastructure.

References

1. Wolpaw, J., Wolpaw, E.W. (eds.): Brain-computer interfaces: principles and practice. OUP, Oxford (2012)
2. Rutkowski, T.M., Mori, H.: Tactile and bone-conduction auditory brain computer interface for vision and hearing impaired users. J. Neurosci. Methods **244**, 45–51 (2015)
3. Mori, H., et al.: Multi-command tactile brain computer interface: a feasibility study. In: Oakley, I., Brewster, S. (eds.) Haptic and Audio Interaction Design. HAID 2013. LNCS, vol. 7989, pp. 50–59, Springer Heidelberg (2013). https://doi.org/10.1007/978-3-642-41068-0_6

4. Mori, H., Matsumoto, Y., Makino, S., Kryssanov, V., Rutkowski, T.M.: Vibrotactile stimulus frequency optimization for the haptic BCI prototype. In: Proceedings of The 6th International Conference on Soft Computing and Intelligent Systems, and The 13th International Symposium on Advanced Intelligent Systems, Kobe, Japan, 20–24 November 2012, pp. 2150–2153 (2012)
5. Lebedev, M.A., Nicolelis, M.A.L.: Brain-machine interfaces: past, present and future. TRENDS Neurosci. **29**(9), 536–546 (2006)
6. Schirner, G., DenizErdogmus, K.C.: TaskinPadir: the future of human-in-the-loop cyberphysical systems. Computer **1**, 36–45 (2013)
7. Gotman, J.: Automatic recognition of epileptic seizures in the EEG. Electroencephalogr. Clin. Neurophysiol. **54**(5), 530–540 (1982)
8. Qu, H., Gotman, J.: A patient-specific algorithm for the detection of seizure onset in longterm EEG monitoring: possible use as a warning device. IEEE Trans. Biomed. Eng. **44**(2), 115–122 (1997)
9. Gigola, S., Ortiz, F., D'attellis, C.E., Silva, W., Kochen, S.: Prediction of epileptic seizures using accumulated energy in a multiresolution framework. J. Neurosci. Methods **138**(1–2), 107–111 (2004)
10. Adeli, H., Ghosh-Dastidar, S.: NahidDadmehr: a wavelet-chaos methodology for analysis of EEGs and EEG subbands to detect seizure and epilepsy. IEEE Trans. Biomed. Eng. **54**(2), 205–211 (2007)
11. Übeyli, E.: Recurrent neural networks employing Lyapunov exponents for analysis of ECG signals. Expert Syst. Appl. **37**(2), 1192–1199 (2010)
12. Übeyli, E.: Analysis of EEG signals using Lyapunov exponents. Neural Network World **16**(3), 257 (2006)
13. Tzallas, A.T., Tsipouras, M.G., Fotiadis, D.I.: Automatic seizure detection based on timefrequency analysis and artificial neural networks. Comput. Intell. Neurosci. **2007** (2007)
14. Ghosh-Dastidar, S., HojjatAdeli, N.: Principal component analysis-enhanced cosine radial basis function neural network for robust epilepsy and seizure detection. IEEE Trans. Biomed. Eng. **55**(2), 512–518 (2008)
15. Weng, W.D., Khorasani, K.: An adaptive structure neural networks with application to EEG automatic seizure detection. Neural Netw. **9**(7), 1223–1240 (1996)
16. Gotman, J., Wang, L.Y.: State-dependent spike detection: concepts and preliminary results. Electroencephalogr. Clin. Neurophysiol. **79**(1), 11–19 (1991)
17. Pradhan, N., Sadasivan, P.K., Arunodaya, G.R.: Detection of seizure activity in EEG by an artificial neural network: a preliminary study. Comput. Biomed. Res. **29**(4), 303–313 (1996)
18. Nigam, V.P., Graupe, D.: A neural-network-based detection of epilepsy. Neurol. Res. **26**(1), 55–60 (2004)
19. Duda, R.O., Hart, P.E., Stork, D.G.: Pattern Recognition, 2nd edn. WILEYINTERSCIENCE, New York (2001)
20. Fukunaga, K.: Statistical Pattern Recognition, 2nd edn. Academic Press Inc., Cambridge (1990)
21. Fukushima, K.: Neocognitron: a self-organizing neural network model for a mechanism of pattern recognition unaffected by shift in position. Biol. Cybern. **36**, 193–202 (1980)
22. Kallenberg, M., et al.: Unsupervised deep learning applied to breast density segmentation and mammographic risk scoring. IEEE Trans. Med. Imaging **35**(5), 1322–1331 (2016)
23. Pereira, S., Pinto, A., Alves, V., Silva, C.A.: Brain tumor segmentation using convolutional neural networks in MRI images. IEEE Trans. Med. Imaging **35**(5), 1240–1251 (2016)
24. Hatipoglu. N., Bilgin. G.: Cell segmentation in histopathological images with deep learning algorithms by utilizing spatial relationships. Med. Biol. Eng. Comput. 1–20 (2017). https://doi.org/10.1007/s11517-017-1630-1

25. Tan, J.H., Acharya, U.R., Bhandary, S.V., Chua, K.C., Sivaprasad, S.: Segmentation of optic disc, fovea, and retinal vasculature using a single convolutional neural network. J. Comput. Sci. (2017). https://doi.org/10.1016/j.jocs.2017.02.006
26. Setio, A.A.A., et al.: Pulmonary nodule detection in CT images: false positive reduction using multi-view convolutional networks. IEEE Trans. Med. Imaging **35**(5), 1160–1169 (2016)

Machine Learning Algorithms for the Diagnosis of Cardiac Arrhythmia in IoT Environment

Samir Yadav$^{(\boxtimes)}$, Vinod Kadam, and Shivajirao Jadhav

Dr. Babasaheb Ambedkar Technological University, Lonere, Raigad, India
{ssyadav,vjkadam,smjadhav}@dbatu.ac.in

Abstract. Cardiac arrhythmia is very harmful heart disease which can cause severe and even potentially deadly symptoms. Early diagnosis of an arrhythmia can save lives. In the modern health care environment, the use of Internet of Things (IoT) technologies brings the suitability for medical professionals and patients, because they are useful in various medical background. An arrhythmia diagnosis system based on IoT sensors helps and monitors automatic transmission and measurement of Electrocardiogram (ECG) signal data, analyzes this data, and alerts medical professional for an urgency. We are developing a new approach for monitoring and diagnosing cardiac arrhythmia using IoT environment. The concerned ECG signal data is generated by using the IoT sensors and standard online (UC Irvine repository) dataset for predicting severity of cardiac arrhythmia. Also, we used a new fuzzy logic-based neural network classifier algorithm for diagnosing cardiac arrhythmia. The experiments are carried out by various machine learning algorithms on standard UCI Repository dataset as well as on real patients data records collected from multiple Indian hospitals. The proposed machine learning algorithm offers the best accuracy than different machine learning approaches with minimum time complexity. Patients can check their heart condition at their home by using this system. The maintenance and development cost of this system is low due to it is a lightweight and small size system.

Keywords: Health care · Machine learning · Internet of Things · UCI repository · Fuzzy logic-based neural network classifier

1 Introduction

Higher rate of patient admission to the hospital causes severe problems in health sectors. Due to the rise in many quick demises caused by chronic heart failure or high blood pressure, it needs to provide continuous health supervising service at home. This inspires the requirement for reasonable, consistent, wearable

Supported by Third phase of Technical Education Quality Improvement Programme (TEQIP-III).

and low-power devices that will increase the quality of life for several patients. The Internet of Things (IoT) is a technique capable of achieving the previously mentioned health services, and could further develop healthcare systems [21,32]. In real time, medical IoT sensors can generate a large volume of patient's data. The cloud environment will be useful to store this data securely which can be accessible for health care application [6,29]. Recently, the patient monitoring systems are one of the significant advancements because of its improved technology [7,24].

Machine learning and data science is an essential areas for health care where it plays a vital role for prediction and diagnosis of many diseases [35]. Many researchers have used different machine learning based technologies in the medical area over the past decades [2,8,14,18,28,31,33,34]. This inspired us to develop a modern online and IoT based secure health care system for monitoring and diagnosing disease like cardiac arrhythmia (CA). CA is one of the deadly heart disease which can affect the human life. It changes rate of heartbeats, i.e., sometimes heartbeats are too fast or too slow or irregular, which can result in heart failure or sudden death [13,22]. There are different types of CA [20], which can affect human body. Electrocardiogram (ECG) signals are usually used for detection of cardiac arrhythmia.

ECG signal used to measures the electrical activity of the heart in a given time. Also, it helps to detect any harm to the heart using size and position of heart chambers. ECG of a normal human contains five features known as intervals or wave, i.e., PR interval, QT interval, QRS interval, and RR interval [12]. The QRS and RR intervals are essential parts of the arrhythmia diagnosis. The crucial features called heart rate variability (HRV) is calculated by using the RR interval [5].

In our system, we measure a patient's ECG signal data by using medical ECG sensor. This information i.e., biometric data which is collected using sensor then sends to the raspberry pi, and further it is transferred to the cloud. We developed a new approach to classify (16 classes) cardiac arrhythmia diseases for which the concerned data has been collected from UCI repository dataset [3,30] and IoT medical sensors used on patients from different hospitals in India.

Further, this work is organized as follows: Sect. 2 discusses advantages and disadvantages of the different works performed by researchers in recent years. Section 3 explains the methodology and the techniques used in this work. Section 4 discusses results and discussion of experimental design in this work and finally conclusion and future scope are given by the Sect. 5.

2   Related Work

At present, many methods for disease diagnosis based on IoT are developed. Table 1 reviews current related studies. It can be seen from this table, researcher developed health-care systems have various features and characteristics which are helpful for medical professionals and biomedical analyst [25–27]. However, these systems are suffered from drawbacks like complex architecture, high cost and not easy to use. Therefore in this work our aim is to develop a IoT based small structure health-care system for diagnosis and real time monitoring of cardiac arrhythmia patients based on machine learning algorithms.

Table 1. Remote healthcare-related research

Sr. No.	Author and year	Summary and comparison
1	Madhyan et al. 2014 [17]	This system measures and monitors the patient's physiological parameter of the body remotely. It shows that the proposed system is energy efficient due to the use of Bluetooth
2	Roonizi et al. 2015 [23]	An enhanced model of signal decomposition based on the Bayesian framework (EKS6) is recommended for the extraction of features and analysis of the ECG signals
3	Gutta et al. 2016 [9]	Multi-class classification of cardiac arrhythmia using multitask learning method based on biometric recognition is given
4	Luz et al. 2016 [16]	A survey of ECG signal features extraction and classification algorithms for cardiac arrhythmia are discussed
5	Jain et al. 2016 [12]	Cardiovascular disease diagnosis using ECG signal parameter detection based on the forwarding search method is proposed
6	Yuehong et al. 2016 [36]	This research highlights different services, technology, and innovations built on the basis of Wearable Medical Sensors (WMSs) and focusses on the opportunities and challenges for development
7	Abdel et al. 2016 [1]	This system provides healthcare applications for effective and supportive services using portable low power IoT devices
8	Raj et al. 2017 [22]	This system provides ECG signal classification using SVM and partical swarm optimization as tuning parameter for accuracy improvement
9	Rad et al. 2018 [20]	Automatic classification of cardiac rhythms using ANN with Bayesian regularization backpropagation is proposed
10	Kaur et al. 2019 [15]	It introduces a random forest classifier and an IoT-based healthcare network. This will improve interactive features among clinicians and patients

3   Methodology

In this section we presents methods we have used in this work for development of IoT based disease diagnosis using different machine learning algorithms.

3.1   System Architecture

The system architecture of our proposed methodology is given in the Fig. 1. Our system have eight essential blocks like UCI Repository Database, Automatic transmission of ECG signal, Medical Records, Feature selection and Extraction, Secure Cloud Storage, Arrhythmia Diagnosing system, and Background Knowledge. The wearable ECG IoT sensor connectors are utilized to gather ECG signal information of patients which are at distant places. The UCI Repository dataset consists of cardiac arrhythmia dataset. These datasets from UCI dataset and

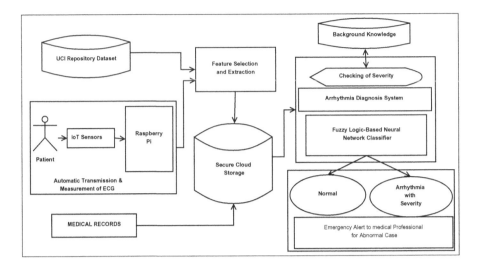

Fig. 1. System architecture

patient's ECG dataset are sent to feature selection and extraction unit for data processing. The medical records consist of the patient's history that is obtained from clinics. All these gathered data then stored in to secure cloud storage. Database management systems steps such as retrieval, partition, aggregation, and merging of data are applied in this data. For classification of arrhythmia based on fuzzy logic-based neural network classifier, the arrhythmia diagnosing system is used. The subpart of this block called severity checking unit that is used for identifying the disease level with the help of proposed smart fuzzy-rules. Further, at the last part of the system, the patient's diagnosed into healthy and type of arrhythmia with seriousness. The understanding of the history works on smart fuzzy rules to make health care decisions records.

3.2 UCI Repository Dataset

The machine learning UCI (University of California, Irvine) Repository contains records of cardiac arrhythmia patients [3]. It contains 452 instances and 279 features, of which 206 are valued linearly and the remaining are insignificant. There are 16 classes associated with these instances. The class one is normal, and the remaining classes are of 15 types arrhythmia. The first four characteristics age, sex, height, and weight, are the patients' basic details. The remainder 276 features are derived through normal ECG recordings of 12 lead [4,10].

3.3 Automatic Transmission and Measurement of ECG Signal

The AD8232 ECG sensor and Raspberry Pi used to obtain ECG signals. AD8232 is an integrated signal conditioning device for an ECG. ECGs are noisy. To remove noise easily from ECG signals, i.e., from PR and QT signals, AD8232 single lead heart rate monitor is used. Raspberry Pi is a low-cost, small-sized device which connects to a TV or LCD screen and utilizes a keyboard and mouse. The Raspberry Pi Model B+ has 512 MB SDRAM with the dual-core ARM11 processor 5 V socket power via usb 3.0. It obtains information from the AD8232 ECG sensor, which further transmits messages through the COM port to the processor. To detect a pulse, the AD8232 sensor has three plates connected to left and right hand and left leg. These plates transform the pulse into the electrical signals.

3.4 Feature Selection and Extraction

Feature selection is used for separating unnecessary or repetitive features/attributes from the dataset. In our system, we used random subspace ensembles classifier for significant features selection [11]. This method reduces the number of features by using feature reduction methods. Feature extraction is the technique of creating a distinct, tinier set of attributes that stills capture most of the valuable data i.e., it establishes new features set. To detect ORS feature, we used a popular Pan-Tompkins algorithm [19]. The algorithm is a number of techniques used to detect R-peaks from ECG signal by derivatives, squaring, integrating, optimized threshold and exploring. For feature extraction, we use Principal component analysis (PCA) which creates linear combinations of the original features. The new feature becomes orthogonal i.e., uncorrelated. These selected and extracted features then stored into the secure cloud storage.

3.5 Secure Cloud Storage

In this system three different types of medical datasets have considered while making the decision. First, data collected from ECG IoT sensor. Second, to compare and to make decisions, we used UCI repository cardiac arrhythmia datasets [3]. And at last, the original clinical history of the patients from different Indian hospitals has been obtained for arrhythmia diseases. All these information then sends to a secured cloud server to make runtime predictions. This data stored securely by applying database management systems steps like retrieval, partition, aggregation, and merging of the data.

3.6 Arrhythmia Diagnosing System

One of the vital units of our system is the arrhythmia prediction unit. Arrhythmia Diagnosis System is responsible for the detection of arrhythmia using our novel algorithm discussed as follows in Algorithm 1. This unit has a subpart called severity checking unit that used for checking the severity of illness with the help of novel developed smart fuzzy-rules. The system sends alert to the physician by sending a short message; hence, the patient detected as affected by arrhythmia disease. To predict arrhythmia and its severity level we have developed a novel Fuzzy logic-based neural network classifier with time consideration algorithm which given in Algorithm 1 as follows:

Algorithm 1: Fuzzy logic-based neural network classifier

Training Phase:

Input: Cardiac arrhythmia record dataset RS_i, Fixed value function Θ.

Output: Set consist of fuzzy rules $\{FL(x), x = 1..., m, t_x = \{TS_y, TE_x, y = 1.., n\}\}$

1 Read the initial record data from RS by period $< p1, p2 >$.
2 Read the set of Attributes ($ASet$) While ($RSi! = EoF$) do
   Start
3 Get the following record data with period $< p1, p2 >$
4 Implement Neural-Tree on the attribute set with period interval $< p1, p2 >$
5 Use weights(W) in the attributes ($FSet$) with a distinct period interval
6 If disease_sign ($Cardiac_arrhythmic$) $== True$ then
7 Call upon the predictive model and apply fuzzy rules to identify and store the severity of the record in R.
8 If disease_sign ($Normal_Class1 = True$) then use fuzzy rules on the data tracking system taking into account the time (p1,p2) for R, calculate the extent of the disease by comparing with records of health history and save this in the data set R2.
9 Else If disease_sign ($CoronaryArteryDisease_class2 == True$) then
   Use fuzzification method on records on the data tracking system taking into account the time $< p1, p2 >$ for R,calculate the extent of the disease by comparing with records of health history and save this in the data set R3
10 Else If the disease_sign ($OldAnteriorMyocardialInfarction_Class3 = True$)then
   Use fuzzification method on records on the data tracking system taking into account the time $< p1, p2 >$ for R,calculate the extent of the disease by comparing with records of health history and save this in the data set R4
11 Else If disease_sign ($OldInferiorMyocardialInfarction_Class4 == True$) then
   Use fuzzification method on records on the data tracking system taking into account the time $< p1, p2 >$ for R,calculate the extent of the disease by comparing with records of health history and save this in the data set R5.
12 Else If disease_sign ($Sinustachycardy_Class5 == True$)then
   Use fuzzification method on records on the data tracking system taking into account the time $< p1, p2 >$ for R,calculate the extent of the disease by comparing with records of health history and save this in the data set R6.
13 Else If disease_sign ($Sinusbradycardy_Class6 == True$) then
   Use fuzzification method on records on the data tracking system taking into account the time $< p1, p2 >$ for R,calculate the extent of the disease by comparing with records of health history and save this in the data set R7.
14 Else If disease_sign ($OldAnteriorMyocardialInfarction_Class10 = True$) then
   Use fuzzification method on records on the data tracking system taking into account the time $< p1, p2 >$ for R,calculate the extent of the disease by comparing with records of health history and save this in the data set R8.

---

(14) Else If disease_sign ($Left ventricule hypertrophy_{C}lass14 = True$) then
Use fuzzification method on records on the data tracking system taking into account
the time $< p1, p2 >$ for R,calculate the extent of the disease by comparing with
records of health history and save this in the data set R9.

(15) Else If disease_sign ($Atrial Fibrillation or Flutter_{C}lass15 == True$) then
Use fuzzification method on records on the data tracking system taking into account
the time $< p1, p2 >$ for R,calculate the extent of the disease by comparing with
records of health history and save this in the data set R10.

(16) Else If disease_sign ($Others_{C}lass16 == True$) then
Use fuzzification method on records on the data tracking system taking into account
the time $< p1, p2 >$ for R,calculate the extent of the disease by comparing with
records of health history and save this in the data set R11.

(17) If ($last\_value(RS) - currant\_value(RS) < \Theta$) then
Go to 6
Else
Go to 3

(18) Return $currant\_value(RS) and Fuzzy\_Rules FL(x)$ with $< p1, p2 >$
End

Tasting Phase:

(1) Read initial query of patient.

(2) Extract the attributes with a period $< p1, p2 >$ and store these attributes into set ASet.

(3) Compare enrollment detail of the user,
if patient has credintials then
Invoke arrhythmia diagnosis for prediction
Else
Invoke the Enrollment method using patient credintials
and other information

(4) Forward the patient info to the diagnosis system in a period $< p1, p2 >$ for resolution.

(5) Take mediacal professional guidance.

(6) Return output data record.

4   Experimental Results and Discussion

We have used various java programming language and Mysql database to develop
the proposed system. The primary task in this work is to categorize data into
the healthy, and unhealthy patients with their level of disease. The experimental
results are carried out on medical datasets collected from various hospitals as
well as the data from UCI repository datasets. Different parameters like accuracy, sensitivity, and specificity (defined in Eqs. 1, 2 & 3) are used to evaluate
the proposed work. Also, we have carried out experiments by using various classification algorithms such as Artificial Neural Network (ANN), Reinforcement
learning (RL), Fuzzy logic (FL), Linear Regression (LR), Decision Tree (DT),
and the proposed Algorithm.

4.1   Parameter for Evaluation

As discussed we used three performance measures of mode i.e., accuracy, sensitivity, and specificity for the evaluation of our models. These parameters value

can be calculated by using true positive (TP), false positive (FP), true negative (TN) and false negative (FN), which is given as follows.

$$Accuracy = \frac{TN + TP}{TN + TP + FP + FN} \tag{1}$$

$$Sensitivity = \frac{TP}{TP + FN} \tag{2}$$

$$Specificity = \frac{TN}{TN + FP} \tag{3}$$

Where, TP = positive patients will be predicted rightly.
TN = Negative patients will be predicted as correctly.
FP = Negative patients will be predicted as positive.
FN = Positive patients will be predicted as negative.

4.2 Results

To calculate the performance of the proposed methodology, the various number of instances has used in our experiments. Here, five different classifications algorithms such as Artificial Neural Network (ANN), Support Vector Machine (RL), K-Nearest Neighbour (KNN), Decision Tree (DT) used for classification of cardiac arrhythmia.

Figure 2 gives the accuracy of the our algorithm and existing methodology.

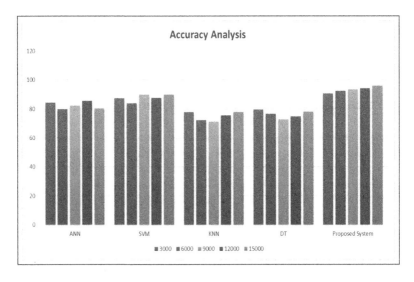

Fig. 2. Analysis of accuracy of models

From Fig. 2, it can be seen that the our proposed method outperforms the existing methods such as ANN, RL, FL, LR, and DT due to various constraints and fuzzy rules applied on classification.

The analysis of specificities of proposed and existing methods like ANN, SVM, KNN, and DT have shown in Fig. 3. We have conducted five different experiments. For these experiments, we have considered 3000, 6000, 9000, 12000 and 15000 various sets of patient's record.

From Fig. 3, it can be seen that the proposed system has higher specificity than the existing methods such as ANN, RL, FL, LR, and DT due to various constraints and fuzzy rules applied on classification.

The comparison of sensitivities of the proposed and existing methods like ANN, SVM, KNN, and DT have shown in Fig. 4. We have conducted five different experiments. For these experiments, we have considered 3000, 6000, 9000, 12000 and 15000 various sets of patient's record.

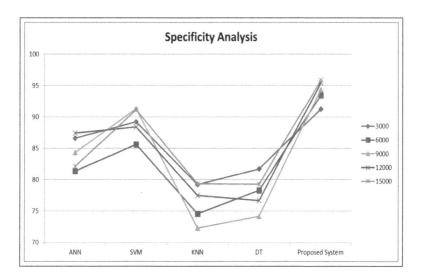

Fig. 3. Analysis of specificity of models

From Fig. 4, it can be seen that the sensitivity of the our model is higher when we compare it with existing methods such as ANN, RL, FL, LR, and DT due to various constraints and fuzzy rules applied on classification.

The analysis of time required to perform classifications also called response time for the proposed model and existing models like ANN, SVM, KNN, and DT have shown in Fig. 5. We have conducted different experiments on various datasets.

From Fig. 5, it can be seen that the time required for classification is lesser for the proposed model as compared to existing methods due to the simplicity of the proposed classifier.

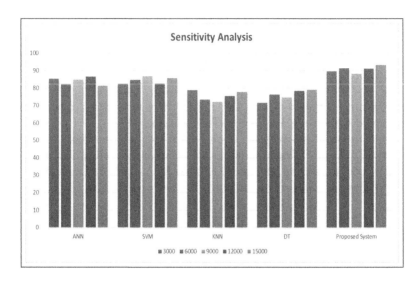

Fig. 4. Analysis of sensitivity of models

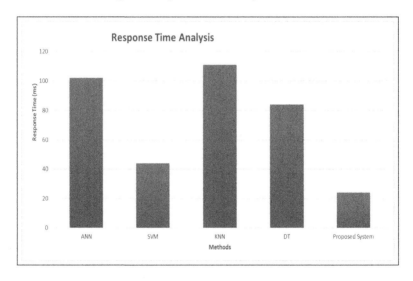

Fig. 5. Response time analysis

5   Conclusion and Future Scope

A novel IoT based health care tracking system using machine learning approach is developed. This system gives real-time monitoring as well as detection and diagnosis of cardiac arrhythmia disease according to the severity using cloud. This system can be useful for real-time as well as off-line stored data. We also have developed a novel Fuzzy logic-based neural network classifier model to classify arrhythmia diseases with it's severity. To compare performance of the model, different experiments are carried out by various machine learning algorithms on standard UCI repository dataset as well as on real patients data records collected from multiple Indian hospitals. Our model has achieved an highest accuracy, sensitivity and specificity and minimum complexity for all number of instances used in the experiments compared to the existing techniques. The alarm is also used in the system to provide an alert to medical professional and patients in the criticalness of patients. The maintenance and development cost of this system is low due to it is a lightweight and small size.

The major limitation of our work is, we have used less number of data and therefore further work focuses on development of such a framework for multi-processing and multi-dimensional data based on Hadoop and new security based algorithm.

Compliance with Ethical Standards

Conflicts of Interest. The authors declare that there is no conflict of interest regarding the publication of this paper.

Ethical Statement. All the subjects provided written informed consent prior to participation. The experimental protocol was approved by the ethics committees of Seth G.S. Medical College and KEM Hospital, Mumbai and was conducted in accordance with the ethical standards outlined in the Declaration of Institutional Ethics Committee (IEC).

References

1. Abdelgawad, A., Yelamarthi, K., Khattab, A.: IoT-based health monitoring system for active and assisted living. In: Gaggi, O., Manzoni, P., Palazzi, C., Bujari, A., Marquez-Barja, J.M. (eds.) GOODTECHS 2016. LNICST, vol. 195, pp. 11–20. Springer, Cham (2017). https://doi.org/10.1007/978-3-319-61949-1_2
2. Ali, M., Guru, D.S., Suhil, M.: Classifying Arabic farmers' complaints based on crops and diseases using machine learning approaches. In: Santosh, K.C., Hegadi, R.S. (eds.) RTIP2R 2018. CCIS, vol. 1037, pp. 416–428. Springer, Singapore (2019). https://doi.org/10.1007/978-981-13-9187-3_38
3. Blake, C., Merz, C.: UCI repository of machine learning databases. Department of Information and Computer Sciences, University of California, Irvine (1998). http://www.ics.uci.edu/mlearn/MLRepository.html
4. Coast, D.A., Stern, R.M., Cano, G.G., Briller, S.A.: An approach to cardiac arrhythmia analysis using hidden Markov models. IEEE Trans. Biomed. Eng. **37**(9), 826–836 (1990)
5. De Chazal, P., O'Dwyer, M., Reilly, R.B.: Automatic classification of heartbeats using ECG morphology and heartbeat interval features. IEEE Trans. Biomed. Eng. **51**(7), 1196–1206 (2004)
6. Dykstra, J., Sherman, A.T.: Design and implementation of Frost: digital forensic tools for the OpenStack cloud computing platform. Digit. Invest. **10**, S87–S95 (2013)

7. Gardner, R.M., Clemmer, T.P., Evans, R.S., Mark, R.G.: Patient monitoring systems. In: Shortliffe, E.H., Cimino, J.J. (eds.) Biomedical Informatics, pp. 561–591. Springer, London (2014). https://doi.org/10.1007/978-1-4471-4474-8_19

8. Ghosh, S., Mukherjee, H., Obaidullah, S.M., Santosh, K.C., Das, N., Roy, K.: A survey on extreme learning machine and evolution of its variants. In: Santosh, K.C., Hegadi, R.S. (eds.) RTIP2R 2018. CCIS, vol. 1035, pp. 572–583. Springer, Singapore (2019). https://doi.org/10.1007/978-981-13-9181-1_50

9. Gutta, S., Cheng, Q.: Joint feature extraction and classifier design for ECG-based biometric recognition. IEEE J. Biomed. Health Inform. **20**(2), 460–468 (2015)

10. Guvenir, H.A., Acar, B., Demiroz, G., Cekin, A.: A supervised machine learning algorithm for arrhythmia analysis. In: Computers in Cardiology 1997, pp. 433–436. IEEE (1997)

11. Jadhav, S., Nalbalwar, S., Ghatol, A.: Feature elimination based random subspace ensembles learning for ECG arrhythmia diagnosis. Soft. Comput. **18**(3), 579–587 (2014). https://doi.org/10.1007/s00500-013-1079-6

12. Jain, S.K., Bhaumik, B.: An energy efficient ECG signal processor detecting cardiovascular diseases on smartphone. IEEE Trans. Biomed. Circ. Syst. **11**(2), 314–323 (2016)

13. Kadam, V., Jadhav, S., Yadav, S.: Bagging based ensemble of support vector machines with improved elitist GA-SVM features selection for cardiac arrhythmia classification. Int. J. Hybrid Intell. Syst. **16**(1), 25–33 (2020)

14. Kadam, V.J., Yadav, S.S., Jadhav, S.M.: Soft-margin SVM incorporating feature selection using improved elitist GA for arrhythmia classification. In: Abraham, A., Cherukuri, A.K., Melin, P., Gandhi, N. (eds.) ISDA 2018. AISC, vol. 941, pp. 965–976. Springer, Cham (2020). https://doi.org/10.1007/978-3-030-16660-1_94

15. Kaur, P., Kumar, R., Kumar, M.: A healthcare monitoring system using random forest and Internet of Things (IoT). Multimed. Tools Appl. **78**(14), 19905–19916 (2019). https://doi.org/10.1007/s11042-019-7327-8

16. Luz, E.J.S., Schwartz, W.R., Cámara-Chávez, G., Menotti, D.: ECG-based heartbeat classification for arrhythmia detection: a survey. Comput. Methods Prog. Biomed. **127**, 144–164 (2016)

17. Madhyan, E., Kadam, M.: A unique health care monitoring system using sensors and ZigBee technology. Int. J. Adv. Res. Comput. Sci. Softw. Eng. **4**(6), 183–189 (2014)

18. Makandar, A., Somshekhar, R.: Image enhancement using filters on Alzheimer's disease. In: Santosh, K.C., Hegadi, R.S. (eds.) RTIP2R 2018. CCIS, vol. 1036, pp. 33–41. Springer, Singapore (2019). https://doi.org/10.1007/978-981-13-9184-2_3

19. Pan, J., Tompkins, W.J.: A real-time QRS detection algorithm. IEEE Trans. Biomed. Eng. **32**(3), 230–236 (1985)

20. Rad, A.B., et al.: ECG-based classification of resuscitation cardiac rhythms for retrospective data analysis. IEEE Trans. Biomed. Eng. **64**(10), 2411–2418 (2017)

21. Rahmani, A.M., et al.: Exploiting smart e-Health gateways at the edge of healthcare Internet-of-Things: a fog computing approach. Future Gener. Comput. Syst. **78**, 641–658 (2018)

22. Raj, S., Ray, K.C.: ECG signal analysis using DCT-based dost and PSO optimized SVM. IEEE Trans. Instrum. Measur. **66**(3), 470–478 (2017)

23. Roonizi, E.K., Sassi, R.: A signal decomposition model-based Bayesian framework for ECG components separation. IEEE Trans. Sig. Process. **64**(3), 665–674 (2015)

24. Ruikar, D.D., Hegadi, R.S., Santosh, K.: A systematic review on orthopedic simulators for psycho-motor skill and surgical procedure training. J. Med. Syst. **42**(9) (2018). Article number: 168. https://doi.org/10.1007/s10916-018-1019-1

25. Ruikar, D.D., Santosh, K., Hegadi, R.S.: Automated fractured bone segmentation and labeling from CT images. J. Med. Syst. **43**(3), 1–13 (2019). https://doi.org/10.1007/s10916-019-1176-x

26. Ruikar, D.D., Santosh, K., Hegadi, R.S.: Segmentation and analysis of CT images for bone fracture detection and labeling. In: Medical Imaging: Artificial Intelligence, Image Recognition, and Machine Learning Techniques, p. 131 (2019)

27. Santosh, K., Antani, S., Guru, D.S., Dey, N.: Medical Imaging: Artificial Intelligence, Image Recognition, and Machine Learning Techniques. CRC Press, Boca Raton (2019)

28. Santosh, K.C., Wendling, L.: Automated chest X-ray image view classification using force histogram. In: Santosh, K.C., Hangarge, M., Bevilacqua, V., Negi, A. (eds.) RTIP2R 2016. CCIS, vol. 709, pp. 333–342. Springer, Singapore (2017). https://doi.org/10.1007/978-981-10-4859-3_30

29. Thota, C., Manogaran, G., Lopez, D., Vijayakumar, V.: Big data security framework for distributed cloud data centers. In: Cybersecurity Breaches and Issues Surrounding Online Threat Protection, pp. 288–310. IGI Global (2017)

30. Dua, D., Graff, C.: UCI machine learning repository (2017). http://archive.ics.uci.edu/ml

31. Yadav, S.S., Jadhav, S.M.: Deep convolutional neural network based medical image classification for disease diagnosis. J. Big Data **6**(1), 1–18 (2019). https://doi.org/10.1186/s40537-019-0276-2

32. Yadav, S.S., Jadhav, S.M.: Machine learning algorithms for disease prediction using IoT environment. Int. J. Eng. Adv. Technol. **8**(6), 4303–4307 (2019)

33. Yadav, S.S., Jadhav, S.M., Bonde, R.G., Chaudhari, S.T.: Automated cardiac disease diagnosis using support vector machine. In: 2020 3rd International Conference on Communication System, Computing and IT Applications (CSCITA), pp. 56–61. IEEE (2020)

34. Yadav, S.S., Jadhav, S.M., Nagrale, S., Patil, N.: Application of machine learning for the detection of heart disease. In: 2020 2nd International Conference on Innovative Mechanisms for Industry Applications (ICIMIA), pp. 165–172. IEEE (2020)

35. Yadav, S.S., Kadam, V.J., Jadhav, S.M.: Comparative analysis of ensemble classifier and single base classifier in medical disease diagnosis. In: Bansal, J.C., Gupta, M.K., Sharma, H., Agarwal, B. (eds.) ICCIS 2019. LNNS, vol. 120, pp. 475–489. Springer, Singapore (2020). https://doi.org/10.1007/978-981-15-3325-9_37

36. Yuehong, Y., Zeng, Y., Chen, X., Fan, Y.: The Internet of Things in healthcare: an overview. J. Ind. Inf. Integr. **1**, 3–13 (2016)

Efficient Method to Extract QRS Complex and ST Segment for Cardiovascular Diseases Prediction

Sanjay Ghodake[1(✉)], Shashikant Ghumbre[2], and Sachin Deshmukh[3]

[1] MIT Academy of Engineering (Alandi (D)), Pune, Maharashtra, India
ghodkesanjay1@gmail.com
[2] Government College of Engineering and Research, Avasari, Pune, Maharashtra, India
[3] Department of CSIT, Dr. Babasaheb Ambedkar Marathwada University, Aurangabad, India

Abstract. For the heart diseases, the early prediction required to save the human being life. There are several ways to perform the early prediction of Cardiovascular Disease (CVD), however the most of the state-of-art approaches are expensive with poor accuracy of prediction. The computerised approach used the Electrocardiogram (ECG) signals to perform the early prediction of CVD. The ECG based approach is simple, effective and inexpensive; hence it gains the significant attention of researchers from last two decades. The Computer Aided Diagnosis (CAD) system introduced the ECG based approach for CVD prediction using the ECG signal of patients on which the algorithms single processing, data mining, and machine learning applied for accurate prediction. ECG based CVD detection framework composed of three main sections that is preprocessing, features extraction, and classification. The steps like preprocessing and features extractions are crucial for the efficiency of CVD detection. In this paper, we proposed the novel framework of CVD detection of Q, R, S, T beats efficiently from the pre-processed ECG signal. From the pre-processed ECG signal, our aim is to extract QRS and ST segments using the dynamic and simple thresholding approach. The segments are used then for the statistical features extraction. The classification is performed by using the Artificial Neural Network (ANN) classifier. The proposed method shows the precision rate, recall rate, detection accuracy and detection time are 0.91, 0.92, 0.91 and 1.51 respectively. It shows the balance between the accuracy and prediction time performance as compared to state of art method.

Keywords: Computer aided diagnosis · Cardiovascular Disease · Electrocardiogram · QRS segment first keyword · ST segment · Hybrid filtering

1 Introduction

During the last decade, the classification and highlight extraction have pulled in critical consideration in numerous territories of sciences, for example, science, prescription, science, and financial matters. Specifically, infection classification and biomarker disclosure become progressively significant in present-day natural and therapeutic research. ECGs are similarly minimal effort and non-intrusive in diagnosing and screening heart

© Springer Nature Singapore Pte Ltd. 2021
K. C. Santosh and B. Gawali (Eds.): RTIP2R 2020, CCIS 1381, pp. 108–121, 2021.
https://doi.org/10.1007/978-981-16-0493-5_10

diseases like CVD. With the improvement of individual ECG screens, a lot of ECGs are recorded and put away; accordingly, quick and proficient algorithms are called for to break down the data and make the diagnosis. The ECG based CVD detection mainly consist of steps like signal pre-processing, extraction, and prediction. Highlights that have been utilized in describing the ECGs incorporate heartbeats interim highlights, frequency-based highlights of ECG signals and Hermite polynomials [1–5]. Past methods of ECG classification incorporate straight discriminants [6], decision tree [7–9], neural systems [1, 10, 11], support vector machine [2–5], and Gaussian mixture model algorithm [12]. A few specialists perform disease recognition utilizing ECG data alongside other clinical estimations [8, 10]. But such techniques which utilized coefficients of different premise functions as highlights for classification, for example, the wavelet coefficients, the coefficients are typically difficult to translate clinically. Rather than applying such method directly on the ECG signals, the extraction of important beats of Q, R, S, P, and T signals, increases the more reliability and accuracy [14, 16–21]. The composition of ECG signals and how the beats may use for the CVD prediction is described below.

During the cardiac cycle the myocardium produces the time-variant voltages and is displayed as the graphical recording. This graphical recording is named as the electrocardiogram. Normally the ECG signal is in the range of ±2 mV. The ECG requires a chronicle bandwidth of 0.05 to 150 Hz. On x-axis, time scale of 0.04 s/mm and on Y-axis voltage sensitivity of 0.1mV/mm are mentioned. The small square of the standard ECG paper speaks to 0.04 s and every huge square speaks to 0.2 s. Figure 1 indicates the normal ECG waveform with its components and the different intervals. The first wave termed as P wave which is generated by the deflection of atrial depolarization. It is a low amplitude wave (50–100 mV). The next biggest segment formed in the ECG curve is the QRS complex which is brought about by the ventricular depolarization. This wave is sharper and high in amplitude. Due to the ventricular repolarization a low amplitude deflection curve is formed and termed as T wave. The length of time starts from the P wave then QRS wave is known as PR interim. The ECG curve has predictable direction, duration and amplitude under normal condition. Hence from the ECG graph various components can be distinguished and deciphered. The typical or irregular function of the heart can be determined (Table 1).

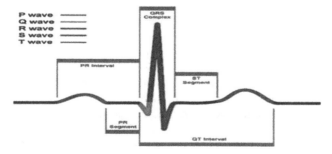

Fig. 1. Components of each ECG signal

Table 1. Normal ECG intervals

ECG waves	Interval
P wave duration	≤0.12 s
Corrected QT	≤0.11 s (male)
	≤0.46 s (*female*)
PR interval	0.12−0.22 s
QRS complex duration	≤0.10 s

In order to understand the heart behavior the most common clinical practice is to record the limb leads or 12 lead ECG. Using the ECG the various kinds of heart disorders such as coronary illness, heart attack etc. may cause heart disappointment for the duration of the day can be detected [13]. The fiducial focuses, for example, P, Q, R, and S in the ECG are located initially by the physicians. From those fiducial points the remaining waves like P, QRS and T are located. Using these fiducial points and the waves many heart abnormalities estimation like ventricular/arterial, arrhythmias, heart rate variations and deviations of ST-segment are conducted. In order to diagnose the ECG data exactly the clinician requires an automatic decision support system because normally more than100000 cardiovascular cycles are recorded per quiet in multi-day with an ECG device. It is a huge task for the physicians to interpret any heart disorders from this large amount of ECG. Thus, it is required to automate approach that extracts the waves of ECG signals efficiently and accurately.

In this paper feature extraction phase is designed in which we first extracted two important segments of QRS and ST segments of each ECG signals using the dynamic thresholding algorithm, further rather than using big size QRS and ST features for the classification we extracted the statistical features from each segment and construct the unique and simple feature vector. Finally in classification phase, the ANN applied. In sections II, brief review of recent ECG features extraction methods presented, in III. The proposed methodology presented, in IV, simulation results and analysis introduced and in V, end and future work described.

2 Related Work

In this section we described various features extraction and waves extraction methods used in CVD detection reviewed.

Feature Extraction Methods. The classification step basically depends on the detection QRS complex and features extraction. Various algorithms dependent on subsidiary [22], digital filters [23], and wavelet change [24] have much of the time been utilized to recognize QRS. With improvement of equipment condition, considerably more techniques embrace wavelet changes. The effectiveness of wavelet change unequivocally relies upon the decision of mother wavelets in wavelet based methods. Other discovery algorithms are proposed in writings scientific morphology [25], concealed Markov model [36], S-change [27], Hilbert change [15], customary sentence structure [28], quadratic filter [29], multiresolution entropy [30], inadequate portrayal [31], and singular value decomposition (SVD) [32]. In spite of the fact that the above recognition techniques gives high precision with their exploratory datasets, this presentation generally relies upon chosen mother wavelets in wavelet change just as fixed parameters and based information in different techniques. In this way, if there should arise an occurrence of ECG designs, which are physiological varieties because of times, people, or a situation, the decision of proper mother wavelets and parameters winds up troublesome. Also, separating hand-created includes physically for QRS complex location may present noteworthy computational intricacy of generally speaking process, particularly in the change spaces. To adjust with different morphologies of ECG signals, a few algorithms embrace versatile limit for a significant parameter in QRS complex recognition. There are two classes of versatile limit, single-level [33, 34, 35, 36] and numerous levels [37, 38]. Be that as it may, the versatile limit improves recognition precision to the detriment of computational intricacy, which makes it hard for ongoing QRS location. Counterfeit neural system based methodologies have been proposed for constant recognition [39]. Further in [40]–[44], some recent works reported on QRS detection from the ECG signal. In [40], support vector machine (SVM) and wavelet energy histogram algorithms are suggested for ECG signal analysis and classification. They structured the cardio-vascular arrhythmia recognition in the ECG signal dependent on three phases including ECG signal pre-processing, highlight extraction utilizing QRS discovery and pulses classification. In [41], a productive and simple to-translate system of heart malady classification proposed dependent on examination of classifiers and component extraction techniques. Author investigated the features extraction method using three classifiers for dimension-reduction such as SVM, SDA and LASSO logistic regression. In [42], the recent work on CVD detection using ECG signal reported. Signal processing and neural networks methods for processing of ECG signals consisting of extracting features from ECG signal in order to detect the types of CVD's. In [43], dynamical ECG recognition framework proposed for CVD's and human detection using the dynamical neural learning component. Said method consist of two stages: a preparation and test stage. In preparation stage, cardiovascular elements inside ECG signals is extricated precisely by utilizing radial basis function (RBF) neural systems. The acquired cardio-vascular system dynamics is spoken to and put away in steady RBF systems. In [44], very recent method proposed for the QRS signals detection from ECG signal using one dimensional convolutional neural network (CNN). It is used for separating ECG morphological parameters. All the separated morphological highlights are utilized by multi-layer perceptron (MLP) for QRS signals recognition. Furthermore, the creator embraced the ECG signal pre-processing method which just contains contrast activity in the worldly domain. However such methods failed to accomplish the exchange off between calculation overhead and prediction accuracy. The work reported in this paper is different in which the lightweight QRS and ST segments extraction designed followed by the statistical features extraction technique.

3 Proposed Methodology

Figure 2 shows proposed ECG signal based CVD recognition framework. The input ECG signal is first processed through the hybrid pre-processing algorithm in which the noises and artefacts are removed from the raw ECG waves. The pre-processing method is suggested to reduce the error rates in ECG signal with minimum computation efforts. After the pre-processing, we applied the proposed wave extraction technique in which we build the QRS and ST complex from the ECG waves for features extraction purpose. This ECG wave extraction technique is very significant for the extraction of particular QRS and ST complex from ECG waves. As ECG wave consist of various other features such as P waves, PR interval, PR segment, QT Interval which are less significant for the prediction of CVD. Hence wave extraction technique is necessary to improve the CVD prediction results. Figure 2 itself justify the need of ECG wave extraction step. Mean, standard deviation, variance, entropy, and smoothness statistical features are computed for extraction of QRS and ST parameters in ECG wave. Feature vector of size 10 is constructed and passed to the classifier for the prediction purpose.

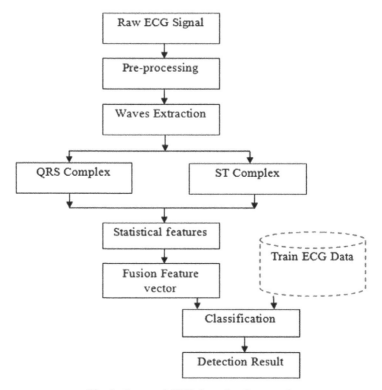

Fig. 2. Proposed CVD detection framework

After successful removal of unwanted information from the ECG signal i.e. pre-processing, we applied the process of wave's extraction and features formation process.

Waves Extraction. This section presents the methodology to build the QRS and ST segmentation from the pre-processing ECG signal. The extraction of beats such as Q, R, S, P, and T performed first and then form the QRS complex and ST segment for the features extraction process. Along with efficiency of wave's detection and extraction, the main objective is that wave's extraction algorithm should take minimum computation efforts. The previous adaptive thresholding methods applied in morphological domain which takes longer computation time, however here we directly applied in time-domain signal so that no time waste in performing any morphological operation. The technique of thresholding is dynamic in which the threshold value is fixed and computed as per the input ECG signal. This method is start from the signal normalization to QRS complex and ST segment extraction as demonstrated in Algorithm 1.

As observed in Algorithm 1, the first step is to perform the normalization of pre-processed signal as:

$$A = \frac{P^2}{\max(|P^2|)} \tag{1}$$

Where, P is the pre-processing ECG signal. The normalized signal further used for the computation of average signaling using the convolution operations as:

$$B = \frac{\text{Ones } (1, 31)}{31} \tag{2}$$

The temporary array of matrix 1's of size 1 * 31 which is used to perform the convolution with normalized signal A as

$$C = conv(A, B) \tag{3}$$

$$C = C(15 + [1{:}N]) \tag{4}$$

The convolution output used to compute the average signal as:

$$AS = \frac{C}{\max(|C|)} \tag{5}$$

The normalization and average signaling of original pre-processed signal helps to minimize the overhead of computation burden while estimating the waves, also improves the accuracy of waves extraction. The dynamic threshold value is computed by taking the mean of AS signal. The computed threshold value applied to select the waves those satisfies the threshold value as:

$$AS = AS > \alpha \tag{6}$$

The left and right waves computed using Differences and Approximate Derivatives (DAD) operator. The DAD computes the differences between adjacent elements of AS along the first array dimension whose size does not equal. This returns the signals of matrix starting from −1 (left) to 1(right). Thus we got the left wave and right wave for the signal. From the left and right waves we perform the subtract delay operations to suppress the low-pass and high-pass filtering.

Algorithm 1: Waves Extraction
Inputs
P: Pre-processed ECG signal
Output
QRS
ST
1. Compute length *N* of *P*
2. *A*: Signal Normalization using Eq. (1)
3. *AS*: Average signalling using Eq. (5)
4. *α*: Mean of Signal AS
5. AS: Apply the threshold on *AS* using Eq. (6)
6. [L, R]: Estimate the left and right waves
7. *Rwave*: Extract *R* wave using Eq. (7)
8. *Qwave*: Extract *Q* wave using Eq. (8)
9. *Swave*: Extract *S* wave using Eq. (9)
10. *Pwave*: Extract *T* wave using Eq. (10)
11. *Twave*: Extract *P* wave using Eq. (11)
12. *QRS*: [*Qwave, Rwave, Swave*]
13. *ST*: [*Swave, Twave*]
14. Return (*QRS, ST*)
15. Stop

After the confinement of left and right waves, we extract the P, Q, R, S, and T waves as using the min and max operations on L (left) and R (right) waves detected.

$$Rwave = \max(P(L{:}R)) \tag{7}$$

$$Qwave = \min(P(L{:}Rwave)) \tag{8}$$

$$Swave = \min(P(L{:}R)) \tag{9}$$

$$Pwave = \max(P(L{:}QwaveQT)) \tag{10}$$

$$Twave = \max(P(Swave{:}R)) \tag{11}$$

The QRS complex and ST segments formed using the extracted waves.

Statistical Features. We extracted total 5 features such as mean, standard deviation, smoothness, entropy, and variance from the QRS and ST complex.

Mean. The Mean estimate the incentive in the QRS/ST waves where focal clustering happens. It can be determined to utilize recipe.

$$\mu = \frac{1}{MN} \sum_{i=1}^{M} \sum_{j=1}^{N} p(i,j) \tag{12}$$

As p(i,j) is data value at (i,j) point of an QRS/ST ECG signal of MxN size.

Standard Deviation. The Standard Deviation (σ) of QRS/ST data at p(i, j) from its mean value is determine by the equation.

$$\sigma \sqrt{\frac{1}{MN} \sum_{i=1}^{M} \sum_{j=1}^{N} (P(i.j) - \mu)^2} \tag{13}$$

Smoothness. Relative smoothness, R is a measure of smoothness of QRS/ST signal that can be utilized to build up descriptors of relative smoothness. The smoothness is resolved to utilize the recipe

$$R = 1 - \frac{1}{1 + \sigma^2} \tag{14}$$

Entropy. Entropy is a factual proportion of haphazardness that can be utilized to describe the surface of the information QRS/ST signal. Entropy, h can likewise be utilized to portray the dispersion variety in a district. Generally speaking Entropy of the QRS/ST can be determined as

$$h = \sum_{k=0}^{L-1} Pr_k (log_2 Pr_k) \tag{15}$$

Variance (var). It is square root of standard deviations formula is:

$$Var = \sqrt{\sigma} \tag{16}$$

The feature vector constructed from the above five statistical measure methods for each QRS and ST complex. The fusion of both complex forms the final feature vector of size 10. For the classification we used the ANN classifier with ratio of 70% training, 15% testing, and 15% validation.

4 Simulation Results

MATLAB is used for CVD detection and Physikalisch-Technische Bundesanstalt (PTB) ECG dataset [45, 46] is used for the experiment. For both heart disease patients and healthy patients ECG signals were recorded by the Professor Michael Oeff, M.D. The 545 ECG signals were collected for different subjects consisting of both CVD and normal. There are total 468 ECG signals affected by CVD and 77 samples are normal.

The hybrid filter was used to achieve the denoising quality and time complexity performance as shown in Table 2 below.

Table 2. Filtering quality evaluation

Method	Mean square error	Signal to noise ratio	Time complexity (s)
Butterworth	0.3545	10.55	0.148
Notch	0.3323	10.78	0.112
Adaptive filtering	0.2566	11.89	0.193
Hybrid	0.1973	13.33	0.156

The exhibition of CVD recognition is estimated as far as precision, recall, and accuracy. The precision, recall, and accuracy rates are calculated as:

$$Precision = \frac{tp}{tp + fp} \tag{17}$$

$$Recall = \frac{tp}{tp + fn} \tag{18}$$

$$Accuracy = \frac{tp + tn}{tp + tn + fp + fn} \tag{19}$$

Where, tp is true positive, tn is true negative, fp is false positive, and fn is false negative. The outcome of proposed features extraction methodology is contrasted and the state-of-art features extraction methods such as discrete wavelet transforms (DWT), Hilbert transform (HT), and SVD with varying number of hidden layers (5 to 15) (Fig. 3).

Table 3. Precision rate evaluation

Layers	SVD	HT	DWT	Proposed
5	0.71	0.73	0.77	0.89
10	0.74	0.75	0.78	0.90
15	0.76	0.76	0.79	0.91

Table 4. Recall rate evaluation

Layers	SVD	HT	DWT	Proposed
5	0.75	0.79	0.81	0.88
10	0.75	0.77	0.82	0.89
15	0.76	0.78	0.8	0.92

Table 5. Detection accuracy evaluation

Layers	SVD	HT	DWT	Proposed
5	0.745	0.775	0.805	0.893
10	0.755	0.782	0.814	0.904
15	0.774	0.789	0.82	0.915

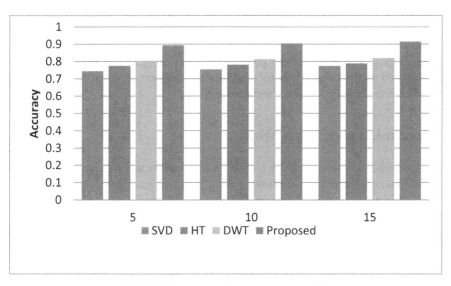

Fig. 3. Performance of accuracy rate analysis

The table number 3, 4 and 5 shows the performance of precision, recall, and accuracy rates for each method with varying the number of hidden layers respectively. We noticed that as the hidden layers increases, the performance is increasing as well for each technique. However, increasing the hidden layer may lead the more computation overhead. The proposed features extraction method approach shows the significant improvement in performances compared to all other methods. The SVD method shows the worst performance among all the techniques. The proposed approach shows the improvement as the detection and extraction of QRS and ST complex performed followed by the statistical analysis which delivered the more unique and reliable set of features for the prediction. Table 6 demonstrate the average performance and comparison with similar methods.

Table 6. Comparative analysis with similar methods

Method	Accuracy (%)	Detection time (s)
SVD	75.8	1.23
HT	78.2	1.74
DWT	81.3	1.95
[26]	88.52	1.72
[27]	82.5	1.59
Proposed	90.4	1.51

As observed in Table 6, the overall accuracy of proposed framework designed in this paper is significantly higher than the previous methods. We investigated the performance of detection time also which shows that the SVD technique takes less time among all the techniques, but its accuracy is very poor which is not acceptable at realist conditions. The proposed method shows the balance between the accuracy and prediction time performance as compared all other techniques.

5   Conclusion and Future Work

This paper exhibited the framework for the CVD discovery using the raw ECG signals for the different subjects under the two main categories such as normal and diseased. We designed the hybrid filtering technique at first to pre-process the raw ECG signal. Then the dynamic thresholding based QRS complex and ST complex extraction algorithm presented. The statistical features extracted from the QRS and ST segmentation before the classification. The simulation results clears that the proposed system significantly superior as compared to state-of-art methods. For future work, we propose to work on design robust features extraction method rather than using statistical features.

References

1. Ghongade, R., Ghatol, A.: A robust and reliable ECG pattern classification using QRS morphological features and ANN. In: Proceedings of the IEEE Region 10 Conference (TENCON 2008), pp. 1–6 (2008)
2. Kallas, M., Francis, C., Kanaan, L., Merheb, D., Honeine, P., Amoud, H.: Multi-class SVM classification combined with kernel PCA feature extraction of ECG signals. In: Proceedings of the 19th International Conference on Telecommunications (ICT 2012), pp. 1–5, Jounieh, Lebanon, April 2012

3. Rabee, A., Barhumi, I.: ECG signal classification using support vector machine based on wavelet multiresolution analysis. In: Proceedings of the 11th International Conference on Information Science, Signal Processing and their Applications (ISSPA 2012), Montreal, Canada, pp. 1319–1323, IEEE, July 2012

4. Shen, M., Wang, L., Zhu, K., Zhu, J.: Multi-lead ECG classification based on independent component analysis and support vector machine. In: Proceedings of the 3rd International Conference on BioMedical Engineering and Informatics (BMEI 2010), Yantai, China, vol. 3, pp. 960–964. IEEE, October 2010

5. Zellmer, E., Shang, F., Zhang, H.: Highly accurate ECG beat classification based on continuous wavelet transformation and multiple support vector machine classifiers. In: Proceedings of the 2nd International Conference on Biomedical Engineering and Informatics (BMEI 2009), Tianjin, China, pp. 1–5. IEEE, October 2009

6. de Chazal, P., O'Dwyer, M., Reilly, R.B.: Automatic classification of heartbeats using ECG morphology and heartbeat interval features. IEEE Trans. Biomed. Eng. **51**(7), 1196–1206 (2004)

7. Fayn, J.: A classification tree approach for cardiac ischemia detection using spatiotemporal information from three standard ECG leads. IEEE Trans. Biomed. Eng. **58**(1), 95–102 (2011)

8. Mair, J., Smidt, J., Lechleitner, P., Dienstl, F., Puschendorf, B.: A decision tree for the early diagnosis of acute myocardial infarction in nontraumatic chest pain patients at hospital admission. Chest **108**(6), 1502–1509 (1995)

9. Tsien, C.L., Fraser, H.S., Long, W.J., Kennedy, R.L.: Using classification tree and logistic regression methods to diagnose myocardial infarction. Stud. Health Technol. Inf. **52**(1), 493–497 (1998)

10. Dorffner, G., Leitgeb, E., Koller, H.: Toward improving exercise ECG for detecting ischemic heart disease with recurrent and feedforward neural nets. In: Proceedings of the 4th IEEE Workshop on Neural Networks for Signal Processing (NNSP 1994), Ermioni, Greece, pp. 499–508, September 1994

11. Hu, Y.H., Tompkins, W.J., Urrusti, J.L., Afonso, V.X.: Applications of artificial neural networks for ECG signal detection and classification. J. Electrocardiol. **26**(supplement), 66–73 (1993)

12. Martis, R.J., Chakraborty, C., Ray, A.K.: A two-stage mechanism for registration and classification of ECG using Gaussian mixture model. Pattern Recogn. **42**(11), 2979–2988 (2009)

13. Zimmerman, T.G., Syeda-Mahmood, T.: Automatic detection of heart disease from twelve channel electrocardiogram waveforms. In: Proceedings of the Computers in Cardiology, Durham, NC, USA, pp. 809–812, October 2007

14. Hadj Slimane, Z.E., Naït-Ali, A.: QRS complex detection using empirical mode decomposition. Digit Signal Process **20**, 1221–1228 (2010)

15. Manikandan, M.S., Soman, K.P.: A novel method for detecting R-peaks in electrocardiogram (ECG) signals. Biomed. Signal Process. Control **7**, 118–128 (2012)

16. Zhu, H., Dong, J.: An R-peak detection method based on peaks of shannon energy envelope. Biomed. Signal Process Control **8**, 466–474 (2013)

17. Bouaziz, F., Boutana, D., Benidir, M.: Multi-resolution wavelet-based QRS complex detection algorithm suited to several abnormal morphologies. IET Signal Process. **8**, 774–782 (2014)

18. Karimipour, A., Homaeinezhad, M.R.: Real-time electrocardiogram P-QRS-T detection–delineation algorithm based on quality-supported analysis of characteristic templates. Comput. Biol. Med. **52**, 153–165 (2014)

19. Chouakri, S.A., Bereksi-Reguig, F., Taleb-Ahmed, A.: QRS complex detection based on multi wavelet packet decomposition. Appl. Math. Comput. **217**, 9508–9525 (2011)

20. Adnane, M., Jiang, Z., Choi, S.: Development of QRS detection algorithm designed for wearable cardiorespiratory system. Comput. Methods Programs Biomed. **93**, 20–31 (2009)

21. Madeiro, J.P., Cortez, P.C., Marques, J.A., Seisdedos, C.R., Sobrinho, C.R.: An innovative approach of QRS segmentation based on first-derivative, Hilbert and Wavelet Transforms. Med. Eng. Phys. **34**, 1236–1246 (2012)
22. Arzeno, N.M., Deng, Z.D., Poon, C.S.: Analysis of first-derivative based QRS detection algorithms. IEEE Trans. Biomed. Eng. **55**, 478–484 (2008)
23. Keselbrener, L., Keselbrener, M., Akselrod, S.: Nonlinear high pass filter for R-wave detection in ECG signal. Med Eng Phys. **19**, 481–484 (1997)
24. Sharma, T., Sharma, K.K.: QRS complex detection in ECG signals using locally adaptive weighted total variation denoising. Comput. Biol. Med. **87**, 187–199 (2017)
25. Trahanias, P.E.: An approach to QRS complex detection using mathematical morphology. IEEE Trans Biomed. Eng. **40**, 201–205 (1993)
26. Coast, D.A., Stern, R.M., Cano, G.G., Briller, S.A.: An approach to cardiac arrhythmia analysis using hidden Markov models. IEEE Trans. Biomed. Eng. **37**, 826–836 (1990)
27. Zidelmal, Z., Amirou, A., Adnane, M., Belouchrani, A.: QRS detection based on wavelet coefficients. Comput. Methods Programs Biomed. **107**, 490–496 (2012)
28. Hamdi, S., Ben, A.A., Bedoui, M.H.: Real time QRS complex detection using DFA and regular grammar. Biomed. Eng. Online **16**, 31 (2017)
29. Phukpattaranont, P.: QRS detection algorithm based on the quadratic filter. Expert Syst. Appl. **42**, 4867–4877 (2015)
30. Farashi, S.: A multiresolution time-dependent entropy method for QRS complex detection. Biomed. Signal Process Control **24**, 63–71 (2016)
31. Zhou, Y., Hu, X., Tang, Z., Ahn, A.C.: Sparse representation-based ECG signal enhancement and QRS detection. Physiol. Meas. **37**, 2093 (2016)
32. Jung, W.H., Lee, S.G.: An R-peak detection method that uses an SVD filter and a search back system. Comput. Methods Programs Biomed. **108**, 1121–1132 (2012)
33. Choi, S., Adnane, M., Lee, G.J., Jang, H., Jiang, Z., Park, H.K.: Development of ECG beat segmentation method by combining lowpass filter and irregular R-R interval checkup strategy. Expert Syst. Appl. **37**, 5208–5218 (2010)
34. Zhang, C.F., Bae, T.W.: VLSI friendly ECG QRS complex detector for body sensor networks. IEEE J. Emerg. Sel. Top Circuits Syst. **2**, 52–59 (2012)
35. Zhang, F., Lian, Y.: QRS detection based on morphological filter and energy envelope for applications in body sensor networks. J Signal Process Syst. **64**, 187–194 (2011). https://doi.org/10.1007/s11265-009-0430-8
36. Kholkhal, M., Reguig, F.B.: Efficient automatic detection of QRS complexes in ECG signal based on reverse biorthogonal wavelet decomposition and nonlinear filtering. Measurement **94**, 663–670 (2016)
37. Lee, J.W., Kim, K.S., Lee, B., Lee, B., Lee, M.H.: A real time QRS detection using delay-coordinate mapping for the microcontroller implementation. Ann. Biomed. Eng. **30**, 1140–1151 (2002). https://doi.org/10.1114/1.1523030
38. Christov, I.I.: Real time electrocardiogram QRS detection using combined adaptive threshold. Biomed. Eng. Online **3**, 1–9 (2004)
39. Vijaya, G., Kumar, V., Verma, H.K.: ANN-based QRS-complex analysis of ECG. J. Med. Eng. Technol. **22**, 160 (1998)
40. Barhatte, A.S., Ghongade, R., Thakare, A.S.: QRS complex detection and arrhythmia classification using SVM. In: 2015 International Conference on Communication, Control and Intelligent Systems (CCIS)
41. Huang, R., Zhou, Y.: Disease classification and biomarker discovery using ECG data. BioMed Res. Int. **2015**, 680381 (2015)
42. Savalia, S., Acosta, E., Emamian, V.: Classification of cardiovascular disease using feature extraction and artificial neural networks. J. Biosci. Med. **5**, 64–79 (2017)

43. Deng, M., Wang, C., Tang, M., Zheng, T.: Extracting cardiac dynamics within ECG signal for human identification and cardiovascular diseases classification. Neural Netw. **100**, 70–83 (2018)

44. Xiang, Y., et al.: Automatic QRS complex detection using two-level convolutional neural network. Biomed. Eng. Online **17**(1), 13 (2018)

45. Goldberger, A.L., Amaral, L.A.N., Glass, L., Hausdorff, J.M., et al.: Components of a new research resource for complex physiologic signals.Circulation **101**(23), e215–e220 (2000)

46. https://www.physionet.org/physiobank/database/ptbdb/

Deep Learning Based Lung Nodules Detection from Computer Tomography Images

Mahender G. Nakrani[1(✉)], Ganesh S. Sable[2], and Ulhas B. Shinde[1]

[1] CSMSS's Chh. Shahu College of Engineering, Aurangabad 431011, Maharastra, India
nakrani.mahender@gmail.com
[2] G S Mandal's Maharashtra Institite of Technology, Aurangabad 431010, Maharastra, India

Abstract. Lung cancer is among the dominant cause of deaths due to cancer. Lung cancer survival largely depends on the stage at which it is diagnosed with early stage diagnosis significantly improves the survival rate of patients. Radiologist diagnoses the Computerized Tomography images by detecting lung nodules from the images. Detection of initial stage lung cancer is very challenging as the sizes of lung nodules are very small and are difficult to locate. Many computer aided detection systems to detect lung nodules were proposed to assist radiologist. Recently, Deep learning neural network has found its way into lung nodule detection system after the success it has exhibited in computer vision tasks. In this paper, we propose a novel deep convolutional neural network based system for lung nodule detection and localization. Our objective is to provide radiologist with a tool to correctly diagnosis Computer Tomography (CT) scans of patients. The system developed was able to detect and localize with the sensitivity of 92.9%. LIDC-IDRI world largest publicly available database for computer tomography scans of human lungs was used for this research.

Keywords: Convolutional neural network · Lung nodules · Deep learning · Computer tomography · Nodule detection

1 Introduction

As per Lancet reinforced Report and World Health Organization's (WHO), there were around 18.1 million cancer cases registered in 2018 worldwide [1]. The report also estimates that the cancer will be cause of 9.7 million deaths in the world in this year. Breast and lung cancers top the list with each being culprit in 12.3% of the total number of new cases registered in 2018 worldwide. Among them, Lung cancer leads the list of mortalities with anticipated to claim one in each five deaths because of cancers. Radiologists detects lung nodules from computer tomography scan of human lungs. The work load of radiologist is rising as the lung cancer has become one of the leading cancers in the world. Computer aided detection (CAD) systems that detect nodules automatically are intended to assists radiologists and reduces their workload [2]. A CAD system for lung nodule detection typically consists of three stages [3], lung CT scan pre-processing, detection of nodule candidate's and reduction of false positives. The first part focuses

© Springer Nature Singapore Pte Ltd. 2021
K. C. Santosh and B. Gawali (Eds.): RTIP2R 2020, CCIS 1381, pp. 122–130, 2021.
https://doi.org/10.1007/978-981-16-0493-5_11

on separating lung lobes from all other body parts. The second part aims for detection of nodule candidates from the segmented scans and the third part focuses on reduction of the number of false positives from the second stage.

The contribution of this paper is to detect nodule candidates. Our main objective is to apply 2D convolutional based deep learning neural network for detection of nodules. The proposed CAD system should be able to detect nodules which are even less than 3 mm in diameter for detection of early stage lung cancer from a 2D lung CT scan of 512 × 512 pixels. A sample of a lung nodule on a 2D CT scan slice is shown in Fig. 1. The proposed CAD system will comprise of two stages ie., CT scan preprocessing and lung nodule detection. Our aim is to achieve high accuracy and sensitivity with just two stages.

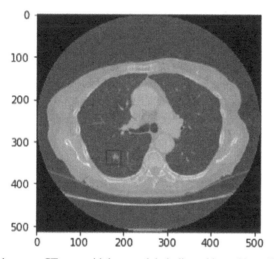

Fig. 1. Slice of pulmonary CT scan with lung nodule indicated by red box (Color figure online)

2 Background

To localize a nodule on a slice means to locate specific position of probable nodules in the lungs. A position is a group of (x, y) coordinate points on a 2D plane of a scan wherein the nodules lies. The proposed system aims is to locate these positions. The success of AlexNet [4], the deep convolutional neural networks (CNN) developed by Alex Krizhevsky in the image classification challenge laid a pathway to solve different image classification and object detection problems. This motivated researchers to apply CNNs to various tasks of computer vision. Deep learning algorithms are these CNNs with large number of layers and are usually referred as deep convolutional neural networks DCNNs. DCNNs also found their way into the medical applications and many CAD systems were developed using DCNNs for classification and detection in medical images over the years. One of these application were DCNNs had promising performance was the detection of lung nodules. In [5] a 2D CNN was implemented which had nodule detection accuracy of 60.2% with 2.1 average false positives per scan. A 3D CNN was implemented in [6] and was able to achieve sensitivity of 80% with false positives of 22.4 per scan. A multi-level 2D MU-Net was developed in [7] which was able to detect lung nodules with sensitivity of 94.4% with average of 215 candidates per scans. In [8], a vanilla 3D CNN was used which was named as DeepMed had a sensitivity of 93.9% for detection of lung nodules with false positives of 4 per scans. A 3D CNN with multi-scale prediction strategy was developed in [9] which was able to detect nodules with sensitivity of 93.4% and 4 false positive per scans.

3   Chest CT Scan Dataset

LIDC-IDRI public archive [13] was used for patient lung CT scan for the experimentation. 8 medical imaging companies and seven academic centers collaborated together to create this public archive. It is a collection of thoracic CT scans of 1018 cases for lung cancer screening and diagnostics. The archive is also linked with marked-up annotated lesions in a XML files. The annotation of the CT scans was done in two phases, i.e., blind phase and open diagnosis phase by four experienced radiologists. The CT scan consists of axial slice of around 100 to 400 images for each patient. Each slice contains scan of 512×512 pixels. All the scans are available in Digital Imaging and Communications in Medicine (DICOM) format.

4   Method

The first stage of the proposed system is to preprocess the CT scan slices to segment lung from other parts of body such as human tissues, organs, bones etc. The segmentation is followed by the splitting of image into 32×32 pixels windows from 512×512 pixels slice giving a total of 256 windows. This windows are utilized to generate heatmaps from the annotation given in XML files to create labels. Figure 2(a) shows a slice of CT scan and the label generated for the slice from XML annotated file is shown in Fig. 2(b). Our initial approach was to generate heatmap of probable lung nodules directly from the input preprocessed slices by applying it to a 2D DCNN, but result were very poor. To overcome this we converted 32×32 pixels windows ($16 \times 16 = 256$ of them) into categorical labels consisting of 256 categories corresponding to 1 for each window. Then we trained the proposed 2D CNN architecture to find the probability of whether a window has nodule or not.

4.1   Preprocessing and Segmentation

The CT scans pixels values are given in Hounsfield Unit (HU) [15–17, 19]. The HU is the measurement of radio density of different objects under the scan. The first step was to transform the voxels values outside the bore of the CT scan into 0. Then a morphological operation based segmentation technique is carried out to segment lung from other parts such as outside air, bones, tissues and body parts that would make the data noisy. The segmentation was done by extracting two largest regions by area from

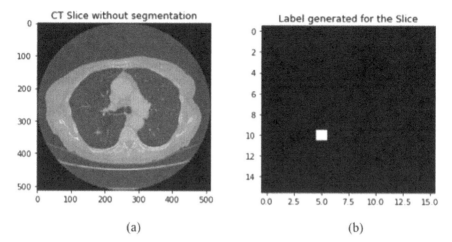

Fig. 2. Example of CT scan slice (a) CT scan slice. (b) Label generated from the XML annotation file.

the image. To these two regions we applied erosion operation and to include nodules attached with the walls of lungs we applied dilation operation to it. Next closing and filling operation was applied to generate mask. Finally the mask was superimposed on the slice to generate lung segmented slice. The original 2D scan slice is shown in Fig. 3(a) and the preprocessed slice is shown in Fig. 3(b).

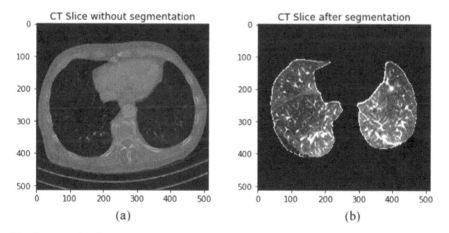

Fig. 3. Example of preprocessing (a) Original slice of CT scan (b) Slice after preprocessing.

4.2 Nodule Detection

The crucial step in any CAD system is candidate detection. The nodule candidate detection method should be able to detect large number of nodule candidate detection with high sensitivity. We propose a novel 2D Deep Convolutional Neural Network (DCNN) architecture for nodule candidates detection from CT scans. Our initial approach was to generate heatmaps of 16×16 (each pixel for subimage of 32×32 mask), but the results were poor. We then decided to flatten the 16×16 giving us 256 categories where the category is 1 if the nodule candidate is present in that mask and all others 0. The proposed DCNN initiates with an input layer of size 512×512. It consists of 13 convolution layer with 8, 16, 32, 64, 128 and 256 batches. The convolution was performed by small kernel of 3×3 and activated by Rectified linear units (ReLU). It consists of 5 Max-pooling layer with kernel of size 2×2 and a fully connected layer which was activated by softmax activation function to giving final output of 256 categories. This DCNN architecture is shown in Fig. 4 and there details are given in Table 1.

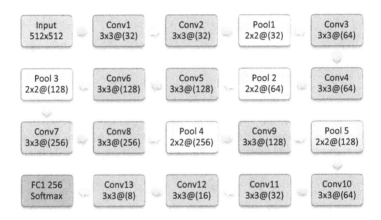

Fig. 4. Deep convolution neural network.

Table 1. Details of deep convolution neural network

Layers	Params	Activation	Output
Input	0		$512 \times 512 \times 1$
Conv1	320	ReLu	$512 \times 512 \times 32$
Conv2	9248	Relu	$512 \times 512 \times 32$
Pool1	0		$256 \times 256 \times 32$
Conv3	18496	Relu	$256 \times 256 \times 64$
Conv4	36928	Relu	$256 \times 256 \times 64$

(*continued*)

Table 1. (*continued*)

Layers	Params	Activation	Output
Pool1	0		128 × 128 × 64
Conv5	73856	ReLu	128 × 128 × 128
Conv6	147584	Relu	128 × 128 × 128
Pool2	0		64 × 64 × 128
Conv7	295168	ReLu	64 × 64 × 256
Conv8	590080	Relu	64 × 64 × 256
Pool3	0		32 × 32 × 256
Conv9	295040	ReLu	32 × 32 × 128
Pool4	0		16 × 16 × 128
Conv10	73792	ReLu	32 × 32 × 64
Conv11	18464	ReLu	32 × 32 × 32
Conv12	4624	ReLu	32 × 32 × 16
Conv13	1160	ReLu	32 × 32 × 8
Flatten	0		2048
Dense	524544	Softmax	256

5 Experimental Results

LIDC-IDRI was used to evaluate the performance of the proposed CAD system. We have chosen slices with nodule candidates only for the experimentation and ignored all slices without nodule candidates. These slices are divided into batch of 80% dataset for training process and 20% dataset for testing. The 80% dataset for training processes if further divided into 70% dataset for training and 30% dataset for validation. The result of top 5 predictions on some samples is shown in Fig. 5. Figure 6 shows the graph of the sensitivity of proposed CAD system against candidates per scan. From the figure it can be noted that the sensitivity of the system becomes almost constant after 5 candidates per scan. Table 2 compares the results of the proposed CAD system with other CAD system illustrated in [18]. The proposed CAD system was able to achieve the sensitivity of 92.9% with only 5 false positives per scan on testing dataset. With simple 20 layer Deep Learning architecture and by converting the problem to categorical classification problem, we were able to produce comparable results without false positive reduction stage.

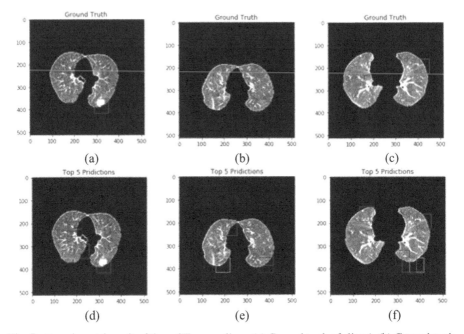

(a)   (b)   (c)

(d)   (e)   (f)

Fig. 5. Experimental result of three CT scans slices. (a) Ground truth of slice 1. (b) Ground truth of slice 2. (c) Ground truth of slice 3. (d) Top 5 predictions slice 1. (e) Top 5 predictions slice 2. (f) Top 5 predictions slice 3.

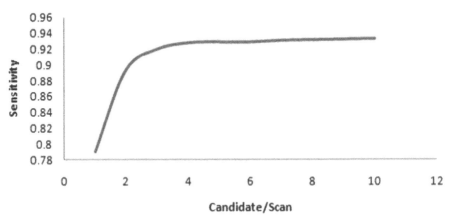

Fig. 6. Sensitivity of the proposed CAD system vs number of candidates per scans.

Table 2. Comparison of DCNN based CAD systems for detection of lung nodules.

System	Sensitivity	Candidates/Scan
ISI-CAD	0.857	335.8
Subsolid CAD	0.369	290.7
Large CAD	0.317	47.5
M5L	0.769	22.3
ETRO-CAD	0.930	333.0
Baseline-CAD (4 anchors)	0.895	25.8
R-CNN	0.946	15.0
Proposed CAD	0.929	5.0

6   Conclusion and Future Scope

In this study, a novel 2D deep convolution neural network for the detection of pulmonary nodule candidate is proposed. The experimentation results carried out on LIDC-IDRI indicates that the proposed CAD system produces comparable result even without implementation of false positive reduction. We will try to implement other deep learning architecture like U-Net, R-CNN to improve the sensitivity of proposed system without effecting false positives in future. We will also try to implement false positive reduction stage to achieve same sensitivity with less number of false positives per scans.

References

1. Latest global cancer data: Cancer burden rises to 18.1 million new cases and 9.6 million cancer deaths in 2018. International agency for research on cancer, World Health Organization, Press Release N° 263, September 2018
2. Ruikar, D.D., Hegadi, R.S., Santosh, K.C.: A systematic review on orthopedic simulators for psycho-motor skill and surgical procedure training. J. Med. Syst. **42**(9), 168 (2018)
3. Nakrani, M.G., Sable, G.S., Shinde, U.B.: Detection of lung nodules in computed tomography image using deep machine learning: a review. In: Proceedings of International Conference on Communication and Information Processing (ICCIP) (2019)
4. Krizhevsky, A., Sutskever, I., Hinton, G.E.: ImageNet classification with deep convolutional neural networks, pp. 1097–1105 (2012)
5. da Nóbrega, R.V.M., Peixoto, S.A., da Silva, S.P.P., Filho, P.P.R.: Lung nodule classification via deep transfer learning in CT lung images. In: IEEE 31st International Symposium on Computer-Based Medical Systems (CBMS), Karlstad, pp. 244–249 (2018)
6. Hamidian, S., Sahiner, B., Petrick, N., Pezeshk, A.: 3D convolutional neural network for automatic detection of lung nodules in chest CT. In: Proceedings of SPIE (2017)
7. Heeneman, T., Hoogendoorn, M.: Lung nodule detection by using deep learning (2018). https://www.beta.vu.nl/nl/Images/werkstuk-heeneman_tcm235-876475.pdf
8. Winkels, M., Cohen, T.S.: 3D G-CNNs for pulmonary nodule detection. arXiv preprint arXiv: 1804.04656 (2018)

9. Gu, Y., Lu, X., Yang, L., Zhang, B., Yu, D., Zhao, Y.: Automatic lung nodule detection using a 3D deep convolutional neural network combined with a multi-scale prediction strategy in chest CTs. Comput. Biol. Med. **103**, 220–231 (2018)
10. Hu, Z., Muhammad, A., Zhu, M.: Pulmonary nodule detection in CT images via deep neural network: nodule candidate detection. In: Proceedings of the 2nd International Conference on Graphics and Signal Processing, ICGSP 2018, pp. 79–83 (2018)
11. Sabari Nathan, D., Saravanan, R., Anbazhagan, J., Koduganty, P.: Comparison of deep feature classification and fine tuning for breast cancer histopathology image classification. In: Santosh, K.C., Hegadi, R.S. (eds.) RTIP2R 2018. CCIS, vol. 1036, pp. 58–68. Springer, Singapore (2019). https://doi.org/10.1007/978-981-13-9184-2_5
12. Tang, H., Kim, D.R., Xie, X.: Automated pulmonary nodule detection using 3D deep convolutional neural networks. In: IEEE 15th International Symposium on Biomedical Imaging (ISBI 2018), Washington, DC, pp. 523–526 (2018)
13. Krishnamurthy, S., Narasimhan, G., Rengasamy, U.: An automatic computerized model for cancerous lung nodule detection from computed tomography images with reduced false positives. In: Santosh, K.C., Hangarge, M., Bevilacqua, V., Negi, A. (eds.) RTIP2R 2016. CCIS, vol. 709, pp. 343–355. Springer, Singapore (2017). https://doi.org/10.1007/978-981-10-4859-3_31
14. Setio, A.A.A., Traverso, A., Bel, T.: Validation, comparison, and combination of algorithms for automatic detection of pulmonary nodules in computed tomography images: the LUNA16 challenge. arXiv:1612.08012 (2016)
15. Ruikar, D.D., Santosh, K.C., Hegadi, R.S.: Automated fractured bone segmentation and labeling from CT images. J. Med. Syst. **43**(3), 1–13 (2019). https://doi.org/10.1007/s10916-019-1176-x
16. Ruikar, D.D., Santosh, K.C., Hegadi, R.S.: Segmentation and analysis of CT images for bone fracture detection and labeling. In: Medical imaging: Artificial Intelligence, Image Recognition, and Machine Learning Techniques. CRC Press (2019). ISBN: 978-0-36713-9612. Chapter 7
17. Ruikar, D.D., Santosh, K.C., Hegadi, R.S.: Contrast stretching-based unwanted artifacts removal from CT images. In: Santosh, K.C., Hegadi, R.S. (eds.) RTIP2R 2018. CCIS, vol. 1036, pp. 3–14. Springer, Singapore (2019). https://doi.org/10.1007/978-981-13-9184-2_1
18. Ding, J., Li, A., Hu, Z., Wang, L.: Accurate pulmonary nodule detection in computed tomography images using deep convolutional neural networks. In: Descoteaux, M., Maier-Hein, L., Franz, A., Jannin, P., Collins, D.L., Duchesne, S. (eds.) MICCAI 2017. LNCS, vol. 10435, pp. 559–567. Springer, Cham (2017). https://doi.org/10.1007/978-3-319-66179-7_64
19. Santosh, K.C., Antani, S., Guru, D.S., Dey, N. (eds.): Medical Imaging: Artificial Intelligence, Image Recognition, and Machine Learning Techniques. CRC Press, Boca Raton (2019)

Enhancement of MRI Brain Images Using Fuzzy Logic Approach

M. Ravikumar[1]([✉]), B. J. Shivaprasad[1], and D. S. Guru[2]

[1] Department of Computer Science, Kuvempu University, Jnanasahyadri, Shimoga, Karnataka, India
ravi2142@yahoo.co.in, shivaprasad1607@gmail.com
[2] Department of Studies in Computer Science, University of Mysore, Manasagangothri, Mysore, Karnataka, India
dsguruji@yahoo.com

Abstract. In this work, fuzzy method is proposed to enhance the contrast of Magnetic Resonance Imaging (MRI) brain images. Negative Image (NI), Log Transform (LT), Gamma Correction (GC), Histogram Equalization (HE), Adaptive Histogram Equalization (AHE) and Dynamic Histogram Equalization (DHE) methods are compared with proposed method. The performance is evaluated by using quantitative measures like Michelon Contrast (MC), Entropy, Peak Signal to Noise Ratio (PSNR), Structure Similarity Index Measurement (SSIM) and Absolute Mean Brightness Error (AMBE) as a parameter on BRATS-2014 dataset. The proposed method gives good results for Entropy, PSNR and AMBE, we need to improve the proposed method for MC and SSIM.

Keywords: Magnetic Resonance Imaging (MRI) · Histogram Equalization (HE) · Adaptive Histogram Equalization (AHE) · Dynamic Histogram Equalization (DHE)

1 Introduction

Human brain is the most complex and essential part of a body which is made up of many neurons. Brain controls and also coordinates all the necessary functions of a body. Diseases will affect the functionality of a brain, diagnosing such diseases is very much essential and is very challenging task. Different image acquisition techniques are available in the literature, such as Magnetic Resonance Imaging (MRI), Computerized Tomography (CT), Positron Emission Tomography (PET), Single Photon Emission Computed Tomography (SPECT) and functional Magnetic Resonance Imaging (fMRI) [1]. MRI is widely used brain imaging techniques, because MRI does not produce radiation to the body. Hence it is not harmful to the brain.

Even-though MRI technique has more advantages, it has a limitation that, MRI technique approach is produces low contrast images. Analysis of these low

© Springer Nature Singapore Pte Ltd. 2021
K. C. Santosh and B. Gawali (Eds.): RTIP2R 2020, CCIS 1381, pp. 131–137, 2021.
https://doi.org/10.1007/978-981-16-0493-5_12

contrast images is a difficult task for doctors and radiologists. Increasing the contrast of an image will ease analyzing the human brain. Thus contrast enhancement is an important stage in MRI brain image analysis and different spatial domain, frequency domain & other learning techniques are used for enhancing of MRI images. Enhancement is a process where in visual quality of an image is improved so that the resultant image is more suitable for specific application [2]. The spatial domain refers to the image plane itself and methods in spatial domain are based on directly modifying the value of the pixels.

In this work, we propose a method i.e, fuzzy logic approach to enhance the MRI brain images and the proposed method is compared with spatial domain methods. Based on the qualitative metrics result is measured and result shows the superiority of the proposed method.

This paper is organized as follows: In Sect. 2, details of different enhancement methods are addressed. In Sect. 3, different quantitative parameters used to measure enhancement is discussed. Section 4, we discuss the experimentation and also result are given.

2   Related Work

To enhance the contrast of MRI brain images, different techniques were proposed like Histogram Equalization (HE) [3,4,6,8,9], Adaptive Histogram Equalization (AHE) [3,7], Contrast Limited Adaptive Histogram Equalization (CLAHE) [3,6], Local Histogram Equalization (LHE) [3], Bi-Histogram Equalization (BBHE) [4,9], Minimum Mean Brightness Error Bi-Histogram Equalization (MMBEBHE) [4,5], Brightness Preserving Dynamic Histogram Equalization (BPDHE) [4,5,7], Recursive Mean Separated Histogram Equalization (RMSHE) [5], Brightness Preserving Dynamic Histogram Equalization (BPDHE) [5], Dualistic Sub-Image Histogram Equalization (DSIHE) [5], Brightness Preserving Dynamic Fuzzy Histogram Equalization (BPDFHE) [6], Different Techniques like Global Histogram Equalization (GHE) [7], Modified BHE, Brightness preserving BHE (BBHE) [9], Fuzzy logic based Adaptive Histogram Equalization (AHE) [4], Multi Scale Retinex (MSR) [8] and Non-subsampled Contourlet Transform (NSCT)- FU [8].

The enhanced image is measured using different performance matrices such as Entropy [4,5,8], Features Similarity Index [4], Contrast Improvement Index [4], Peak Signal Noise Ratio (PSNR) [5,6,8], Absolute Mean Brightness Error (AMBE) [5,7,9], Mean Squared Error (MSE) [6], Root Mean Square Error (RMSE) [6,9], Weber contrast [7], Michelson contrast [7,9], Mean Absolute Error (MAE) [8], Structure Similarity Index Measurement (SSIM) [8] and Pixel distance [9].

From the related work, it is observed that, most of the works on enhancement of brain MRI images used histogram equalization, adaptive equalization and Contrast Limited Adaptive Histogram Equalization methods. To evaluate the transformed images commonly Peak Signal Noise Ratio, Root Mean Square Error and Absolute Mean Brightness Error techniques are used to measure the performance.

3 Proposed Method

In this section, we discuss the proposed method for enhancing brain tumor images and the block diagram of the proposed method is shown in the Fig. 1.

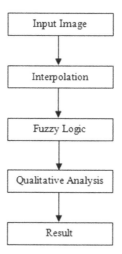

Fig. 1. Block diagram of the proposed method.

Initially, we take a gray scale brain images, in order to improve visual appearance of the image i.e., quality of the image, an interpolation method is applied. Using the bi-linear interpolation method, the visual appearance of the image is maintained while resizing of an image [10]. Once the interpolation is completed, we use the fuzzy logic approach for enhancement purpose. In fuzzy logic image is partitioned and each partitioned will be considered as fuzzy window, using mean and variance fuzzy window is enhanced. İn same way all the fuzzy windows are enhanced and finally summing up all fuzzy windows. Fuzzy logic mainly includes fuzzification, inference engine and defuzzification [11], which are given in the Eq. 1 and 2.

Here fuzzification is required to map the input image with fuzzy plane, vice versa for defuzzification i.e., the membership of a point $P_{ij}(x, y) \in D$ to the window $W_{ij}(x, y)$ are given by the Eq. 1.

$$W_{ij} = \frac{(P_{ij}(x, y))^{\gamma}}{\sum_{i=1}^{n} \sum_{j=1}^{m} (P_{ij}(x, y))^{\gamma}} \tag{1}$$

Where, $W_{ij} : D \rightarrow [0, 1]$
W_{ij} describe the membershop and $P_{ij}(x, y)$ describe the pixel value.
$\gamma \in (0, \infty)$ and γ controls the fuzzification and defuzzification.

The transform ψ_{enh} is built as a sum of the transformed W_{ij} weights with degree of membership ψ_{ij} . The enhanced image is given by Eq. 2.

$$\psi_{enh}(f) = \sum_{i=1}^{n}\sum_{j=1}^{m}W_{ij} \times \psi_{ij}(f) \qquad (2)$$

Where, $\psi_{ij}(f)$ is image (f) before enhancement.
$\psi_{enh}(f)$ is image (f) after enhancement.

After the fuzzy logic process is completed, we compute the quality of an images by using quantitative measure like Michelon Contrast (MC), Entropy, Peak Signal to Noise Ratio (PSNR), Structure Similarity Index Measurement (SSIM) and Absolute Mean Brightness Error (AMBE) as a parameter. Obtained results are tabulated in Table 1 and the images are shown in the Fig. 2.

In the next section, experimentation is discussed.

4   Experimentation

In subsequent sections, result and discussion are given.

4.1   Result

For the purpose of experimentation, we have used 200 MRI brain images collected from the dataset- BRATS-2014 repository. To enhance the brain images, the different spatial domain enhancement methods are used like Negative Image (NI), Log Transform (LT), Gamma Correction (GC), Histogram Equalization (HE), Adaptive Histogram Equalization (AHE) and Dynamic Histogram Equalization (DHE) are used. After the experimentation, the results are shown in the Fig. 2.

Using different c values, image is enhanced using Log transform. The results are shown in the Fig. 4 and PSNR value is tabulated in Table 2. When c value is 1.5, it gives good result, result is measured based on PSNR as a parameter.

Gamma value is considered randomly (0.89) for the purpose of enhancement.

4.2   Discussion

To select the best enhancement method quantitative analysis parameters are used, they are MC, PNSR, SSIM and AMBE. Table 1 gives the different quantitative measure results.

From Table 1, it is observed that the proposed method gives good result for Entropy, PSNR and AMBE. Qualitative matrices values are shown in the Fig. 3, where the lower AMBE and higher Entropy, MC, SSIM PSNR values indicates good quality image. Further, we need extend our proposed method to improve MC and SSIM matrices.

Table 1. Comparison of different quantitative methods.

	Entropy	MC	PSNR	SSIM	AMBE
HE	3.42	1.00	7.22	0.123	74.89
AHE	0.005	1.00	13.42	0.62	28.45
DHE	3.47	0	15.95	0.829	30.921
Negative	2.00	0.30	1.09	−0.11	198.08
Log Transform	1.82	1.00	190.8	0.84	36.86
Gamma Correction	0	0	28.45	0.62	13.42
Proposed (Fuzzy)	**3.69**	**0.99**	**23.91**	**0.61**	**8.33**

Table 2. Different c values of PSNR.

Type	C = 0.1	C = 0.5	C = 1.0	C = 1.5	C = 2.0	C = 2.5	C = 3.0	C = 3.5	C = 4	C = 4.5
PSNR	17.63	18.00	18.45	**19.08**	15.00	14.40	13.00	12.49	12.10	11.94

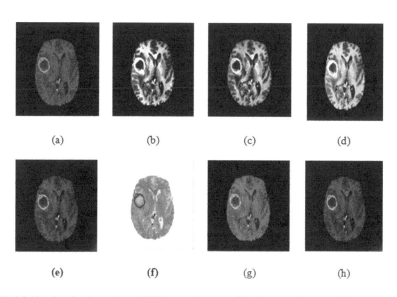

(a)   (b)   (c)   (d)

(e)   (f)   (g)   (h)

Fig. 2. Methods of enhancing MRI brain images. (a) Original Image (b) HE (c) AHE (d) DHE (e) Fuzzy Logic (f) Negative Image (g) Log Transform and (h) Gamma Correction.

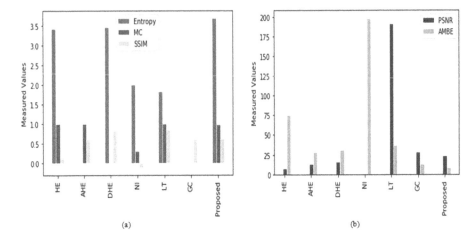

Fig. 3. Qualitative performance Metric Values (a) Represent the results for Entropy, MC and SSIM (b) Represent the results for PSNR and AMBE.

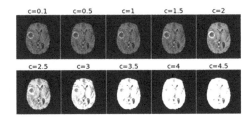

Fig. 4. Different values of c to enhance MRI brain image using Log Transform.

5   Conclusion

In this work, we have proposed a Fuzzy logic approach to enhance MRI brain images. The proposed method is compared with HE, AHE, DHE, Negative, Log Transform and Gamma correction methods. These methods are verified using performance matrices like, MC, PSNR, SSIM and AMBE as parameter. The proposed method works good for entropy, PNSR and AMBE, we need to improve the proposed method for MC and SSIM. Further, we plan to extend our work towards frequency domain and machine learning approaches to improve the performance.

References

1. Ruikar, D.D., Santosh, K.C., Hegadi, R.S.: Contrast stretching-based unwanted artifacts removal from CT images. In: International Conference on Recent Trends in Image Processing and Pattern Recognition, pp. 3–14. Springer (2018)
2. Bedil, S.S., Khandelwal, R.: Various image enhancement techniques - a critical review. Int. J. Adv. Res. Comput. Commun. Eng. **2**(3) (2013)

3. Agravat, R.R., Raval, M.S.: Deep learning for automated brain tumor segmentation in MRI images, pp. 183–201 (2018)
4. Kaur, H., Rani, J.: MRI brain image enhancement using Histogram Equalization techniques. In: International Conference on Wireless Communications, Signal Processing and Networking (WiSPNET), Chennai, pp. 770–773 (2016)
5. Subramani, B., Velucham, M.: MRI brain image enhancement using brightness preserving adaptive fuzzy histogram equalization. Int. J. Image Syst. Technol. **28**(3), 1–6 (2018)
6. Senthilkumaran, N., Thimmiaraja, J.: Study on histogram equalization for MRI brain image enhancement. In: Proceedings of International Conference on Recent Trends in Signal Processing, Image Processing and VLSI, ICrtSIV 2014, pp. 317–325 (2014)
7. Suryavamsi, R.V., Reddy, L.S.T., Saladi, S., Karuna, Y.: Comparative analysis of various enhancement methods for astrocytoma MRI images. In: International Conference on Communication and Signal Processing (ICCSP), Chennai, pp. 0812–0816 (2018)
8. Senthilkumaran, N., Thimmiaraja, J.: Histogram equalization for image enhancement using MRI brain images. In: 2014 World Congress on Computing and Communication Technologies, Trichirappalli, pp. 80–83 (2014)
9. Zhou, F., Jia, Z., Yang, J., Kasabov, N.: Method of improved fuzzy contrast combined adaptive threshold in NSCT for medical image enhancement. Hindawi BioMed Res. Int. **2017**, 1–10 (2017)
10. Oak, P.V., Kamathe, R.S.: Contrast enhancement of brain MRI images using histogram based techniques. Int. J. Innov. Res. Electr. Electron. Instrum. Control Eng. **1**(3), 90–94 (2013)
11. Han, D.: Comparison of commonly used image interpolation methods. In: Proceedings of the 2nd International Conference on Computer Science and Electronics Engineering (ICCSEE 2013), pp. 1556–1559 (2013)
12. Thakur, A., Mishra, D.: Fuzzy contrast mapping for image enhancement. In: Proceedings of the 2nd International Conference on Signal Processing and Integrated Networks (SPIN), pp. 549–552 (2015)

Image Analysis and Recognition

Exploiting Radon Features for Image Retrieval

S. A. Angadi[1(✉)] and Hemavati C. Purad[2]

[1] Visvesvaraya Technological University, "Jnana Sangama", Belagavi 590018, Karnataka, India
saangadi@vtu.ac.in
[2] Tontadarya College of Engineering, Gadag, Karnataka, India
hemacpurad@gmail.com

Abstract. Radon transform is one of the features used in image reconstruction from the early days of image processing, later it has been used in many applications of astronomy and other fields, such feature is explored for the process of matching and retrieval of images. The proposed system has 4 phases, pre-processing, feature extraction, knowledgebase construction and image retrieval. The local features such as sum, mean, standard deviation, & the eigen values are extracted for each radon transformed image and stored in a knowledgebase. Cityblock distance measure and query-by-image method are employed for matching and retrieval of images respectively and the accuracy of 87% is achieved.

Keywords: CBIR · Radon transform · Feature extraction

1 Introduction

The rapid growth of technologies and increased usage of internet which has exploded online storage/retrieval demands fast and efficient storage, search and retrieval process. Most of the image retrieval systems lack the understanding of the user's perception behind the query image, discussed in Mussarat Yasmin et al. [1] and is most challenging problem known as semantic gap problem. The best feature descriptors are in need to fill the semantic gap between low level to high level features as described in Olfa Allani et al. [2] which builds pattern based ontology module to fill the semantic gap. However due to images' variability, ambiguity and the wide range of illumination, image characterization is challenging. The most popular way to tackle such problems is to utilize the low level features such as color, texture histograms, object shapes etc. In recent years, a variety of feature representation methods have been proposed to solve how to describe the visual objects in different images, some methods focus on the local information, such as Scale-Invariant Feature Transform (SIFT), Speeded-Up Robust Feature Transform (SURF), Histogram of Oriented Gradients (HOG) etc. and others are holistic descriptors such as GIST mentioned in Xiao Cai et al. [3].

Expressing an image content with only one feature is difficult, explained in Abir Baöazaoui et al. [4], Hailong Liu et al. [5], Wei Yu et al. [6], Peizhong Liu et al. [7], to obtain the relevant information from the image data two or more different features are extracted, resulting in two or more feature descriptors at each image point. Most of

© Springer Nature Singapore Pte Ltd. 2021
K. C. Santosh and B. Gawali (Eds.): RTIP2R 2020, CCIS 1381, pp. 141–151, 2021.
https://doi.org/10.1007/978-981-16-0493-5_13

the literature review reported that, fusion of one or more features may result in efficient retrieval of images in the CBIR system. Some of the literature review related to this is summarized here. The papers Guoyong Duan et al. [8], Jun Yue et al. [9], Anu Bala et al. [10], Soumya Prakash Rana et al. [11], and L.K. Pavithra et al. [12], Mahmood Sotoodeh et al. [13], Fahimeh Alaei et al. [14] have proposed a combination of color, texture and shape features for image retrieval. The fusion of low level global and local features described in Jufeng Yang et al. [15] uses local texture information of a medical images, which does not possess any colour, by Self-Organising Map algorithm to extract the features of region of interest (ROI). In the proposed method initially radon transform which is scale and translation invariant, global feature is employed and local features such as sum mean eigen values are computed.

The Radon transform and its spectral properties are largely used for image reconstruction, illustrated in F. Boschen et al., Hanchuang Wang et al. [16, 17] for medical imaging, and is robust against compression, filtering, blurring, sharpening, and noise etc. and improves the low frequency components by Dattatray V. Jadhao et al. [18] and has the advantage of dimensionality reduction, can capture the directional features of the pattern image by projecting the pattern onto different orientation slices suggested by G. Y. Chen et al.; Elouedi Ines et al.; Minghao Piao et al. [19–21] and allows to characterize easily the features of geometrical transforms. It also permits an easy extraction of an indexing vector of the image by Frederic Lefebvre et al. [22]. Radon transform is translation invariant, rotation invariant and scaling invariant as claimed in Yan Chen et al. and Yudong Zhang et al. [23, 24]. These properties of the radon transform motivated us to consider the radon transform as feature and explored in retrieval of images.

Proposed method uses both the global and local features. The radon transform is used as a global feature and sum of the radon coefficients in each row, mean of standard deviation of radon transformed image, and eigen values by Nikolaos V Boulgouris et al. [25] are computed on radon transformed image and all the values are concatenated horizontally to form a feature vector of size 122-by-1 for each image and stored in a knowledgebase. Query-by-Image method is used for querying, Cityblock distance measure is employed to compare and sort the distance values. According to their similarity the top ranked images are retrieved and displayed. The system achieved the accuracy of 87%.

The rest of the paper is organized as follows, Sect. 2 describes the proposed methodology, results and discussion is presented in Sect. 3 and Sect. 4 concludes the work.

2 Proposed Methodology

A new methodology, which is based on scale and translation invariant features of radon transform for content based image retrieval is proposed in this paper. The low level features such as sum, mean, standard deviation, and eigen values are computed on radon transformed image. The method comprises of 4 phases, pre-processing, feature extraction, knowledgebase construction and image retrieval. The block-diagram of the proposed work is given in Fig. 1, which depicts all the processing steps used in this work.

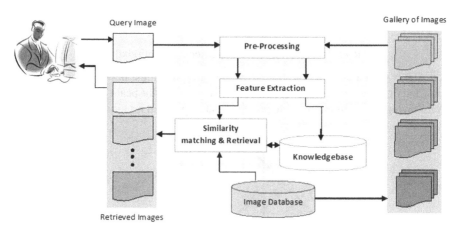

Fig. 1. Block diagram of the proposed system

2.1 Preprocessing

Pre-processing is a common name for operations with images at lowest level of abstraction. The aim of pre-processing is an improvement of the image data that suppresses unwanted distortions or enhances some image features important for further processing [26–28]. In this proposed method, the pre-processing is performed to make the image suitable for extracting the features needed. Initially Input Image of size <256X384X3 uint8> is loaded from the image database, and resized to <80X80X3 uint8> pixels and RGB to grayscale conversion is applied to obtain a grayscale <80X80 uint8> image; finally converted to double type <80X80 double>.

2.2 Features

Headings Feature is a piece of information in general and/or a specific structure or may also be the result of a general neighborhood operation or feature detection applied to the image [29, 30]. The proposed method is deployed by exploiting the Radon Transform Feature-which is scale and translation invariant, and local features such as mean, standard deviation and eigen-values are computed over the radon transformed image. The detailed description of features employed are discussed in below sections.

Radon Transform. The Radon Transform is defined as a collection of 1D projections around an object at angle intervals Θ(theta). A projection of a two-dimensional function f(x, y) is a set of line integrals. The radon function computes the line integrals or projections of an image matrix from multiple sources along parallel paths, or beams, in a certain direction (or along specified directions). Beams are spaced 1 pixel unit apart. To represent an image, the radon function takes multiple, parallel-beam projections of the image from different angles by rotating the source around the center of the image. In general, the RT of a 2-D image function f(x, y) is denoted as R(θ, r), which is defined as follows,

$$R(\Theta,r)=R\{f(x,y)\}=\iint f(x,y)\delta(r\text{-}x\cos\Theta\text{-}y\sin\Theta)dxdy \qquad (1)$$

where $\delta(.)$ is the Dirac function. $\theta \in [0, \pi)$ denoting the angle between the beam and x-axis, and $r \in (-\infty, \infty)$ is the perpendicular distance from the beam crossing the origin. The following Fig. 2 shows a single projection at a specified rotation angle.

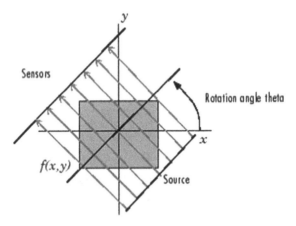

Fig. 2. Parallel-beam projection at rotation angle theta

The Radon transform of the grayscale image, I, is computed as mentioned in Eq. (1) for a specific set of angles, theta, as shown in Fig. 2. The resultant of Eq. (1) is the radon image R, which contains radon transform in each column for each angle in theta (default is 0:179). The centre pixel of I is defined to be floor $((size(I) + 1)/2)$ which is the pixel on the x-axis corresponding to $x' = 0$.

Eigen Values. Eigenvalues and eigenvector features are prominently used in the analysis of linear transformations. Originally utilized to study principal axes of the rotational motion of rigid bodies, eigenvalues and eigenvectors have a wide range of applications, for example in stability analysis, vibration analysis, atomic orbitals, facial recognition, and matrix diagonalization. In the language of linear algebra, the Radon transform is a linear transformation, maps a linear combination of functions to the same linear combination of the Radon transforms of the functions separately. Diagonal matrix of eigenvalues is obtained on radon transformed image to find important subsystems or patterns inside noisy image.

Sum, Mean and Standard Deviation. In mathematics, summation (denoted with an enlarged capital Greek sigma symbol $\sum$) is the addition of a sequence of numbers; the result is their sum or total. The proposed method computes the sum of all the pixels in each row of radon image and stored in a column vector as shown in the Eq. (2),

$$row(p, 1) = \sum_{p=1 \text{ to } m} \sum_{q=1 \text{ to } n} R(p, q) \qquad (2)$$

Where m and n are number of rows and columns of a radon transformed image, p and q are indices and the 2D array 'R' gives the radon coefficients of radon transformed image. Equation (2) computes the sum of all the pixel values of each row of radon transformed image R, and the sum is stored in a column vector called row(p, 1), where p is the number of rows and 1 is the number of columns. The column wise mean is calculated over the coefficients of radon transformed image. The resultant values are same for all the columns, so mean is computed again over previously obtained mean values. Further the standard deviation is computed for each column of radon transformed image and it is observed that all the values are same for each column, so mean of those values are computed and stored in a variable STD_MEAN. A low standard deviation indicates that the data points tend to be very close to the mean, whereas high standard deviation indicates that the data points are spread out over a large range of values.

Sum, mean and standard deviation are first order statistics and concerned with properties of individual pixels. These statistics are considered as one of the image features on radon transformed image R, and the detailed steps for constructing the knowledgebase by extracting the features from the radon transformed images are given in the Algorithm 1.

The proposed method employs Radon transform as a global feature on the pre-processed input image of size 80-by-80 with default theta (0:179) values; and Eq. (1) results in radon image R of size 117-by-180, in which the columns contain the Radon transform for each angle in theta (0:179). The sum of all pixel values in each row of radon image R is computed and stored in a column vector of size 117-by-1 and reshaped to 1-by-117, mean of column wise mean (of size 1-by-1), mean of standard deviations (of size 1-by-1) and mean of means obtained for the diagonal matrix of eigen values (of size 1-by-1) are computed as local features over the radon image and all these features along with class and image names are concatenated horizontally to form the feature vector of size (1-by-122) for each image. The FEATURE_EXTRACTION algorithm detailed in Algorithm 1, iteratively builds the knowledgebase FDB_W2.mat containing 500 feature vectors of each image from the database named WANG-SET-2. The database WANG-SET-2 is built by choosing 5 classes of 500 images from the original Wang database.

Algorithm 1: Feature_Extraction(I)
Input: Pre-processed Image Iij, where j ∈ images in i, where i ∈ subfolders of database D.
Output: Knowledgebase, FDB_W2.mat containing 120 columns of radon coefficients of radon image, class name (at 121th column) and an input image name (at 122nd column) for each image Iij.
Begin
Step-1: For each Iij, in j∈i∈D
Do
    Step-1a: Compute the radon transform R for the pre-processed image Iij by applying equation (1) with default theta values 0:179.
    Step-1b: Compute the sum of all the pixel values in each row of radon transformed image as in equation (2) and store in a column vector variable "row" and finally reshape it to row vector.
    Step-1c: Compute the mean of means obtained for Radon transformed image RIij.
    Step-1d: Compute the mean of standard deviations of the radon transformed image RIij.
    Step-1e: Compute the Diagonal matrix of eigen values, on row-by-row sized radon transformed image.
    Step-1f: Compute the mean of means obtained for Diagonal matrix of Eigen values.
    Step-1g: Form the feature vector of each input image Iij by concatenating horizontally all the values computed via steps 1a to 1f.
    Step-1h: Store the Feature vector for Iij in the knowledgebase FDB_W2.mat
End (of for loop)
Step-2: Continue to construct the knowledgebase, FDB_W2.mat until there are no more images left without extracting the features, in each subfolders of the image database D.
End (of begin)

2.3 Similarity Matching and Image Retrieval

Image Retrieval is performed by employing query-by-image method, in which, an image is submitted from the image database as a query image rather than giving the text description or a keyword. The proposed method pre-processes the given query image in the form suitable for further processing as explained in Sect. 2.1. The algorithm FEA-TURE_EXTRACTION is applied in order to extract the described features as mentioned in the Sect. 2.2. and feature vector for the query image is obtained.

Similarity Matching is performed using cityblock distance measure. The City block distance between two points, a and b, with k dimensions is calculated as given in the Eq. (3),

$$\sum\nolimits_{j=1 \text{ to } k} |a_j - b_j| \tag{3}$$

The City block distance is always greater than or equal to zero. The measurement would be zero for identical points and high for points that show little similarity. For high dimensional vectors the City block works better than the Euclidean distance. In most cases, this distance measure yields results similar to the Euclidean distance.

Initially the feature vector set of the pre-processed images from the knowledgebase, which are stored in a mat file called FDB_W2.mat are loaded for similarity matching. And cityblock distance between feature vector of query image to feature vectors of each image in knowledgebase is computed using Eq. (3). The cityblock distance yielded is normalized. The normalized city-block distances are sorted according to most similar to least similar. Finally top ranked images are retrieved and displayed from the gallery.

Algorithm 2: Retrieval of Images
Input: Knowledgebase FDB_W2.mat (pre-computed using Algorithm-1) and query Image's Feature vector "queryImageFeature".
Output: Retrieves top-N images from the gallery of images. Where N is the number of images to be retrieved.
Begin
Step 1: Compute Cityblock distance between queryImageFeature vector to the feature vectors of each image in the knowledgebase FDB_W2.mat using equation (3).
Step 2: Sort the obtained cityblock distances, corresponding classes and names of input images (where all the three values are concatenated horizontally) from most similar to least similar values.
Step 3: Finally display top-N images, whose features are close to the query image based on the sorted cityblock distance values.
End (of begin)

3   Results and Discussion

An experimental test has been carried out on Wang Database (also called as Corel database) images in order to test the efficacy of the proposed method. For experimental purpose 5 classes of 500 images, 100 images per class are selected out of 10,000 images from Wang database and saved as WANG-SET-2. The classes selected are dinosaurs, flowers, horses, mountains, and sportsmen. City-block distance measure is used for matching and Query-by-Image method is employed to retrieve the images as mentioned in Algorithm 2. An example query image and the retrieval results are shown in Figs. 3 and 4 respectively. The query image 'dinosaur' is selected, for that the system retrieves top 10 similar images from the gallery, as shown in Fig. 3. Similarly the Fig. 4 shows the result of query image containing 'Horse'.

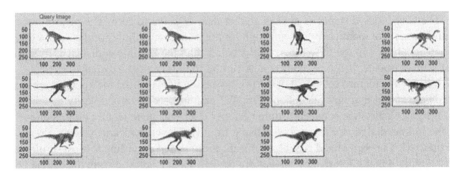

Fig. 3. Snapshot of the retrieved images for query image containing the 'dinosaur'

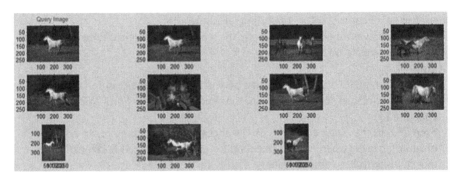

Fig. 4. Snapshot of retrieval result obtained for query image containing 'Horse'

Image retrieval systems are evaluated using rank based measures and interested in analyzing only the top results of images retrieved. We have evaluated the retrieval performance of our system using Precision and recall. For a retrieval system, precision and recall values are calculated based on the Eqs. (4) and (5),

$$\text{Precision} = \text{Number of relevant images retrieved}/\text{Total number of retrieved image} \quad (4)$$

$$\text{Recall} = \text{Number of relevant images retrieved}/\text{Total number of relevant images} \quad (5)$$

Equations (4) and (5) comprise the calculation of the precision and recall for the query image. The performance of the system is measured by retrieving top 10 relevant images those that have smallest city block distance to the query image and the experiment is repeated for each class, ten times and average of those ten experiments for each class is obtained and finally the precision is calculated by averaging the average precision of all the classes. Recall is computed for retrieval of total (100) images of each class, and experiment is repeated for ten times for each class and average recall is reported and shown in the Fig. 5. The accuracy of 87% is obtained.

The results of the proposed CBIR system show that the class 'Horse' is having small recall value because the proposed feature is very robust in finding color features as it has distinguished dinosaurs and sportsman classes properly. The Fig. 3 explains how

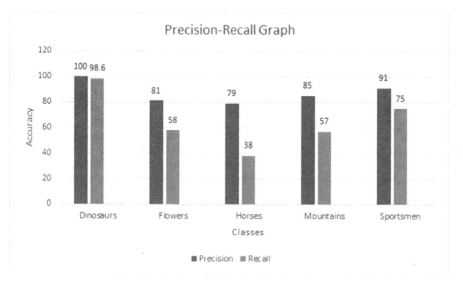

Fig. 5. The precision and recall of our proposed system

the anorexic dinosaurs are retrieved by giving anorexic dinosaur image as query image. Similarly, the images with the green background such as flowers and horses are retrieved by querying flower image with the green background. White horses and white mountains are retrieved by giving white horse as a query image.

4   Conclusion

In this paper, the Radon transform, a global feature is employed for image retrieval. The feature extracted here is the fusion of both the local and global features. Initially colour input image is grayscaled and resized to 80 × 80 pixels and Wang database is used, 5 classes of 500 images are selected for our system. The radon transform is employed and sum, mean, standard deviation, the eigen values are extracted as local features and concatenated horizontally to form the feature vector. Query-by-image method is used for querying, Cityblock distance is used to compute the distance between query image feature to the image feature in the knowledgebase for each class and resultant values are sorted to retrieve the top ranked images. The performance of our system is evaluated by measuring average precision and recall values, and the accuracy of 87% is achieved.

References

1. Yasmin, M., Mohsin, S., Sharif, M.: Intelligent image retrieval techniques: a survey. J. Appl. Res. Technol. **12**, 87–103 (2014)
2. Allani, O., Zghal, H.B., Mellouli, N., Akdag, H.: A knowledge-based image retrieval system integrating semantic and visual features. In: 20th International Conference on Knowledge Based and Intelligent Information and Engineering Systems, KES 2016, York, United Kingdom, 5–7 September 2016 (2016). Procedia Computer Science **96**, 1428–1436

3. Cai, X., Nie, F., Huang, H., Kamangar, F.: heterogeneous image feature integration via multimodal spectral clustering. In: CVPR (2011)
4. Baöazaoui, A., Barhoumi, W., Ahmed, A., Zagrouba, E.: Modeling clinician medical-knowledge in terms of med-level features for semantic content-based mammogram retrieval. Expert Syst. Appl. **94**, 11–20 (2017)
5. Liu, H., Li, B., Lv, X., Huang, Y.: Image retrieval using fused deep convolutional features. International congress of information and communication technology (ICICT). Procedia Comput. Sci. **107**, 749–754 (2017)
6. Yu, W., Yang, K., Yao, H., Sun, X., Xu, P.: Exploiting the complementary strengths of multilayer cnn features for image retrieval. Neurocomputing **237**, 235–241 (2016)
7. Liu, P., Guo, J.-M., Wu, C.-Y., Cai, D.: Fusion of deep learning and compressed domain features for content-based image retrieval. IEEE Trans. Image Process. **26**(12), 5706–5717 (2017)
8. Duan, G., Yang, J., Yang, Y.: Content-based image retrieval research. In: 2011 International Conference on Physics Science and Technology (ICPST) (2011). Physics Procedia 22, 471–477
9. Yue, J., Zhenbo Li, L., Liu, Z.F.: Content-based image retrieval using color and texture fused features. Math. Comput. Model. **54**, 1121–1127 (2011)
10. Bala, A., Kaur, T.: Local Texton Xor patterns: a new feature descriptor for content based image retrieval. Eng. Sci. Technol. Int. J. **19**, 101–112 (2016)
11. Rana, S.P., Dey, M., Siarry, P.: Boosting content based image retrieval performance through integration of parametric and nonparametric approaches. J. Vis. Commun. Image R. **58**, 205–219 (2019)
12. Pavithra, L.K., Sree Sharmila, T.: An efficient framework for image retrieval using color, texture and edge features. Comput. Electr. Eng. **70**, 580–593 (2018)
13. Sotoodeh, M., Moosavi, M.R., Boostani, R.: A novel adaptive LBP-based descriptor for color image retrieval. Expert Syst. Appl. **127**, 342–352 (2019)
14. Alaei, F., Alaei, A., Pal, U., Blumenstein, M.: A comparative study of different texture features for document image retrieval. Expert Syst. Appl. **121**, 97–114 (2019)
15. Yang, J., Liang, J., Shen, H., Wang, K., Rosin, P.L., Yang, M.-H.: Dynamic match Kernel with deep convolutional features for image retrieval. IEEE Trans. Image Process. **27**(11), 5288–5302 (2018)
16. Boschen, F., Kummert, A.: Discretized radon transform and its spectral properties. In: 2000 10th European Signal Processing Conference, pp. 1–4 (2000)
17. Wang, H., et al.: Highprecision seismic data reconstruction with multi-domain sparsity constraints based on curvelet and high-resolution radon transforms. Appgeo (2018). https://doi.org/10.1016/J.Jappgeo.2018.12.003
18. Jadhao, D.V., Holambe, R.S.: Feature extraction and dimensionality reduction using radon and fourier transforms with application to face recognition. In: International Conference on Computational Intelligence and Multimedia Applications (2007)
19. Chen, G.Y., Kégl, B.: Feature extraction using radon, wavelet and Fourier transform (2007)
20. Ines, E., Dhikra, H., Régis, F., Amine, N.-A., Atef, H.: Fingerprint recognition using polynomial discrete radon transform. In: Image Processing Theory, Tools And Applications. IEEE (2014)
21. Piao, M., Jin, C.H., Lee, J.Y., Byun, J.-Y.: Decision tree ensemble-based wafer map failure pattern recognition based on radon transform-based features. IEEE Trans. Semicond. Manuf. **31**(2), 250–257 (2018)
22. Lefebvre, F., Macq, B., Legat, l.-D.: RASH:RAdon soft hash algorithm. In: 11th European Signal Processing Conference, pp. 1–4 (2002)
23. Chen, Y., Wu, Q., HE, X.: Human action recognition by radon transform. In: IEEE International Conference on Data Mining Workshops (2008)

24. Zhang, Y., Wu, L.: A rotation invariant image descriptor based on radon transform. Int. J. Digit. Content Technol. Appl. **5**(4), 209–217 (2011)
25. Boulgouris, N.V., Chi, Z.X.: Gait representation and recognition based on radon transform. IEEE (2006)
26. Ruikar, D.D., Santosh, K.C., Hegadi, Ravindra S.: Contrast stretching-based unwanted artifacts removal from CT images. In: Santosh, K.C., Hegadi, Ravindra S. (eds.) RTIP2R 2018. CCIS, vol. 1036, pp. 3–14. Springer, Singapore (2019). https://doi.org/10.1007/978-981-13-9184-2_1
27. Ruikar, D.D., Santosh, K.C., Hegadi, R.S.: Automated fractured bone segmentation and labeling from CT images. J. Med. Syst. **43**(3), 1–13 (2019)
28. Ruikar, D.D., Santosh, K.C., Hegadi, R.S.: Segmentation and analysis of CT images for bone fracture detection and labeling. Medical Imaging: Artificial Intelligence, Image Recognition, and Machine Learning Techniques, p. 131 (2019)
29. Santosh, K.C., Lamiroy, B., Wendling, L.: DTW-Radon-based shape descriptor for pattern recognition. Int. J. Pattern Recognit. Artif. Intell. **27**(3), 1350008 (2013)
30. Santhosh, K.C., Lamiroy, B., Wendling, L.: DTW for matching radon features: a pattern recognition and retrieval method. In: Blanc-Talon, J., Kleihorst, R., Philips, W., Popescu, D., Scheunders, P. (eds.) ACIVS 2011. LNCS, vol. 6915, pp. 249–260. Springer, Heidelberg (2011). https://doi.org/10.1007/978-3-642-23687-7_23

A Contrast Optimal Visual Cryptography Scheme for Half-Tone Images

D. R. Somwanshi[1]([✉]) [ID] and Vikas T. Humbe[2]

[1] Department of Computer Science, College of Computer Science and Information Technology (COCSIT), Latur, Maharashtra, India
somwanshi1234@gmail.com
[2] School of Technology, Swami Ramanand Teerth Marathwada University Nanded, Sub-Center, Latur, Maharashtra, India
vikashumbe@gmail.com

Abstract. In this paper, we propose a new contrast optimal visual Cryptography Scheme for Half-tone images. This scheme can share Secrete Half-tone Image among n members and all n number of members can reveal the secrete by superimposing the n share together and $n - 1$ members can not reveal the secrete. The proposed method generate the noise like shares but share size and the image recovered after superimposition is of same size which is equal to original secrete image. It reduces the pixel expansion based on the designed codebook and matrix transposition. It also reduces the transmission speed, complexity computation, security issue and restore the secrete image without any distortion. The scheme proposed can share the black& white and gray level image and produces the more accurate results and is easy to implement.

Keywords: Half-tone images · Pixel expansion · Optimal contrast · Secure secrete sharing scheme

1 Introduction

Visual Cryptography (VC) is a technique of encrypting the written material such as handwritten notes, images, printed text, etc. in a seamlessly secured way and which can easily be deciphered by visual systems of human being without using complex calculation [1]. It means decryption is not required or huge amount of calculation is not required to decrypt the text. This allows everybody to make use of the system, even if they don't have sufficient knowledge about encryption or decryption of cryptographic system and for those who fails to perform any of the computations required.

This technique is first introduced by Naor et al. [1], in 1994. In this technique the message or image data to be protected or encrypted is divided into n different parts or share. Each part or share is distributed among the n different users, and then we can specify as a minimum of any k from the n different shares are stacked together or required to get the original information. The $k - 1$ shares cannot be used to generate the original message. For example, suppose there are six thieves wants to share a bank account, but

K. C. Santosh and B. Gawali (Eds.): RTIP2R 2020, CCIS 1381, pp. 152–162, 2021.
https://doi.org/10.1007/978-981-16-0493-5_14

the problem is that they do not faith on each other, and they divided the password for the bank account in such a manner that, any three of them or more should be working with can have access to their account, nevertheless not less than three thieves can access the account.

In Visual Cryptography, the visual information that is to be encrypted is taken as input to the system and then every pixels (white or black) of the images is broken down into smallest parts. There should be the same number of white and black parts. If a pixel of the image is broken down into two parts, then there will be one white and one black part. Similarly if the pixel of the original image is broken down into four parts, then there should be two white and two black parts of pixel [1, 3, 4].

For encryption by using (2, 2) scheme secret image information is broken down into two parts. Here each pixel in secrete image that is represented by block of 4 or 2 sub-pixel in each share of the image. After combining two shares the original secrete image can be revealed. With one share secrete image cannot be revealed [1].

For encrypting each pixels of the original secret image several technique are used one of the method, in (2, 2) secrete sharing method every pixel in the secret image is showed as 2 sub-pixel in both the shares, for this reading every pixel in secret image is needed and if the pixel that is read is white then one row from first two rows of Table 1 is taken and the probability of this is 0.5 and 2 pixel blocks are assigned to the shares as represented in the third and fourth columns of Table 1. Similar to that, if the pixel that is read is black, then one of the row from the last two rows from Table 1 is taken and the probability of this is 0.5. The sub-pixel block is assigned to each share from that. When two shares of image are stacked or combined together, and if the two white pixels are overlapped on one another, then the output pixel obtained should be white similar to that if the black pixel in one share of image overlapped with either a white or black pixel in another share of the image, then the output pixel obtained will be black. This indicates that the stacking or combining operation of the shares signifies the Boolean OR operation. In Table 1, last column shows the obtained sub-pixel, if the sub-pixels of all two shares in the third and fourth columns of Table 1 are stacked together [1].

Table 1. (2, 2) Secrete sharing scheme.

Original Pixel	Proba-bility	Share1 Sub- Pixel	Share2 Sub-Pixel	Share1 || Share 2
▢	0.5	◼▢	◼▢	◼▢
▢	0.5	▢◼	▢◼	▢◼
◼	0.5	◼▢	▢◼	◼◼
◼	0.5	▢◼	◼▢	◼◼

While creating and combining the shares some problems such as pixel expansion, alignment problem, and extensive codebook design, flipping Issues, and share distortion may arise [23]. In pixel expansion numbers of pixel in share are increased, due to that size of share increases and alignment problems may occurs. Due to alignment problems combined secrete image looks different. If image is not superimposed in proper direction flipping issue may arise. Proposed method provides the new contrast optimal secrete sharing based scheme using half-tone images that overcomes the above problems.

2 Related Work

Naor and Shamir generalized basic secrete sharing scheme into k out of n visual cryptography scheme [1]. In this type scheme of visual cryptography, total of n shares of the original image are generated and provided to n participant. Minimum k of those n participant have to provide theirs share for reveling the secrete image, secrete image cannot be reveled if less than k share are presented for reveling process. The scheme provides the convenience to user that secrete can be reveled even if some of the shares from n share loses and only k share are required but the. Contrast of the recovered image is poor and pixel expansion is double in this scheme. To provide the security to this scheme, G. Atenieseet et al. further modified (k, n) model to general access structure model of visual cryptography [10]. According to them number of share n, created are divided into two parts or subsets as per the importance and need. First part of the subset is called qualified subset and second is forbidden subset. From qualified subset and k shares can be used to generate the original secret, but then again less numbers of shares than the k shares cannot be used to generate the original secret image. Again k or more parts or shares from the forbidden set, are cannot be used recover the original secret image. So, the general access structure of the visual cryptography increases the security of the system. Abhishek Parakhet et al. [11] has illustrated the basic idea of visual cryptography which is based on concept of Recursive threshold scheme. Through which they represented that recursive hiding of slighter secrets in shares of superior secrets with secret sizes expanding at every step and thereby aggregate the information, here the each bits of shares transfers to $(n − 1)/n$ bit of secret which is approximately equal to 100%. Again to maintain the good contrast and improve the security, Zhi Zhou et al. proposed the new scheme, based on called halftone visual cryptography [5, 6], in which the binary pixel of secrete image is converted into array of sub pixels, which are called halftone cells, in every n shares create from secrete image. Chang-Chou et al. recommended new scheme of visual cryptography for gray level secrete images [12]. The scheme uses the dithering technique for conversion of secretes gray level image into the imprecise binary image. Then they have applied existing algorithms which used binary image for creation of the shares.

To reduces the pixel expansion F. Liu et al. recommended a new method of visual cryptography for color images [14]. The method proposed by them used the three different techniques, which can be used for color image illustration using these three techniques they extracts Red, Green and Blue channels from the original image.

Table 2. Comparative review of different Visual cryptography Techniques

Author	Title and Year	Image type	Technique
Mahmoud E. Hodeish, Vikas T. Humbe	An Optimized Half tone Visual Cryptography Scheme Using Error Diffusion 2018 [23]	Binary and Gray Scale Image	Improves pixel expansion, Eliminate explicit requirement of codebook, reduce the random pattern of the shared images and Performance analysis through statistical analysis
Shivendra Shivani	Verifiable Multi-tone Visual Cryptography 2017 [22]	Gray scale and Color image	Cheating prevention by adding pixels in a share that contains a self-embedding verifiable bit for integrity test of that pixel and improves the constraints like random shares, codebook requirement, and contrast loss
Mahmoud E. Hodeish, Linas Bukauska, Vikas T. Humbe	An Optimal (k,n) Visual Secret Sharing Scheme for Information Security 2016 [20]	Binary image	(k, n) scheme based on codebook design, transport of matrices, Boolean n-Vector, and XORing operation
Angel Rose, A Sabu, M Thampi	A secure verifiable scheme for secret image sharing 2015 [19]	Binary image, gray scale image	Arnold transformation technique, Bit-Plane Complexity Steganography, structural similarity index value, mean square error (MSE) and test
Souvik Roy and P. Venkateswaran	Online payment system using steganography and visual cryptography 2014 [18]	Binary image, gray scale image	Text based steganography, and visual cryptography
Rajendra A B and Sheshadri H S	Visual cryptography in internet voting system 2013 [21]	Gray level image	2-out-of-2 visual cryptography

Anyone channel can be used in half toning process but because of this half-toning original superiority of image are decreases.Multi secrete visual cryptography is first introduces by Wu and Chen [15], the scheme allows us to combine two secret information images into two parts. The Share named A and B can be created using two secret binary images. Two secrete images can be hidden in to these two shares in such a way that if the two secrete shared are stacked together then first secret can be revealed, and the two secrete shares are rotated by 90° anti-clockwise direction then the second secret image can be revealed. To extends this scheme Shyu et al. [16, 17] proposed a new scheme for Multi-Secret Visual Cryptography, through which two or more secret information images can be secured and revealed at a time using two shares of image.

Comparative review of different Visual cryptography Techniques is presented in Table 2.

3 Preliminary Concepts

3.1 HVC Scheme

Half tone visual cryptography process the halftone image, an image is made up of a series of dots instead of continuous tone is called halftone image. Series of dots in halftone image can be of different colors, different sizes, and different shapes. Larger dots in image are used to represent darker more dense areas of the image, whereas smaller dots are used for lighter areas in image [4].

3.2 Error Diffusion

Error diffusion This is the technique of converting gray level image to binary image form in such a way that picture in binary image form looks similar to gray level image with somewhat better quality image. This is the efficiency and simplicity of binary image. The process of error diffusion diffuses or minimizes the error in binary image. The Error diffusion at pixel level is performed by using filtering the quantization error and by using the feedback to the input. Filtering the quantization error process diffuses the quantization error on one pixel away to the neighboring gray pixels. The pleasing halftone images is displayed in nature because of the error diffusion process and noise and blue noise removal process using high frequency [19]. To diffuse the error up to the 12 neighboring pixels instead of 4 pixel in their previous algorithms, Jarvis et al. [6] have introduced another error diffusion algorithm for error diffusion.

3.3 Code Matrix's

For the simulation of method we develop the code block for black and white pixel for the creation of shares which is shown in Table 3.

Table 3. Code blocks

Code Block For Black Pixel   Code Block For White Pixel

$$
\begin{bmatrix} 0 & 1 & 1 & 1 \\ 1 & 0 & 1 & 1 \\ 1 & 1 & 0 & 1 \\ 1 & 1 & 1 & 0 \end{bmatrix} \text{C1}
\qquad\qquad
\begin{bmatrix} 0 & 0 & 0 & 0 \\ 1 & 1 & 1 & 1 \\ 1 & 1 & 1 & 1 \\ 1 & 1 & 1 & 1 \end{bmatrix} \text{C1}
$$

$$
\begin{bmatrix} 1 & 0 & 1 & 1 \\ 1 & 1 & 0 & 1 \\ 1 & 1 & 1 & 0 \\ 0 & 1 & 1 & 1 \end{bmatrix} \text{C2}
\qquad\qquad
\begin{bmatrix} 1 & 1 & 1 & 1 \\ 0 & 0 & 0 & 0 \\ 1 & 1 & 1 & 1 \\ 1 & 1 & 1 & 1 \end{bmatrix} \text{C2}
$$

$$
\begin{bmatrix} 1 & 1 & 0 & 1 \\ 1 & 1 & 1 & 0 \\ 0 & 1 & 1 & 1 \\ 1 & 0 & 1 & 1 \end{bmatrix} \text{C3}
\qquad\qquad
\begin{bmatrix} 1 & 1 & 1 & 1 \\ 1 & 1 & 1 & 1 \\ 0 & 0 & 0 & 0 \\ 1 & 1 & 1 & 1 \end{bmatrix} \text{C3}
$$

$$
\begin{bmatrix} 1 & 1 & 1 & 0 \\ 0 & 1 & 1 & 1 \\ 1 & 0 & 1 & 1 \\ 1 & 1 & 0 & 1 \end{bmatrix} \text{C4}
\qquad\qquad
\begin{bmatrix} 1 & 1 & 1 & 1 \\ 1 & 1 & 1 & 1 \\ 1 & 1 & 1 & 1 \\ 0 & 0 & 0 & 0 \end{bmatrix} \text{C4}
$$

4 Proposed Methodology

4.1 Algorithm

1. Input any gray level image
2. Apply the Jarvis's Halftone algorithms for converting image into half tone image.
3. Calculate the size width and height of image
4. Create the n shares as per the requirement and fill zeros in it
5. Define the code matrix for black pixel and white pixel input image (for *4* shares considered in this research) as per the Table 1
6. Traverse from each pixel in input image and perform the following steps.
   a. For $i = 1$ to width
   b. For $j = 1$ to height
   c. If pixel in input image is black then
      (1) Randomly select one code block from black code block
      (2) Select one row randomly from the selected code block and then assign this to row vector V
   d. End
   e. If pixel in input image is white then
      (1) Randomly select one code block from white code block
      (2) Select one row from the selected code block randomly and assign this to row vector V

   f. End

   g. Assign each pixel in row vector V to different shares *S1, S2, S3, S4* sequentially (First pixel to share one, second pixel to share 2 etc.) for share construction.

   h. End

   i. End

7. Combine the shares with XOR Operator for getting the secrete image as below *Result* $= S1 \oplus S2 \oplus S3 \oplus + S4$

5 Experimental Results and Discussion

The results of the proposed method obtained, are presented in Fig. 1:

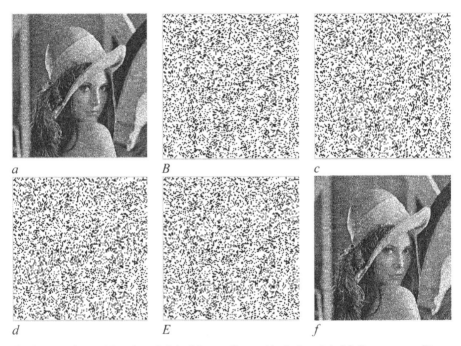

Fig. 1. Experimental Result. a) Original Secrete Image. b), c), d) and e). f) is Reconstructed Image (Size: *512 × 512*)

In Fig. 1 the original secrete image and the image obtained after stacking of all the shares are exactly of same quality and size means there is no pixel expansion. Code block designed here creates all shares of equal size of the original secrete image and also the reconstructed image.

6 Discussion and Performance Analysis

6.1 Pixel Expansion

The proposed scheme reduces the problem of the Pixel expansion 100%, because the share size and the generated image size after superimposing the share is exactly equal which is shown in Fig. 1 (a) original image, Share Images (b), (c), (d), (e) and recovered image (f).

6.2 Contrast and Statistical Analysis

As the output displayed in Figs. 1(a), and (f) the recovered image is obtained without any pixel distortion. To evaluate the quality of recovered secret image and to prove that the recovered secret image is of same quality of the original image, there are variety of statistical metrics of image restoration are presented as below.

Mean Square Error (MSE). Mean Square Error (MSE) [24] can be mathematically computed using the formula.

$$MSE = \frac{1}{MXN} \sum_{i=1}^{M} \sum_{j=1}^{N} \left(h_{ij} - h'_{ij} \right)^2 \tag{1}$$

Where h_{ij} h'_{ij} are the pixel values of original image and recovered image, respectively

Peak-Signal-to-Noise Ratio (PSNR). Peak-Signal-to-Noise Ratio (PSNR) [25] is also and mathematical or engineering formulations calculated using *MSE* with the help of following formula.

$$PSNR = 10 * \log \frac{R^2}{MSE} \tag{2}$$

Statistically, when the value of PSNR = 1, it indicates that the scheme proposed delivers the extreme visual quality.

Universal Index Quality (UIQ). To calculate the Universal Index Quality (UIQ) [26] following equation can be used.

$$UIQ = \frac{4\sigma_{zy}\overline{xy}}{\sigma_x^2 + \sigma_y^2 \left[(\bar{x})^2 + (\bar{y})^2 \right]} \tag{3}$$

This is used to modeling the image distortion as combination of three factors these are

1. Loss of correlation,
2. Luminance distortion, and
3. Contrast distortion

The UIQ of the image among the two images ranges from -1 to $+1$. There is a strong positive linear correlation between the two images X and Y, if the value of UIQ is nearer to $+1$. The value -1 of the UIQ represents the negative relationship among the two images and finally the zero value represents that there is no relationship among the two images [26].

Maximum Difference (MD). Maximum Difference (MD) measure is used to calculate the error between original image and reconstructed image. MD is directly proportional to the contrast giving an image dynamic range and which can be calculated with the expression [27].

$$MD = max|x_{ij} - y_{ij}| \tag{4}$$

Average Difference (AD). The Average difference is method of calculating the difference in two images; original image and recovered image. Average Difference (AD) among original secrete image and recovered secrete image and is calculated with the average difference metrics as below [28].

$$AD = \frac{1}{MXN} \sum_{i=1}^{M} \sum_{j=1}^{N} (X_{ij} - Y_{ij}) \tag{5}$$

Where, the value of X, Y represents the original secrete image and calculated recovered secrete image. The values of the all the metrics discussed above are presented in Table 1.

MSE, MD and AD value is zero, PSNR value is infinite ∞ and UIQ value is 1 represents and assures that the original secrete images have been completely recovered without any damage or loss of any meaningful information of the recovered image. Table 4 shows those values.

Table 4. Value of difference statistical metrics obtained in experiment

Statistical metrics	Value obtained in experiments
MSE	0
PSNR	∞
UIQ	1
MD	0
AD	0

7   Conclusion

This paper proposed the contrast optimal visual cryptography scheme for Half-tone images, and eliminate the problem of pixel expansion, explicit requirement of code book, low transmission and poor quality of reconstructed image after combing shares. We work on half-tone image which are directly converted using Jarvis's Halftone algorithms from gray scale image. Four different shares created using codebook designed for black and whiter pixel and all four are needed for reconstruction. Selection of code book and row vector from selected code book is performed randomly depending upon the pixel in input image. Shares are combined using XOR Operation and better quality output image is obtained by completing the reconstructed share. Perform and evaluation of our method is tested with existing method and using some statistical metrics and we found better results in terms of quality and zero pixel expansion.

References

1. Naor, M., Shamir, A.: Visual Cryptography. Eurocrypt (1994)
2. Hou, Y.-C.: Visual cryptography for color images. Pattern Recogn. J. Pattern Recogn. Soc. **36**, 1619–1629 (2002)
3. Weir, J., WeiQi, Y.: Visual Cryptography and its Application, Ventus Publishing Aps eBook, pp. 1–144 (2012)
4. Ecaterina Moraru (Valica), "Visual Cryptography", Published in: Technology, Art & Photos on Slide share, pp. 1–38 (2008)
5. Zhou, Z., Arce, G.R., Di Crescenzo, G.: Halftone Visual Cryptography. IEEE Trans. Image Process. **15**(8), 2241–2453, August 2006
6. Wang, Z., Arce, G.R., Di Crescenzo, G.: Halftone visual cryptography via error diffusion. IEEE Trans. Inf. Forensics Secur. **4**(3), 383–396, September 2009
7. Purushothaman, V., Sreedhar, S.: An improved secret sharing using XOR-based visual cryptography, green engineering and technologies (IC-GET). In: 2016 Online International Conference, November 2016
8. Jin, D., Yan, W.Q., Kankanhalli, M.S.: Progressive color visual cryptography. J. Electron. Imag. **14**(3), 1–13 (2005)
9. Hou, Y.-C., Quan, Z.-Y.: Progressive visual cryptography with unexpanded shares. IEEE Trans. Circuits Syst. Video Technol. **21**(11), 1760–1764 (2011)
10. Ateniese, G., Blundo, C., DeSantis, A., Stinson, D.R.: Visual cryptography for general access structures. In: Proceedings ICAL 1996, pp. 416–428, Springer, Berlin (1996)
11. Parakh, A., Kak, S.: A Recursive Threshold Visual Cryptography Scheme. CoRR abs/0902.2487 (2009)
12. Lin, C.-C., Tsai, W.-H.: Visual cryptography for gray level images by dithering techniques. Pattern Recogn. Lett. **24**(1–3) (2003)
13. Verheul, E., Tilborg, H.V.: Constructions and properties of K out of N visual secret sharing schemes. Des. Codes Cryptogr. **11**(2), 179–196 (1997)
14. Liu, F., Wu, C.K., Lin, X.J.: Colour visual cryptography schemes. IET Inf. Secur. **2**(4), 151–165 (2009)
15. Wu, C.C., Chen, L.H.: A Study On Visual Cryptography, Master Thesis, Institute of Computer and Information Science, National Chiao Tung University, Taiwan, Republic of China (1998)
16. Shyu, S.J., Huanga, S.Y., Lee, Y.K., Wang, R.Z., Chen, K.: Sharing multiple secrets in visual cryptography. Pattern Recogn. **40**(12), 3633–3651 (2007)
17. Chen, T.-H., Wu, C.-S.: Efficient multi-secret image sharing based on Boolean operations. Sig. Process. **91**(1), 90–97, January (2011)
18. Roy, S., Venkateswaran, P.: Online payment system using steganography and visual cryptography. In: IEEE Students' Conference on Electrical, Electronics and Computer Science, pp. 1–5 (2014)

19. Rose, A.A., Thampi, S.M.: A secure verifiable scheme for secret image sharing 2015. Procedia Comput. Sci. **58**, 140–150 (2015)
20. Hodeish, M.E., Bukauska, L., Humbe, V.T.: An optimal (k, n) visual secret sharing scheme for information security. Elsevier-Procedia Comput. Sci. **93**, 760–767 (2016)
21. Rajendra, A.B., Sheshadri, H.S.: Visual Cryptography in Internet Voting System, pp. 60–64. IEEE (2013)
22. Shivani, S.: VMVC: Verifiable multi-tone visual cryptography. Multimed Tools Appl. **77**, 1–20 (2017). https://doi.org/10.1007/s11042-017-4422-6
23. Hodeish, M.E., Humbe, V.T.: An optimized half tone visual cryptography scheme using error diffusion. Multimed Tools Appl. 1–17, Springer, January 2018
24. Chen, C.Y., Chen, C.H., Chen, C.H., Lin, K.P.: An automatic filtering convergence method for iterative impulse noise filters based on PSNR checking and filtered pixels detection. Expert Syst. Appl. **63**, 198–207 (2016)
25. Shankar, K., Eswaran, P.: RGB based multiple share creation in visual cryptography with aid of elliptic curve cryptography. China Commun. **14**(2), 118–130 (2017). https://doi.org/10.1109/CC.2017.7868160
26. Wang, Z., Bovik, A.C.: A universal image quality index. IEEE Signal Process. Lett. **9**(3), 81–84 (2002)
27. Rajkumar, S., Malathi, G.: A comparative analysis on image quality assessment for real time satellite images. Indian J. Sci. Technol. **9**, 1–11 (2016)
28. Ece, C., Mullana, M.M.U.: Image quality assessment techniques in spatial domain. IJCST **2**(3) (2011)

Mineralogical Study of Lunar South Pole Region Using Chandrayaan-1 Hyperspectral (HySI) Data

R. Mohammed Zeeshan[1]([✉]), B. Sayyad Shafiyoddin[1], R. R. Deshmukh[2], and Ajit Yadav[1]

[1] Department of Computer Science, Milliya Arts, Science and Management Science College, Beed 431122, Maharashtra, India
zeeeshan.shaikh@gmail.com, syedsb@rediffmail.com
[2] Department of Computer Science and IT, Dr. Babasaheb Ambedkar Marathwada University, Auranagabad, India
rrdeshmukh.csit@bamu.ac.in

Abstract. The main focus of the presented work was to better predict the surface mineralogy from the Chandrayaan-1 hyperspectral data set covering the area from South Pole region. To address the space weathering effect and to quantify mineralogy the Bi-directional reflectance function have been implemented. The implemented model was tested against two standard lunar laboratory mixtures and with the Apollo 10084 bulk soil sample. About 85 spectra were initially selected from varying locations and only active spectra with significant absorption were used for modeling. The minerals like plagioclase and Clinopyroxene were identified. Many spectra exhibits more iron content simulating mature area. Model result show no olivine content and very low Orthopyroxene content may be because of more crustal thickness, no impact would have penetrated to the lower mantle. Study reveals the potential of hyperspectral data multiplexed with mathematical model for not only mineral quantification but also helps to predict other associated parameters like grain size, iron fraction, phase function, however the spectra from mature soil and the limited HySI coverage acts as challenge for modelling process, modeling the data at longer wavelengths will be an advantage to improve the accurate mineral prediction.

Keywords: Hyperspectral · Absorption coefficient · Bi-directional reflectance

1 Introduction

The mineral analysis of the lunar surface using remotely sensed data from various lunar missions helps to understand the evolution and the composition of the moon [1].

The mineralogical study can be done by means of spatial analysis and spectral profile analysis. Creating the false color composite based on different band shaping algorithms like band strength band curve and band tilt is considered as standard spatial analysis technique [2, 3]. These band shaping algorithms was originally devised for UV-VIS

© Springer Nature Singapore Pte Ltd. 2021
K. C. Santosh and B. Gawali (Eds.): RTIP2R 2020, CCIS 1381, pp. 163–175, 2021.
https://doi.org/10.1007/978-981-16-0493-5_15

164   R. M. Zeeshan et al.

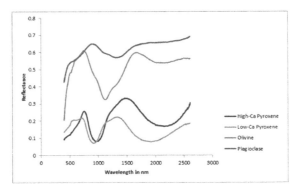

Fig. 1. Spectral signatures of common lunar minerals. (RELAB)

data to determine the surface mineralogy and later modified for HySI data as per HySI specifications [4]. The compositional variation using band parameters to identify different common lunar minerals like orthopyroxene, Clinopyroxene, olivine and Plagioclase were analyzed [5–8]. On the other hand the moon minerals can be well distinguished in the visible and near infrared part of the spectrum by its characteristic absorption features [9]. The chemical structure of the mineral causes change in shape, position and strength of the absorption band across the spectrum [10, 11]. The spectral profile analysis helps to study and identify different mineral from the lunar surface. Figure 1 shows the common major lunar minerals derived from RELAB database. Olivine, pyroxene and plagioclase are the common lunar minerals. The highland is Anorthositic in nature and it is mainly consists of plagioclase type of mineral with characteristic absorption at 1200 nm around ten percent of mafic component. From the Lunar return samples it is clear that for highland plagioclase is in between 15 to 75%. Whereas the low land area contains 10 to 35% of plagioclase. The second common mineral is from pyroxene group having characteristic absorption is at 1000 and 2000 nm with increasing amount of calcium the absorption shifts at the longer wavelengths. The shifts can clearly observe in Low-Ca pyroxene (Orthopyroxene) and High-Ca pyroxene (Clinopyroxene). The lunar soil contains 10 to 60% of pyroxene. The olivine type of mineral is having triplet absorption at 900, 1050 and 1250 nm. The lunar soil contains 2 to 15% of olivine.

2   Literature Review

The hyperspectral remotely sensed data provides a great opportunity to assess the mineralogy but the derived spectral signatures has great influence of an active mechanism that is consistently effects the lunar surface is termed as space weathering process. Space weathering involves two mechanisms firstly solar wind ion implantation and micro meteoritic bombardment [12, 13]. These two processes create a thin layer of submicroscopic iron (SMFe) which alters the spectral properties of soil and reflects strong influence on the spectra [14, 15]. Reduction in the overall reflectance, shallow absorption bands and red sloped continuum are the effects of space weathering on the spectra and can be observed in (Fig. 2). The hyperspectral moon mineralogy mapper (M3) data set onboard

Chandrayaan-1 with product id m3g20090104t161446_v01_rfl downloaded from PDS geosciences node lunar orbital data explorer (www.ode.rel.wustl.edu/moon) the extent of the area covers the mare crisium that is used to demonstrate the influence of space weathering. The red spectra derived from relatively fresh area appearing bright in the image and the blue spectra from mare area appearing dark. The red spectral profile has good spectral contrast as compared to blue spectra. As surface becomes mature it experiences more degree of space weathering accumulating more iron and hence reduces the spectral contrast which complicates the understanding of the surface mineralogy [16–19]. The space weathering effect acts as barrier for accurately assessing the surface mineralogy and compositional analysis. These effects can be address by means of a model based on theory of radiative transfer that best explains the space weathering effect on remotely sensed spectra. Three models are commonly used for modeling the remotely acquired spectra from airless bodies like moon are available. These models suitably describes the behavior of the electromagnetic radiation with the host material. The modified Gaussian model [20], Hapke's Bi-directional reflectance function based on fundamental equation of radiative transfer [21] and Shkuratov model [22]. The MGM model considers the absorption width, center and strength but as Chandrayaan-1 hyperspectral imager (HySI) data is used in this work, the spectral coverage of HySI act as constraint to use MGM model, whereas in Shkuratov model the viewing geometry is not included which greatly effects the reflectance pattern with change in geometry. On the other hand the Hapke's Bi-directional reflectance function that considers the host material as an intimate mixture and accommodates the effect of space weathering with other associated parameters [23–27]. The Hapke's model commonly used for modeling the remote spectra for mineral quantification [28–33].

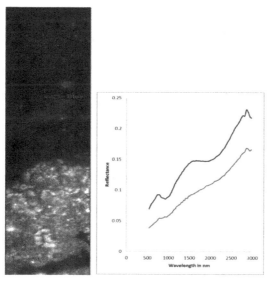

Fig. 2. Moon Mineralogy Mapper data set covering southern part of Mare Crisium and spectral profile from the highlighted areas.

3 Methodology

Modeled spectra or the hypothetical spectra is created for modeling the reflectance spectra derived from Chandrayaan-1 hyperspectral (HySI) data, the Bi-directional reflectance function is used and given by following Eq. (1). The five pure end member spectra of common lunar minerals are selected from RELAB database given (Fig. 3). Considering the spectral coverage of HySI from 0.43 μm to 0.96 μm the five end member is sufficient because within this spectral range most of the lunar mineral does not exhibit the absorptions and if we have data at longer wavelengths can only sense to take more end members for modeling process.

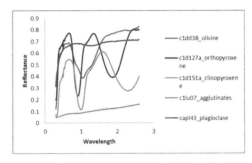

Fig. 3. The end member spectra used from RELAB database.

$$r_B = K \frac{w}{4\pi} \frac{\mu_0}{\mu_0 + \mu} \left(\left[1 + B(g)P(g) + H(\mu_0)H(\mu) - 1 \right] \right) \tag{1}$$

$P(g)$ is phase function defines the angular scattering pattern of particles. g denotes phase angle, $B(g)$ is opposition which is define as sudden bright spots that appears usually with small phase angles because of shadow-hiding and Chandrasekhar's H function accounts for multiple scattering part of the light. w is the single scattering albedo is the ratio of the scattering and extinction coefficient and need to be calculated first by Eq. (2) from [34]. The absorption coefficient of each end member is then calculated from Eq. 23(b) of [27] and it requires the real part of complex refractive index [35] as shown in (Fig. 4). The absorption coefficient of iron is derived using method from [27] which requires the complex refractive indices of iron taken from [36]. For introducing the space weathering effect into the model the absorption coefficient of end member and iron is added together and single scattering albedo was calculated.

The mixture is calculated by adding single scattering albedo of space weathered material given in Eq. (3) and requires specific density of each material end member taken from [35]. Finally the w_{im} substituted in Eq. (1) and remaining parameters like phase function for different angles, porosity parameter for different filling factors, and Chandrasekhar's parameters are calculated and substituted. Parameter study by considering each parameter and in combination of different parameters is carried and found that every parameter is working according to model definitions. The parameters like grain size, porosity, and phase function, iron fraction with different mass fractions of minerals used to increase or decrease the spectral contrast of the modeled spectra. To increase the spectral contrast of modeled spectra, decrease grain size, decrease porosity, increase value of phase function, increase high albedo mineral like plagioclase, decrease iron content will result in overall increased spectral contrast and adding low albedo mineral like agglutinate, increasing grain size, increase porosity and adding more iron fraction which has the effect of space weathering results in decreased spectral contrast of the modeled spectra. After parameter study model validation is done using standard laboratory mixtures and with the Apollo 10084 bulk soil sample spectra.

$$W = 1 - \left(\frac{-27.856R + \sqrt{4.029R^2 + 602.932R + 268.696R}}{51.71R + 16.392} \right)^2 \tag{2}$$

$$w_{im} = \sum_{i=1}^{n} \frac{M(i)}{\rho(i)d(i)} w_{sw}(i) \bigg/ \sum_{i=1}^{n} \frac{M(i)}{\rho(i)d(i)} \tag{3}$$

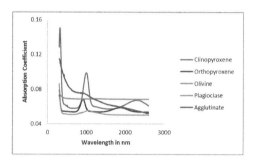

Fig. 4. The Derived Absorption coefficient of Selected end member.

3.1 Testing the Modeled Spectra

The artificial modeled spectra tested against the three components standard lunar mixtures by setting other parameters like grain size, phase function as average at 30° phase angle, 65% porosity and zero iron content means no space weathering effect for lab samples. Initially all mineral mass fractions and other parameters are set to lab specified values and it is found that the resultant modeled spectra is having close resemblance with the laboratory mixtures as shown in Fig. 5. For finding out the difference between measured and modeled spectra the mineral mass fractions is adjusted and it is observed that there is about 10% difference between measured lab spectra and modeled spectra. For the next test case the Apollo 10084 soil sample is used where the effect of space weathering is considered by adding iron fractions to the model and the mass fraction is set to the lab obtained values for the sample [37] and it is clearly observed from Fig. 5 that the model is successful in creating the actual trend in the spectra with 0.0005 of iron fraction. The laboratory mixture composition and composition of 10084 soil sample is given in Table 1. After model evaluation the spectra from the Chandrayaan-1 hyperspectral data set is derived and through modeling process mineralogy of the study area with particle size, iron fraction, porosity and function can be better predicted.

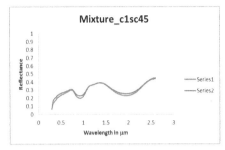
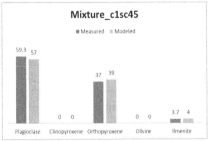

(a)

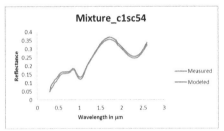
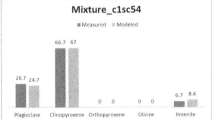

(b)

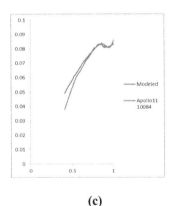

(c)

Fig. 5. (a), Measured and calculated modeled reflectance spectra of mixture c1sc45 and difference between measured and modeled composition. (b), Measured and Calculated Modeled reflectance spectra of mixture c1sc54 and Difference between Measured and Modeled Composition. (c), Measured and Modeled spectra of Apollo 10084 soil sample.

Table 1. Test mixture compositions

Mixture Id	Plagioclase	Clinopyroxene	Orthopyroxene	Ilmenite	Agglutinates	Source
C1sc45	59.3	0	37	3.7	0	RELAB
C1sc54	26.7	66.7	0	6.7	0	RELAB
10084	16.8	8.4	8.4	6.4	53.9	Morris 1978

4 Results and Discussion

The level-4 data product acquired by Chandrayaan-1 HySI instrument is used for analysing the surface minerals. The image cube is downloaded from the official website with product Id ending with 20090605T225556626. The image covering the area from south pole region close to Boguslawsky crater and it is also closed to proposed Chandrayaan-2 landing site near Mazinus and Simpelus craters is used. Figure 6 shows the extent of the study area and the actual HySI coverage with product id. The Hyper spectral imager (HySI) sensor acquires the data in pushbroom mode covering 430–964 nm having 64 continuous bands with the spectral resolution of 15 nm and spatial resolution of 80 m with 20 km swath [38].

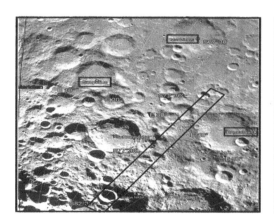

Fig. 6. South Pole map on google moon tool and extent of study area (left). Actual HySI South Pole coverage and highlighted study area.

The original data set is in radiance format and it is converted into reflectance image and divided into two parts. The Fig. 7 shows the subset of the study area with sampling site locations. The image is corrected and cross calibrated using 10084 Apollo bulk soil sample. Further the geometrical correction is performed using the geometry information provided with the data set. The region seems to be rugged and heavily craterd. For spectral profile analysis around 40 spectra from varying locations were selected. Most of the spectra has no significant absorption and having red sloped continuum which indicates the high maturity of the area. However some 16 spectra selected from area appearing bright show adequate absorptions that can be suitable for modeling. These spectra are selected from the small young craters and from the crater walls of the large craters. The spectra from crater wall showing good reflectance with suitable band strength may be because of the gravitational slumping the material underneath getting exposed and these spectra are used for modeling. The spectral resolution of modeled spectra is 5 nm where as the Hysi spectral resolution is 15 nm so before modeling each Hysi spectra is interpolated to 5 nm and imported to the model routine.

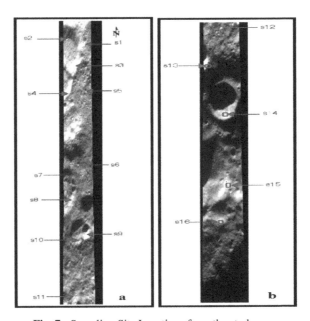

Fig. 7. Sampling Site Locations from the study area.

The Fig. 8 shows the reflectance spectra from the highlighted areas and the normalized spectra for better representation. The HySi spectral coverage is up to 964 nm so the modeled spectra are also clipped to 964 nm; both spectra are plotted into the same plot window. The spectra with no significant absorption and having very low reflectance is discarded from further process. Each spectra is modeled individually and the mineral mass fraction for each spectra is given in Table 2 and the parameter like grain size, porosity, iron fraction is given in Table 3. The spectra s7 shows the maximum reflectance around 21% with good absorption strength similarly spectra s4 and s13 is also showing good reflectance as they are extracted from the crater walls. The model results obtained for these spectra showing low iron mass fractions around 0.00015 with smaller grain size and showing 35 to 45% of Clinopyroxene with low percent of agglutinates with around 40% of plagioclase. The spectra s1 derived from northern part of the part (a) from Fig. 7 with low reflectance around 10% with complete absorption at 900 nm with 48% of orthopyroxene. About 11% low reflectance for spectra s16 shows mature soil with 0.00032 iron fraction with 30% plagioclase and 37% Clinopyroxene and extracted from the crater wall. From modeling results no spectra shows olivine mineralogy may be it is lower mantle material so no impact would have penetrated through the crust to the lower mantle but here all representative spectra shows high values of Clinopyroxene may be the cratering event have penetrated to the upper mantle or the huge south pole Aitken event causes the material to spread around the south pole region. There are prominent rays scattered around the South Pole region form south pole Aitken. Figure 9 shows the sample measured and modeled spectra. The root mean square error and correlation coefficient used as a metric to measure the close resemblance of the measured and modeled spectra.

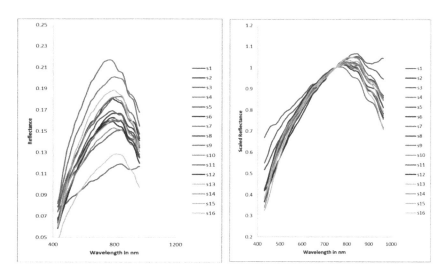

Fig. 8. Raw and Scaled reflectance spectra derived from the study area

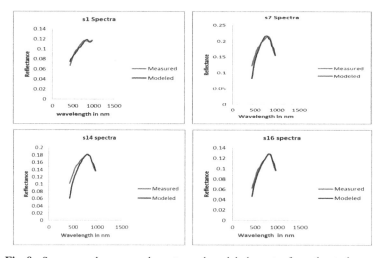

Fig. 9. Some sample measured spectra and modeled spectra from the study area.

Table 2. Mass fractions of each mineral obtained from the modeling process.

Spectra Id	Plagioclase	Clinopyroxene	Orthopyroxene	Olivine	Agglutinates
s1	31	0	48	0	21
s2	36	40	04	0	20
s3	36	36	0	0	28
s4	38	35	09	0	18
s5	33	35	12	0	20
s6	30	39	10	0	21
s7	42	44	0	0	14
s8	38	40	0	0	22
s9	37	39	03	0	21
s10	33	33	11	0	23
s11	30	38	05	0	16
s12	35	37	10	0	18
s13	33	40	10	0	15
s14	34	41	10	0	15
s15	33	35	07	5	24
s16	30	37	07	0	24

Table 3. The model parameters of each modeled spectra.

Spectra Id	Iron volume fraction	Average grain size (μm)	Porosity (%)	Phase function	Rmse	Correlation coefficient
s1	0.00026	63	60	average	0.0023	0.9874
s2	0.00026	99	60	average	0.0031	0.9903
s3	0.00031	108	60	average	0.0018	0.9967
s4	0.00016	86	60	average	0.0036	0.9923
s5	0.00025	113	60	average	0.0039	0.9920
s6	0.00025	98	60	average	0.0048	0.9843
s7	0.00015	51	60	average	0.0041	0.9832
s8	0.00029	94	60	average	0.0022	0.9981
s9	0.00029	85	60	average	0.0029	0.9953
s10	0.00029	87	60	average	0.0021	0.9935

(*continued*)

Table 3. (*continued*)

Spectra Id	Iron volume fraction	Average grain size (μm)	Porosity (%)	Phase function	Rmse	Correlation coefficient
s11	0.00023	88	60	average	0.0032	0.9921
s12	0.00023	88	60	average	0.0021	0.9982
s13	0.00015	92	60	average	0.0062	0.9482
s14	0.00018	93	60	average	0.0051	0.9676
s15	0.00023	78	59	average	0.0029	0.9949
z16	0.00032	75	66	average	0.0028	0.9938

5   Conclusion

The Chandrayaan-1 Hyperspectral data from south pole region near Chandrayaan-2 landing site is used for mineral analysis. The spectral profile analysis considering color composites based on different band parameters and spectral profile analysis is a standard technique for mineral mapping but for quantifying the mineralogy the radiative transfer model has been implemented for better prediction of mineralogical variation on the lunar surface. Moreover the space weathering process on an airless body like moon complicates and impedes the ability to get meaningful information hence the model also helps to determine the degree of space weathering in terms of iron fraction content. The results obtained from modeling process shows the high mass fraction of Clinopyroxne and low Orthopyroxene content for small fresh craters spread across the study area. The highland spectra shows high mass fraction for plagioclase. The mature spectra shows overall low reflectance and model return high iron content for such spectra with high agglutinates. The selected spectra shows around 40% of Clinopyroxene as spectra is selected from small young craters or either from crater wall having steep slopes. Spectra derived except the selected spectra showing very low reflectance with almost no absorption indicates very mature entire area hence difficult to model. At south pole region there is a strong chance of getting hydroxyl absorptions but the limited coverage of HySI dataset is constraint however data sets from other sensor with larger spectral coverage will gives an opportunity to model the spectra more accurately.

References

1. Jin, S., Arivazhagan, S., Araki, H.: New results and questions of lunar exploration from SELENE, Chang'E-1, Chandrayaan-1 and LRO/LCROSS. Adv. Space Res. **52**(2), 285–305 (2013)
2. McCord, T.B., Adams, J.B.: Progress in remote optical analysis of lunar surface composition. The Moon **7**(3–4), 453–474 (1973). https://doi.org/10.1007/BF00564646
3. Isaacson, P.J., Pieters, C.M.: Northern imbrium noritic anomaly. J. Geophys. Res. **114**, E09007 (2009). https://doi.org/10.1029/2008JE003293

4. Bhattacharya, S., Chauhan, P., Rajawat, A.S., Kumar, A.S.K.: Lithological mapping of central part of Mare Moscoviense using Chandrayaan-1 Hyperspectral Imager (HySI) data. Icarus **212**(2), 470–479 (2011). https://doi.org/10.1016/j.icarus.2011.02.006
5. Borst, A.M., Foing, B.H., Davies, G.R., van Westrenen, W.: Surface mineralogy and stratigraphy of the lunar south pole-Aitken basin determined from Clementine UV/VIS and NIR data. Planetary Space Sci. **68**(1), 76–85 (2012). https://doi.org/10.1016/j.pss.2011.07.020
6. Sayyad, S.B., Mohammed, Z.R., Deshmukh, R.R.: Mineral mapping of mare cresium using Chandrayaan-1 Hyperspectral (HySI) data. J. Appl. Sci. Comput. **5**(7), 88–95 (2018)
7. Sivakumar, V., Neelakantan, R.: Mineral mapping of lunar highland region using Moon Mineralogy Mapper (M3) hyperspectral data. J. Geol. Soc. India **86**(5), 513–518 (2015). https://doi.org/10.1007/s12594-015-0341-1
8. Sivakumar, V., Neelakantan, R., Santosh, M.: Lunar surface mineralogy using hyperspectral data: implications for primordial crust in the Earth-Moon system. Geosci. Front. **8**(3), 457–465 (2017). https://doi.org/10.1016/j.gsf.2016.03.005
9. Burns, R.G.: Mineralogical Applications of Crystal Field Theory. Cambridge University Press, New York (1970)
10. Anbazhagan, S., Arivazhagan, S.: Reflectance spectra of analog basalts; implications for remote sensing of lunar geology. Planet. Space Sci. **57**(12), 1346–1358 (2009)
11. Anbazhagan, S., Arivazhagan, S.: Reflectance spectra of analog anorthosites: Implications for lunar highland mapping. Planet. Space Sci. **58**(5), 752–760 (2010)
12. Hapke, B.: Effects of a simulated solar wind on the photometric properties of rocks and powders. Ann. N. Y. Acad. Sci. **123**, 711–721 (1965). https://doi.org/10.1111/j.1749-6632.1965.tb20395.x
13. Hapke, B.: Inferences from the optical properties of the moon concerning the nature and evolution of the lunar surface. Radio Sci. **5**, 293–299 (1970). https://doi.org/10.1029/RS005i002p00293
14. Hapke, B.: Darkening of silicate rock powders by solar wind sputtering. The Moon **7**, 342–355 (1973). https://doi.org/10.1007/BF00564639
15. Hapke, B., Cassidy, W., Well, E.: Effects of vapor-phase deposition processes on the optical, chemical, and magnetic properties of the lunar regolith. The Moon **13**,339–353 (1975). https://doi.org/10.1007/BF00567525
16. Keller, L.P., Mckay, D.S.: Discovery of vapor deposits in the lunar regolith. Science **261**, 1305–1307 (1993)
17. Keller, L.P., McKay, D.S.: the nature and origin of rims on lunar soil grains. Geochim. Cosmochim. Ac. **61**, 2331–2341 (1997)
18. Taylor, L.A., Pieters, C.M., Keller, L.P., Morris, R.V., McKay, D.S.: Lunar mare soils: space weathering and the major effects of surface-correlated nanophase Fe. J. Geophys. Res.-Planets **106**, 27985–27999 (2001)
19. Taylor, L. A., et al.: Mineralogical and chemical characterization of lunar highland soils: insights into the space weathering of soils on airless bodies, J. Geophys. Res. Planets **115**, E02002 (2010)
20. Sunshine, J.M., Pieters, C.M., Prait, S.F.: Deconvolution of mineral absorption bands: an improved approach. J. Geophys. Res. **95**(B5), 6955–6966 (1990). https://doi.org/10.1029/JB095iB05p06955
21. Hapke, B.: Bidirectional reflectance spectroscopy: I. Theory. J. Geophys. Res. Solid Earth **86**(B4), 3039–3054 (1981). https://doi.org/10.1029/JB086iB04p03039
22. Shkuratov, Y.G., Starukhina, L., Huffmann, H., Arnold, G.: A model of spectral albedo of particulate surfaces: implications for optical properties of the Moon. Icarus **137**(2), 235–246 (1999). https://doi.org/10.1006/icar.1998.6035
23. Hapke, B., Wells, E.: Bidirectional reflectance spectroscopy. II Experiments and observations. J. Geophys. Res. **86**, 3055–3060 (1981)

24. Hapke, B.: Bidirectional reflectance spectroscopy. III - Correction for Macroscopicroughness, Icarus **59**, 41–59 (1984)

25. Hapke, B.: Bidirectional reflectance spectroscopy IV - The extinction coefficient and the opposition effect. Icarus **67**(2), 264–280 (1986). https://doi.org/10.1016/0019-1035(86)901 08-9

26. Hapke, B.: Theory of reflectance and emittance spectroscopy. Topics in Remote Sensing, Cambridge. Cambridge University Press, UK (1993)

27. Hapke, B.: Space weathering from Mercury to the asteroid belt. J. Geophys. Res. Planets **106**(E5), 10039–10073 (2001). https://doi.org/10.1029/2000JE001338

28. Clark, R.N., Roush, T.L.: Reflectance spectroscopy: quantitative analysis techniques for remote sensing applications. J. Geophys. Res. **89**, 6329–6340 (1984). https://doi.org/10.1029/JB089iB07p06329

29. Mustard, J.F., Pieters, C.M.: Quantitative abundance estimates from bidirectional reflectance measurements. In: Proceedings of the 17th Lunar and Planetary Science Conference, Part 2 (1987). J. Geophys. Res. **92**, E617–E626 (1987). https://doi.org/10.1029/JB092iB04p0E617

30. Lucey, P.G.: Mineral maps of the Moon. Geophys. Res. Lett. **31**, L08701 (2004). https://doi.org/10.1029/2003GL019406

31. Lawrence, S.J., Lucey, P.G.: Radiative transfer mixing models of meteoritic assemblages. J. Geophys. Res. **112**, E07005 (2007). https://doi.org/10.1029/2006JE002765

32. Cahill, J.T.S., Lucey, P.G., Wieczorek, M.A.: Compositional variations of the lunar crust: results from radiative transfer modeling of central peak spectra. J. Geophys. Res. **114**, E09001 (2009). https://doi.org/10.1029/2008JE003282

33. Cahill, J.T.S., Lucey, P.G., Stockstill-Cahill, K.R., Hawke, B.R.: Radiative transfer modeling of near-infrared reflectance of lunarhighland and mare soils. J. Geophys. Res. **115**, E12013 (2010). https://doi.org/10.1029/2009JE003500

34. Yan, B., Wang, R., Gan, F., Wang, Z.: Minerals mapping of the lunar surface with Clementine UVVIS/NIR data based on spectra un mixing method and Hapke model. Icarus **208**, 11–19 (2010)

35. Hiroi, T., Pieters, C.M.: Estimation of grain sizes and mixing ratios of fine powder mixtures of common geologic minerals. J. Geophys. Res. **99**(E5), 10867 (1994). https://doi.org/10.1029/94JE00841

36. Johnson, P.B., Cristy, R.W.: Optical constants of metals: Ti, V, Cr, Mn, Fe, Co, Ni, and Pd, prb. Phys. Rev. **9**, 5056–5070 (1974)

37. Morris, R.V.: The surface exposure /maturity/ of lunar soils - some concepts and Is/FeO compilation. In: Proceedings of the 9th Lunar and Planetary Science Conference, Houston, TX, 13–17 March 1987, vol. 2, pp. 2287–2297. Pergamon Press, Inc., New York (1978)

38. Kumar, A.: Hyper spectral imager for lunar mineral mapping in visible and near infrared band. Curr. Sci. **96**, 496 (2009)

Confusion Matrix-Based Supervised Classification Using Microwave SIR-C SAR Satellite Dataset

Shafiyoddin Sayyad[1]([⊠]), Mudassar Shaikh[2], Anand Pandit[2], Dattatraya Sonawane[2], and Sandip Anpat[3]

[1] Department of Physics, Milliya Arts, Science and Management College, Beed, Maharashtra, India
syedsb@rediffmail.com
[2] Department of Electronic Science, New Arts, Commerce and Science College, Ahmednagar, Maharashtra, India
mudassarshaikh333@gmail.com, appandit2007@gmail.com, dattasonawane02@gmail.com
[3] Department of Computer Science, Marathwada Mitramandal College of Commerce, Pune, Maharashtra, India
sandipanpat2@gmail.com

Abstract. The microwave Synthetic Aperture Radar (SAR) is an active type of remote sensing. The classification analysis has become one of the very important task, after the availability of microwave SAR datasets from the satellite. The one of the major challenges faced is the accuracy regarding classification analysis. In the present paper the two supervised classification techniques used, i.e., Wishart and Support Vector Machine (SVM). The accuracy results for both classifiers are analyzed on the basis of confusion matrix and omission, commission error. The overall process is carried out on SIR-C L-band SAR dataset of Kolkata (W.B.) India. The four major classes studied is Water, Trees and Mangrove, Paddy, Settlement. The present work focus on the agricultural application. From the overall work it is found that the accuracy of the Wishart supervised classifier is 92.18% and for SVM supervised classifier it is 99.58%. There is very huge difference of 07.40% between these two classifiers. Hence the SVM classifier has better accuracy compare to Wishart classifier. The overall work done by using software tool PolSARPro Ver. 5.0 and NEST Ver. 5.0.16.

Keywords: SAR · Wishart · SVM · Confusion matrix

1 Introduction

In remote sensing the synthetic aperture radar (SAR) is an active microwave type, which assimilated very high resolution images of the earth. It has ability to sense the objects during the day as well as at night, though there is change in environmental conditions, also it penetrate through clouds, smoke, fog, etc. [1, 2]. The classification is important tasks after availability of microwave sar datasets from satellite [3].

© Springer Nature Singapore Pte Ltd. 2021
K. C. Santosh and B. Gawali (Eds.): RTIP2R 2020, CCIS 1381, pp. 176–187, 2021.
https://doi.org/10.1007/978-981-16-0493-5_16

Since, last few decades, many researcher works on accuracy assessment of the microwave SAR image using classification techniques. the supervised algorithm based on complex Wishart distribution proposed by (Lee et al., 1994) [4]. The L Band ALOS PALSAR L-1.1 fully polarimetric dataset for land cover classification has been used by (Mishra et al., 2011) [5]. They analyzed and compared supervised classification results using five classes of land like minimum distance, maximum likelihood, parallelepiped, based on pauli decomposition and wishart classification based on eigen value decomposition by using ENVI and PolSARpro software [5]. The overall study suggests that the most of class having classification accuracy less than 85%. In the present paper work two supervised classification technique used, i.e., Wishart and Support Vector Machine (SVM). The microwave SAR dataset selected is the SIR-C Satellite SAR dataset. The aim of proposed work is to classify microwave SAR image using the above said supervised classification techniques and accuracy is estimated on the basis of error calculation from the classified image. The overall work done by using software tool PolSARPro ver. 5.0 and NEST ver. 5.0.16. The complete study need ground truth study, climate data, survey of india and more importantly the expertise in the training samples selection and processing of data which can be significantly done by all authors.

2 Classification Techniques

The classification is a technique used to identify the different objects in the remotely sensed imagery. It helps to understand image information in more depth. The class is nothing but the group of pixels, or Digital Numbers (DNs) having same spectral properties. Hence, in any multispectral image the pixel having its characteristics is expressed in the form of vectors, i.e., spectral properties of the pixel. These numbers of vector determine the specific class in the objects. The number of classes can be made by using certain indices. The process involves labelling of each class entity using DNs. The spectral pattern recognition is one of the key points in the classification that can be utilized pixel by pixel spectral information. Hence, with the help of spatial and temporal pattern recognition the features of the land cover are identified [6, 7]. The classification technique is based upon Entropy (H), Anisotropy (A) and Alpha (α) polarimetric decomposition parameters defined by Cloude et al. 1996 [8]. In classification assessment the coherency matrix is calculated on the basis of eigenvalue and eigenvector [T]. Hence, the arbitrary coherency matrix is written as,

$$\langle [T] \rangle = [U_3]\left[\sum\right][U_3]^{-1} \tag{1}$$

$$= \sum_{i=1}^{i=3} \lambda_i u_i u_i^{*T} \tag{2}$$

where $[\sum]$ is a 3×3 diagonal matrix with nonnegative real elements and $[U_3]$ is an unitary matrix [9].

2.1 Supervised Classification

The supervised classification technique needs prior knowledge of the field area. The results of computer generated unsupervised classification are also helpful in this study. In a study region the each area is known as a training site. The each training site having different spectral characteristics of DNs. In this classification, basic three steps are involved. First in the training stage identifies the training under areas and develop a numerical description of each land cover type in the selected study region. In the second step, each land cover class in a specific class and labeled it with specific land cover type name. While making class takes care of selecting same pixels in the group. If pixel are not able to understand then make an unknown class for that. In these ways after entire study area is categories, then it is forwarded towards the output stage. After successful selection of training area and categories into different class, at last run the whole process. The accuracy of creating class depends upon selecting the same pixel training area. Then at last classified output image get for further analysis [10].

2.1.1 Wishart Classifier

The Wishart H-A Alpha classification is special type of H A Alpha classification. Here the coherency matrix $\langle T_i \rangle$ of a pixel i of a multilook image knowing the class ωi, the Wishart complex distribution is given by,

$$p\big(\langle T_i \rangle \big/ \omega_m\big) = \frac{N^{-qN} \exp\Big(-tr\Big(N\big[\sum_m\big]^{-1} \langle T_i \rangle\Big)\Big)}{K(N, q)\big|\sum_m\big|^N} \tag{3}$$

Since, $\sum_m = E(\langle T_i \rangle \,|\langle T_i \rangle \; \varepsilon \omega_m)$

$$\sum_m = \frac{1}{N_m} \sum_{i=1}^{N_m} \langle T_i \rangle \tag{4}$$

where Nm pixel number of ωm, $K(N, q)$ which is factor of standardization [11]. Using Wishart classification method there is significant improvement in each iteration. After applying Wishart method the original class boundaries in H and the alpha become less distinct with considerable overlap. The advantage of using ishart method is its effectiveness in automated classification [12–14].

2.1.2 SVM Classifier

The Support Vector Machines (SVMs) is one of the powerful tools for performing supervised classification. It have more ability to classify nonlinear data. Particularly, the dataset like fully Polarimetric SAR dataset [15].

Suppose the N training samples represented by a set of pairs $\{(yi,xi), i = 1,.....,N\}$, where $yi \pm 1$ and $x_i \varepsilon R^n$ are the feature vectors with n components. The classifier is represented by a function, such as where α indicates the parameters of the classifier. The SVM method consists of hyperplane with labels $y = \pm 1$ are located on each side. These closest vectors are the support vectors, and the distance between them is the optimal margin shown in Fig. 1.

The vectors that are not on this hyperplane lead to $w.x + b \neq 0$. The support vectors lie on two hyperplanes that are parallel to the optimal hyperplane with the equation $w.x + b = \pm 1$.

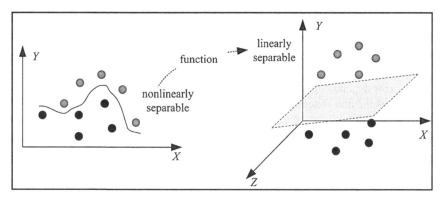

Fig. 1. Support Vectors Machine, (a) nonlinearly separable case, (b) linearly separable case

The maximization of the margin with two support vector hyperplanes leads to the following [16–19]:

$$\min\left\{\frac{1}{2}\|w\|^2\right\} \tag{5}$$

With $y_i(w.x + b) >= 1_i, i = 1, \ldots, k$

If the training samples are not linearly separable (Fig. 1 a), a regularization parameter is introduced and initialized with a large value corresponding to assigning a higher penalty for errors. Therefore, the constrained optimization problem becomes [20, 21],

$$\min\left\{\frac{1}{2}\|w\|^2 + C\sum_{i=1}^{k}\xi_i\right\} \tag{6}$$

With $y_i(w.x + b) >= 1 - \xi_i i$,
where C is the regularization parameter.

3 Error and Accuracy Assessment

The error and accuracy of classification are determined empirically by corresponding reference ground data.

3.1 Confusion Matrix

The results are tabulated in the form of a square matrix known as confusion matrix. This matrix helps in summarizing the performance of classification and corresponding categorized pixels to each land cover type class. The results help to summarized sample results. This is also called as error matrix. With the help of this the omission and commission error is calculated as below.

3.2 Commission Error

It is refers as to the samples of a certain class of the reference data that were not classified [22]. The error depends upon selection of pixel class from the area which can be under studied.It uses the non-diagonal row elements. It can be calculated as,

$$\left(\frac{a+b+c}{T_R}\right) \tag{7}$$

3.3 Omission Error

It is refers as to the samples of a certain class of the classified data that were wrongly classified [23]. It uses the non-diagonal column elements. The error depends upon the proper selection of certain area which can be classified. It can be calculated as,

$$\left(\frac{d+e+f}{T_C}\right) \tag{8}$$

It is organized as a two dimensional matrix where columns representing the reference data and rows representing the classification data.

Table 1. Confusion matrix for four class

Classified data	Reference data				Row total	Commission error (%)
	C1	C2	C3	C4		
C1	w	a	b	c	T_R	
C2	d	x				
C3	e		y			
C4	f			z		
Column Total	T_C					
Omission Error (%)						

Table 1 shows four classes are labelled as C1, C2, C3, C4. The matrix column shows reference data class. Matrix row shows classified data class. The diagonal values w, x, y, z shows the proportion of correctly classified pixels. The row total (T_R) get by adding non-diagonal elements a, b and c except w in the first row. Similarly the column total (T_C) calculated by adding the values of non-diagonal elements d, e and f except w. The same procedure obeyed for all rows and columns. Using this value T_C and T_R omission and commission error is calculated [22–24].

3.4 Accuracy Assessment

Accuracy assessment has two types of calculations, i.e., user's accuracy and producer's accuracy. The user's accuracy provides the user information about the accuracy of the land cover data. It can be calculated as the number of correctly classified samples divided by total row (T_R).

The producer's accuracy indicates the percentage samples of correctly classified reference class. It can be calculated as dividing the number of correctly classified samples by the total column (T_C). The both accuracy depends upon the amount of omission error and commission error. The relation between accuracy and error,

$$\text{User's accuracy } (\%) = (100 - \text{commision error})\% \qquad (9)$$

$$\text{Producer's accuracy } (\%) = (100 - \text{omission error})\% \qquad (10)$$

The overall accuracy can be calculated by dividing the number of correctly classified samples positioned in a diagonal of confusion matrix by the total number of reference pixels checked.

4 Study Area

The study area is in Kolkata (W.B.) India with latitude of 21^0 52' 58.80" N to 22^0 45' 07.20" N and longitude 87^0 51' 57.60" E to 88^0 35' 06.00'' E. The study area is one of the metro city of India. The huge amount of trees and paddy is present in the available area. Only a small amount of the settlement area is present in the SAR image. The major area covered by Trees and Vegetation. The river present in the study area is connected to Java Sea. The study area is used for interpretation of agricultural applications. The SAR dataset outline map and the region of the study area is shown in Fig. 2.

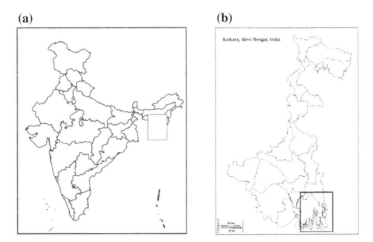

Fig. 2. Study area (a) Map of India (b) Selected area of Kolkata

5 Result and Discussion

In the study area the four classes like Water, Trees + Mangroves, Paddy and Settlement are made using classification techniques. In supervised classification the prior knowledge of the object present on the field area essential. This can be obtained by manual class generation with the help of the computer generated results called unsupervised classification and the scattering effect profile of the object present in the selected study region. In supervised classification two classification techniques, i.e., Wishart and SVM classification is used. Based on the ground truth data number samples were collected out of that, the four major classes are selected as mentioned above for studying the present paper work. In the PolSARPro software, using graphics editor the number of training samples are given. In both the classification 4 testing samples are selected and for every testing samples 5 training areas are given. This selected area is chosen on the basis of ground truth data. Once these training samples are selected, then it runs and the results of both supervised Wishart and SVM classifier shown in the Fig. 3 (a) and (b).

The coherency matrix given by Eq. (1) generates the confusion matrix. This matrix helps to calculate the accuracy of correct class made by users. It shows the accuracy of four classes and the class population. The class population generated from the number of pixels, or DN's contain in each class. It obtained from the supervised classified results and the number of training sample selections.

The Table 2 shows the class population for both the Wishart and SVM supervised classification. The confusion matrix for Wishart and SVM classification is shown Table 3 and 4 respectively, where row and column indicates user's and producer's defined classes.

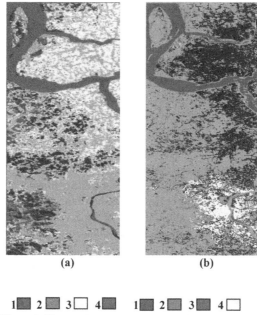

(a)   (b)

1■ 2■ 3☐ 4■   1■ 2■ 3■ 4☐
1. Water 2. Trees + Mangrove 3. Paddy 4. Settlement

Fig. 3. Supervised classification for Kolkata, India SIR-C SAR (a) Wishart (e) SVM Classifier

Table 2. Class population for wishart and SVM supervised classification

Class	Name of class	Class population Wishart	Class population SVM
C1	Water	4526	11810
C2	Trees + Mangrove	2701	03113
C3	Paddy	1825	02557
C4	Settlement	4841	02793

Table 3. Confusion matrix for wishart classification

Class	C1	C2	C3	C4
C1	98.48	00.31	00.00	01.22
C2	01.26	95.63	00.04	03.07
C3	05.86	01.53	89.48	03.12
C4	00.33	14.52	00.00	85.15

Table 4. Confusion matrix for SVM classification

Class	C1	C2	C3	C4
C1	100.00	00.00	00.00	00.00
C2	000.03	99.29	00.55	00.13
C3	000.00	00.04	98.83	01.13
C4	000.00	00.25	00.90	98.85

The accuracy of each class is calculated with the help of error finding in each class. The two error calculation is necessary, i.e. omission and commission error which is calculated using Eq. (10) and (11). The Table 5 and 6 shows the error and accuracy assessment for Wishart supervised classification and SVM supervised classification respectively. Using this the accuracy for each class and overall accuracy is calculated. The comparison of error and accuracy between Wishart and SVM classification is shown by graphical representation. The Fig. 4 shows omission error, Fig. 5 shows commission error and the Fig. 6 shows the accuracy assessment for both the classification.

Table 5. Error and accuracy assessment for wishart supervised classification

Class	Total population	Omission error (%)	Total population	Commission error (%)	Accuracy (%)
C1	069/4526	0.015	157/4614	0.03400	98.48
C2	118/2701	0.043	745/3328	0.22300	95.63
C3	192/1825	0.105	001/1634	0.00061	89.48
C4	719/4841	0.148	195/4317	0.04500	85.15
Overall Accuracy					**92.18**

Table 6. Error and accuracy assessment for SVM supervised classification

Class	Total population	Omission error (%)	Total population	Comm. error (%)	Accuracy (%)
C1	0/11810	0.0000	0.94/11810.94	0.000079	100.0
C2	23/3113	0.0073	08.01/3098.01	0.002580	99.29
C3	30/2557	0.0117	42.13/2569.13	0.016370	98.83
C4	32/2793	0.0110	33.84/2794.84	0.012100	98.85
Overall Accuracy					**99.58**

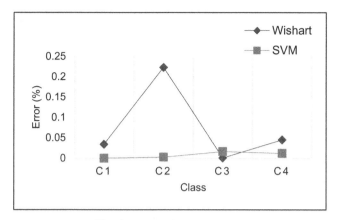

Fig. 4. Graph of omission error

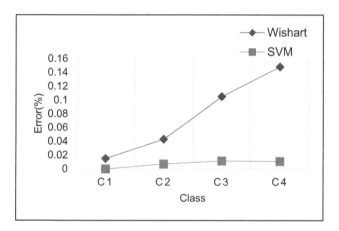

Fig. 5. Graph of commission error

From the simulation result, it is found that more errors present in the Wishart supervised classification in all the four classes as compare with the SVM supervised classification. The reason behind the more error present is the presence of mix classification and the resolution of the DN's. Hence, using these errors calculation the accuracy can be calculated and found as, in case of the Wishart supervised classifier, it is 92.18% and for SVM supervised classifier it is 99.58%. There is a difference of 7.40% between these two classifiers. Hence, from the overall study it is found that the confusion matrix based accuracy assessment for microwave SIR-C SAR dataset is best suited.

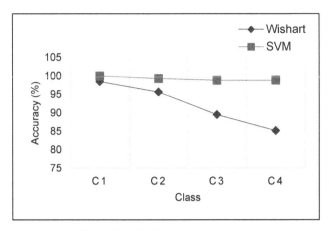

Fig. 6. Graph of accuracy assessment

6 Conclusion

The present paper works confusion matrix based for microwave L band SIR-C dataset is successfully analyzed and accessed. The accuracy assessment based on an error calculation is found to be the realistic method for microwave SAR data analysis. In the two types of supervised classifier, Wishart and SVM results the accuracy for classes like Trees and Mangroves and Paddy are found to be better accuracy. Both the supervised classifiers shows more that 90% accuracy. The majority area in the study site is the agricultural area, the dataset is L band so it is said that the microwave L band dataset is better results for agricultural application. Hence, from the overall study it is found that the confusion matrix based accuracy assessment for microwave SIR-C SAR dataset is best suited.

References

1. Jensen, R.J.: Remote Sensing of the Environment an Earth Resource Perspective, 2nd edn., pp. 12–35. Pearson, London (2014)
2. Calla, O.P.N.: Microwave Remote Sensing, pp. 1–25. Delhi, Director, DESIDOC, Metcalfe House (2009)
3. Shaikh, M.A., Khirade, P.W., Sayyad, S.B.: Classification of polarimetric SAR (PolSAR) image analysis using decomposition techniques. Int. J. Comput. Appl. (IJCA) Proc. Nat. Conf. Digit. Image Sig. Process. (NCDISP 2016) **1**, 20–23 (2016a)
4. Lee, J.S., Grunes, M.R., Kwok, R.: Classification of multi-look polarimetric SAR imagery based on the complex wishart distribution. Int. J. Remote Sens. **15**(11), 2299–2311 (1994)
5. Mishra, P., Singh, D.: Land cover classification of PALSAR images by knowledge based decision tree classifier and supervised classifiers based on SAR observables. Progress Electromagnetics Res. B **30**, 47–70 (2011)
6. Shaikh, M.A., Khirade, P.W., Sayyad, S.B.: Unsupervised and supervised classification of polsar image using decomposition techniques: an analysis from L- band SIR-C data. Asian J. Multidisciplinary Stud. (AJMS) **4**(8), 140–144 (2016)
7. Turkar, V., Rao, Y.S.: Supervised and unsupervised classification of PolSAR images from SIR-C and ALOS/PALSAR using PolSARPro. In: Proceeding of Conference (2009)

8. Cloude, S.R., Eric, P.: A review of target decomposition theorems in radar polarimetry. IEEE Trans. Geosci. Remote Sens. **34**(2), 498–518 (1996)

9. Ouarzeddine, M., Souissi, B., Belhadj-Aissa, A.: Classification of polarimetric SAR images based on scattering mechanisms. Spatial Data Quality (2007)

10. Cloude, S.R., Pottier, E.: An entropy based classification scheme for land applications of polarimetric SAR. IEEE IGRS **35**(1), 68–78 (1997)

11. Ouarzeddine, M., Souissi, B.: Unsupervised Classification using Wishart Classifier. USTHB, F.E.I, BP No 32 EI Alia Bab Ezzouar, Alger

12. Lee, J.S., Grunes, M.R., Ainsworth, T.L., Du, L.J., Schuler, D.L., Cloude. S.R.: Unsupervised classification using polarimetric decomposition and complex wishart distribution. IEEE Trans. Geosci. Remote Sens. **37/1**(5), 2249–2259 (1999)

13. Shenglong, G., Yurun, T., Yang, L., Shiqiang, C., Wen, H.: Unsupervised classification based on H/Alpha decomposition and Wishart classifier for compact Polarimetric SAR. IEEE IGARSS **2015**, 1614–1617 (2015)

14. Lee, J.S., Grunes, M.R., Pottier, E., Ferro-Famil, L.: Unsupervised terrain classification preserving polarimetric scattering characteristics. IEEE Trans. Geosci. Remote Sens. **42**(4), 722–731 (2004)

15. Hosseini, R.S., Entezari, I., Homayouni, S., Motagh, M., Mansouri, B.: Classification of polarimetric SAR images using support vector machines. Can. J. Remote Sens. **37**(2), 220–233 (2011)

16. Lardeux, C., Frison, P.L., Rudant, J.P., Souyris, J.C., Tison, C., Stol, B.: Use of the SVM classification with polarimetric SAR data for land use cartography. In: Proceeding of IEEE, pp. 497–500 (2006)

17. Fukuda, S., Hirosawa, H.: Polarimetric SAR image classification using support vector machines. IEICE Trans. Electron. E84-C **12**, 1939–1945 (2001a)

18. Fukuda, S., Hirosawa, H.: Support vector machine classification of land cover: application to polarimetric SAR data. In: Geoscience and Remote Sensing Symposium (IGARSS), pp. 187–189. IEEE (2001b)

19. Lardeux, C., Frison, P.L., Rudant, J.P. , Souyris, J.C., Tison, C., Stoll, B.: Use of the SVM classification with polarimetric SAR data for land use cartography. In: IGARSS 2006, Denver, Colorado, pp. 497–500 (2006)

20. Ramakalavathi, M., James, A., Nicolas, H., Younan, L., Bruce, M.: Supervised classification using polarimetric SAR decomposition parameters to detect anomalies on earthen Levees. In: IEEE Geoscience and Remote Sensing Symposium (IGARSS), pp. 983–986 (2016)

21. Zhao, Q., Principe, J.C.: Support vector machines for SAR automatic target recognition. IEEE Trans. Aerosp. Electron. Syst. **37**(2), 643–654 (2001)

22. Janssen, L.F., Frans, J.M., Wel, V.D.: Accuracy assessment of satellite derived land-cover data: a review. Photogram. Eng. Remote Sens. **60**(4), 419–426 (1994)

23. Story, M., Congalton, R.G.: Accuracy assessment at: user's perspective. Photogram. Eng. Remote Sens. **52**(3), 397–399 (1986)

24. Kumar, N.: Lecture Notes on Remote Sensing-Digital Image Processing Information Extraction. IISc Banglore, MSL2, 1

25. SIR-C Dataset. https://earthexplorer.usgs.gov/

Forensic Identification of Birds from Feathers Using Hue and Saturation Histogram

Vini Kale[(✉)] and Rajesh Kumar

Government Institute of Forensic Science, Aurangabad, Maharashtra, India
kale_vini@rediffmail.com, rajeshkumar512@gmail.com

Abstract. The planet earth is house of a variety of species of bird. Many of these birds are now getting extinct. Moreover, a lot of wildlife crimes are committed against birds including shooting, trapping, poisoning, and illegal sale of rare species. Feathers may be good evidence in such cases to identify species of birds. In this study, we propose a pattern recognition based technique for identification of birds from feathers. Our literature survey could not reveal a systematic study on identification of birds from the images of feathers. We have made a digital database of 60 feathers from 15 different species of birds. Hue and saturation histogram, which yields a feature vector of 46 dimensions, is extracted from the images of feathers. Nearest Neighbor (NN) algorithm is utilized for identification of birds using various distance metrics, which resulted into a maximum accuracy of 95.46%.

Keywords: Bird identification · Feather · Hue-saturation histogram · Nearest neighbor

1 Introduction

The animal kingdom on the planet earth is very rich. Out of the existing species of animals and birds on the planet, many of them are getting extinct, mainly due to climate change, pollution and illegal hunting. To save these birds from further extinction, governments of various countries around the globe have made facilities in terms of bird/animal sanctuary and at the same time made stringent regulations to stop illegal hunting, shooting, trapping and purchase/sale of birds. Even caging of birds has been made punishable. In forensic scenario, whenever an illegal activity pertaining to birds takes place, many a times at scene of incidence only feathers of birds are found. In such scenario, the most obvious question is whether the birds can be identified from the given feathers. In current scenario, birds are identified from feathers from their morphological similarities from a given atlas or database. But the process are mostly manual, however, a few articles reported the identification of birds from feathers using image analysis approach.

Feathers are complex epidermal organ much like hairs that forms plumage of birds. Feathers of various colors, sizes, shapes and form are arranged on the body of a bird in a specific pattern. The bird's plumage plays an important role in keeping the body of the

© Springer Nature Singapore Pte Ltd. 2021
K. C. Santosh and B. Gawali (Eds.): RTIP2R 2020, CCIS 1381, pp. 188–195, 2021.
https://doi.org/10.1007/978-981-16-0493-5_17

bird warm, supporting the flight, and sending communicating signals to other animals [1]. The shape and color of the feathers of the birds also differ according to their gender and age. A typical feather from the bird from the database has been shown in Fig. 1.

Fig. 1. The figure shows a typical feather, essentially consists of two parts: shaft and barbs (hair like structure). The shaft further contains two parts: rachis and calamus.

As our literature survey could not retrieve a systematic pattern recognition approach for identification of birds from feathers, we propose a preliminary study for automatic identification of birds from the images of their feathers. For identification, hue-saturation histogram is considered as features and classification is done using simple nearest neighbor algorithm.

We have organized rest of the paper as follows: Sect. 2 describes related work in this domain, Sect. 3 discusses proposed method including creation of database, results and discussion is elaborated in Sect. 4 followed by conclusion and future directions, which are placed in Sect. 5.

2 Related Work

In forensic set-up, Identification of birds from feathers is mostly done by merely comparing anatomical structure of unknown feather from a database of feathers. As far as pattern recognition framework for identification is concerned, much effort has not been put by the researchers. In [2], authors have addressed problem of subspecies identification of red tailed hawk based on color and texture analysis of feathers. Classification was done using simple ANOVA (Analysis of variance). The authors could successfully identify the subspecies of red tailed hawk. Not much was found in our literature survey.

3 Proposed Method

The problem of identification of bird from feather has been summarized in Fig. 2. As shown in the figure, hue-saturation histogram (color features) is extracted from the pre-processed query feather image, which is matched against the pre-trained database of feathers from 15 species. The detailed description of the proposed methodology has been further described in following sections.

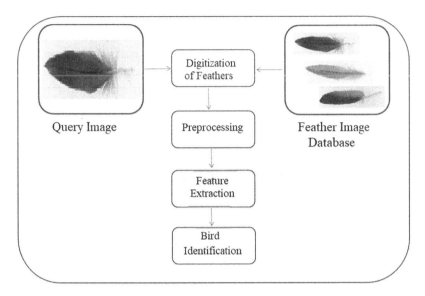

Fig. 2. Proposed methodology for bird identification

3.1   Creation of Database.

70 feathers from 15 species of birds, ranging from two to eight feathers from each species, are collected from Siddhartha Zoological Garden, Aurangabad, Maharashtra and Nirmal Agro Tourism, Nashik, Maharashtra. The name of the species has been given in Table 1. These feathers were digitized using a flatbed scanner at a spatial resolution of 500dpi and 8 bit quantization. The images are further cropped to get only feather region to avoid any kind of disturbances due to background.

Table 1. Species of bird included in the database.

Species	Species	Species
White Necked Stork	Green Pheasant	Emu
Red Shoulder Macaw	Red Golden Pheasant	Black Swan
African Love Birds	Pigeon	Hanh's Macaw
Grey Salon	Love Birds	Fantail Pigeon
Silver Pheasant	Duck	Rose Breasted Cockatoo

The images in the database have been shown in Fig. 3. From the database, 60% images were used for training while rest of the 40% was used for testing. Figure 3 shows typical feather images incorporated in the database.

Fig. 3. Typical feather images incorporated in the database. Feathers of birds: (a) White Necked Stork (b) Emu (c) Red Golden Pheasant (d) Hanh's Macaw (e) Duck (f) Grey Salon (g) Black Swan (h) Green Pheasant (i) Love Birds

3.2 Hue-Saturation Histogram

The color features used for the targeted task was extracted from hue-saturation histogram of feather image. Visualization of color depends on the wavelength of electromagnetic radiation and surface of object, where any of the interaction of electromagnetic spectrum, such as, absorption, transmission, fluorescence and reflection takes place. Any color image consists of two parts: luminance and chrominance. Luminance represents the intensity information in the image while chrominance represents color information. In R-, G-, B- representation of a color image, these two information are not separated. Thus, the image is converted into HSV format. Here, hue is represented by H, saturation by S and value by V. Hue and saturation are chrominance part of the image while value is luminance part of the image.

i. **Hue:** Hue is the color portion of the model, expressed as a number from 0 to 360° [3]:

$$H = \theta \ if \ B \leq G$$

$$H = 360° - \theta \ if \ B > G$$

Where,

$$\theta = \cos^{-1} \frac{1/2[(R - G) + (R - B)]}{[(R - G)^2 + (R - B)(G - B)]^{1/2}}$$

$$R = Red \ component \ of \ the \ image$$

$$G = Green \ component \ of \ the \ image$$

$$B = Blue \ component \ of \ the \ image$$

ii. **Saturation:** Saturation is a measure of the degree by which white light is mixed into a pure color [3].

$$S = 1 - \frac{3}{R + G + B} \times [\min(R, G, B)]$$

Where,

$$R = Red \ component \ of \ the \ image$$

$$G = Green \ component \ of \ the \ image$$

$$B = Blue \ component \ of \ the \ image$$

iii. **Value:** The brightness or intensity of the color in a scale of 0–100% is described as value, where 0 represent completely black and 100 is the brightest one [3].

$$V = (R + G + B)/3$$

Where,

$$R = Red\ component\ of\ the\ image$$

$$G = Green\ component\ of\ the\ image$$

$$B = Blue\ component\ of\ the\ image$$

3.3 Identification Protocol

The features are extracted from the newly created feather image dataset, where 60% of the images are used for training and 40% is used for testing. For training, the cropped images are first converted into HSV format. As features have to be extracted as hue-saturation histogram, we have used only hue and saturation components for feature extraction. Further, hue and saturation histograms are created with total 36 bins for hue (as the hue lies in 0–360°) and 10 bins for saturation. Thus, a total of 46 dimension feature vector is obtained and normalized to make the histograms in the range of 0 and 1. The same process is followed to obtain feature vectors from the test images.

We have made use of the nearest neighbor (NN) algorithm for identification of the test image with the database of feather images. Nearest Neighbor is a non-linear classification technique, where the classification task is done by comparing the distance of the unknown object to the known ones. The object can be classified by the majority votes of its neighbors, and the object is assigned to the class having the majority of minimal distances.

4 Results and Discussion

For the present work, nearest neighbor was utilized for identification of birds from the feather. As nearest neighbor classifier is based on calculating distance from unknown to known objects, various distance measures, namely, Euclidean, City block, Chebychev, Minkowski and Cosine were utilized for better recognition in HSV color space. The results have been shown in Table 2.

As shown in the Table 2 above, the highest accuracy of 95.46% was achieved using city block distance and nearest neighbor classifier with value of K = 1. Moreover, it has also been observed that Euclidean, Minkowski and Cosine metrics are giving similar kind of accuracy. As far as Chebychev distance is concerned, it has performed poorly amongst the considered distance metrics with an accuracy of 81.82%.

Table 2. Performance of nearest neighbor classifier on bird identification from feathers.

Distance metric	Accuracy (%)
Euclidean	86.38
City block	95.46
Chebychev	81.82
Minkowski	86.38
Cosine	86.38

5 Conclusion

The current study presents a pattern recognition approach for bird identification from feathers. The feather images taken from 15 species of birds were converted into HSV color space and hue and saturation histograms were calculated. For identification task, nearest neighbor was utilized. The best accuracy of 95.46% was achieved using nearest neighbor with city block distance. Euclidean distance performed poorer than the city block distance and yielded an accuracy of 86.38%. As our literature survey could not retrieve a systematic effort for an automated bird identification system for feather, it is encouraging to get a decent accuracy utilizing only color features. Thus, the current study makes the pattern recognition approach an efficient method for bird identification from feathers. The current study, being the first systematic pattern recognition approach for bird identification from feathers, exploring the efficacy of texture and shape features and usage of larger database are the obvious extension of the present work.

References

1. Yu, M., et al.: The developmental biology of feather follicles. Int. J. Dev. Biol. **48**, 181–191 (2004)
2. Solís, A.Y., Chanona, J.J., Vergara, M., Alonso, M.L., Blasco, J.: Image analysis: a tool for identification and classification of feathers from birds of prey (2019). https://www2.atb-pot sdam.de/cigr-imageanalysis/images/images12/tabla_137_C0270.pdf
3. Gonzalez, R.C.; Woods, R.E.: Digital Image Processing, Third Edition. Pearson Prentice Hall, Upper Saddle River (2013)
4. Sural, S., Qian, G., Pramanik, S.: Segmentation and histogram generation using the HSV color space for image retrieval. In: IEEE ICIP 2002, pp. 589–592 (2002)
5. Yue, J., Li, Z., Liu, L., Fu, Z.: Content-based image retrieval using color and texture fused features. Math. Comput. Model. **54**(3–4), 1121–1127 (2011)
6. Liu, G.H., Yang, J.Y.: Content-based image retrieval using color difference histogram. Pattern Recogn. **46**(1), 188–198 (2013)
7. Yu, H., Li, M., Zhang, H.J., Feng, J.: Color texture moments for content-based image retrieval. In: Proceedings International Conference on Image Processing, vol. 3, pp. 929–932. IEEE (2002)
8. Kekre, H.B., Thepade, S.D., Sarode, T.K., Suryawanshi, V.: Image retrieval using texture features extracted from GLCM, LBG and KPE. Int. J. Comput. Theory Eng. **2**(5), 695–700 (2010)

9. Sutojo, T., Tirajani, P.S., Setiadi, D.R., Sari, C.A., Rachmawanto, E.H.: CBIR for classification of cow types using GLCM and color features extraction. In: 2nd International Conferences on Information Technology, Information Systems and Electrical Engineering (ICITISEE), pp. 182–187. IEEE (2017)
10. Shah, M.P., Singha, S., Awate, S.P.: Leaf classification using marginalized shape context and shape+texture dual-path convolutional neural network. In: 2017 IEEE International Conference on Image Processing (ICIP), Beijing, pp. 860–864 (2017)
11. Adinugroho, S., Sari, Y.A.: Leaves classification using neural network based on ensemble features. In: 5th International Conference on Electrical and Electronics Engineering, IEEE (2018)
12. Du, J.X., Wang, X.F., Zhang, G.J.: Leaf shape based plant species recognition. Appl. Math. Comput. **185**(2), 883–893 (2007)
13. Haralick, R.M., Shanmugam, K., Dinstein, I.: Textural features for image classification. IEEE Trans. Syst. Man Cybern. **SMC-3**(6), 610–621 (1973)
14. Porebski, A., Vandenbroucke, N., Macaire, L.: Haralick feature extraction from LBP images for color texture classification. In: Image Processing Theory, Tools & Applications. IEEE (2008)
15. Aksoy, S., Haralick, R.M.: Textural features for image database retrieval. In: Proceedings of IEEE Workshop on Content-Based Access of Image and Video Libraries, pp. 45–49. IEEE (1998)

Transformation of Voice Signals to Spatial Domain for Code Optimization in Digital Image Processing

Akram Alsubari[1]([⊠]) [iD], Ghanshyam D. Ramteke[2] [iD], and Rakesh J. Ramteke[1] [iD]

[1] School of Computer Sciences, KBC North Maharashtra University, Jalgaon, India
akram.alsubari87@gmail.com, rakeshj.ramteke@gmail.com
[2] Department of Computer Science, KCES's, Institute of Management and Research, Jalgaon, India
shyam.ramteke@gmail.com

Abstract. The paper is intended to transform the voice-signal from the frequency domain into a spatial domain in form of grayscale image and applied the image processing techniques. To satisfy our hypothesis, two models of signal processing were carried out in this research: Speaker Recognition and Signal Segmentation. For applying the image processing techniques on the voice-signal, two methodologies were proposed to convert the signal into grayscale-image: signal-range based and fuzzy-based. The signal-range based is to convert the signal range from (-1 $\leftrightarrow$ 1) into (0 $\leftrightarrow$ 256). The second method of conversion, Fuzzy Gaussian Membership Function is applied to convert the signal range into (0 $\leftrightarrow$ 1), then multiply them by 255 to be in the range of grayscale image. In the Speaker Recognition, the LBP is used as pre-processing for filtering the intensity of the signal image. The HOG is used to extract the features of signal-image. So, the total length of features-vector is 324. The classification learner tool in MATLAB was used for classifying the feature-vectors and the results were found to be satisfactory. The automatic word segmentation was proposed based on thresholding and morphology operators. The segmentation accuracy is 93.67% in the Marathi-language. The highest recognition rate in speaker identification system is 96.9%.

Keywords: Voice recognition · Signal segmentation · VSI algorithm · Fuzzy · LBP · HOG

1 Introduction

In the cognition process, the pattern recognition is used to classify the data on the basis of knowledge which is already gained. The process to understand the pattern in which class belongs is known as pattern recognition. Those objects/patterns can be one-dimensional for instance: voice signal or image of two dimensional/three dimensional [1–3]. Pattern recognition domain leads to many applications such as speech recognition, face recognition, biometric system and so on. The image in pattern recognition system is represented as matrix, where the signal is represented as vector. Thus, this

© Springer Nature Singapore Pte Ltd. 2021
K. C. Santosh and B. Gawali (Eds.): RTIP2R 2020, CCIS 1381, pp. 196–209, 2021.
https://doi.org/10.1007/978-981-16-0493-5_18

research was carried out to understand the signal in the form of matrix or grayscale image. The voice signal is represented as a vector and the frequency range of signal is between $+1$ and -1, a couple of two values are the minimum and maximum range of the signal respectively. The measurement unit of frequency is Hertz. For the present work, the sampling frequency is used as 22050 Hz which is one of the most sampling frequencies: 8000 Hz, 11025 Hz, 32000 Hz, 44100 Hz and 48000 Hz. While, the image intensities are represented as matrix in the range between 0 to 255. The image is based on the reflected signals from the object. Those reflected signals will represent the intensity of each pixels.

The main research question during this research work "Is it possible to implement the signal processing system through the computer vision techniques?". Then, the hypothesis was prepared according to the research question. Therefore, all the voice signals in this experiment were converted to a grayscale-images by proposing two methods: Voice Signal to Image (VSI) algorithm and fuzzy-based method. The frequency range of the voice signal become as grayscale image 0 to 255. Then applied many computer vision techniques for proving our hypothesis whether is accepted or rejected. Two systems were proposed in this research: speaker recognition and word segmentation. The speaker recognition system can be categorized into two types: first, text-dependent system which is the training and testing voice are having the same text. Second, the text-independent is more popular in the speaker recognition, the training and testing voice are different text. In this paper, the text-independent system is used.

The rest of the section describes the hypothesis of this research. The literature survey (the traditional applications) of speaker recognition describes in the section two. The voice to image conversation methods discuss in the section three. Speaker recognition and word segmentation systems based on computer vision techniques are described in the section four and five, respectively. The section six discusses the results and implementation. The conclusion of the research describes in the last section.

1.1 Hypothesis

H0: the voice signal can be implemented in a form of image.
H1: the voice signal cannot be implemented in a form of image.

2 Literature Survey

Several researchers have been developing for increasing recognition accuracy based on various sophisticated techniques since last recent years. Zhaofeng Zhang et al. [4], have developed the method of blind dereverberation based on generalization spectral subtraction (GSS). In a real environment, the proposed method has been applied to the speaker recognition of the distant-talking (far-field). It achieved a recognition rate of 92.2%. Daniel Ramos-Castro et al. [5], have dealt with a novel method for score normalization scheme. It was the use of a piece of equipment to recognize those persons from their tone. It used for speaker verification, which was based on test-normalization method. It was another application to improve the biometric system.

Surbhi Mathur et al. [6], has implemented how to recognize the anonymous speaker by extracting the speaker unambiguous details included in the speech wave of human. In the forensic stream, a sound expert rendered its opinion in terms of different two evidences: prosecution hypothesis and defence hypothesis. In the hypothesis of prosecution, the unknown test sample was originated from the provided source. The unfamiliar text sample was revealed from a potential suspect population.

R. I. Damper et al. [7], investigated speaker identification using sub-band processing based on hidden markov models. The output of each sub-band was used to train and test individual HMM. Many potential applications can be seen in commerce and business, security, surveillance. Kevin R. Farrell et. al. [8], have presented the speaker recognition of text-independent. It used the TIMIT database with the help of 38 dissimilar speakers, which were from the same dialect region. The modified neural tree network (MNTN) was a hierarchical classifier. The MNTN provided a logarithmic saving from retrieval.

Douglas A. Reynolds [9] has exposed the verification system and the speaker identification on the basis of the model of Gaussian mixture speaker. The system of identification was a maximum likelihood classifier as well as the verification system was a likelihood ratio hypothesis. Four speech databases were available: TIMIT, TIMIT switchboard and YOHO. The closed-set identification accuracies with the help of 630 speakers in the databases of TIMIT and NTIMIT have been 99.5% and 60.70% respectively. The overall accuracy of closed-set identification system was outperformed as 99.50%.

3 Voice to Image Conversion

Two methods were adopted for converting the voice signal into gray scale image: signal-range based and fuzzy-based.

3.1 Signal-Range Based Method

The main process in this research is the conversation of the signal into grayscale image. The proposed algorithm for this conversation which is names as Voice Signal to Image (VSI):-

Step-1: Read the voice signal.

Step-2: Add one to the voice vector for making the range of signal from zero to two.

Step-3: Multiply each elements of the signal vector by 128 for changing the range from zero to 256 (see the Eq. 1).

Step-4: Compute the square root of the vector length.

Step-5: Convert the vector into matrix based on the squared value.

The following equation is proposed for the voice signal into image conversation:-

$$\sum_{i=1}^{n} (A_{i \to n} + 1) \times 128 \tag{1}$$

Where, $A_{i \to n}$ is the voice signal and the range of the signal is between -1 and 1 on the form of vector. Therefore, $+1$ is added to remove the negative values from the vector. With the help of original signal in Fig. 1, the constant number of 128 is to make the range of signal in the range of grayscale image 0–256 as shown in the Fig. 2. As shown in Fig. 3, signal image has been generated.

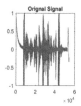

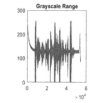

Fig. 1. A Sample of Original Signal.

Fig. 2. Sample of Grayscale Range

Fig. 3. Sample of Signal Image

3.2 Fuzzy-Based Method

The function of Gaussian fuzzy membership is usually used in classification phase for the computer vision systems [10], so in this paper, the fuzzy function will be applied to converted voice signal into grayscale image. The fuzzy conversation steps as follows:-

Step-1: Read the voice signal and it will be represented as vector.

Step-2: Computed the mean and standard deviation for voice vector.

$$\mu(H) = \frac{\sum H_i}{n} \tag{2}$$

$$\sigma(H) = \sqrt{\frac{\sum |H_i - \mu|^2}{n}} \tag{3}$$

Where, H is the vector of voice and n is the length of the H vector. The μ is the mean or the average of H. Ultimately, σ is the standard deviation.

Step-3: Applied the Gaussian fuzzy membership function to the voice vector on the basis of the mean and standard deviation.

$$F(H_w) = e^{\frac{-(H_i-\mu)^2}{2\sigma^2}} \tag{4}$$

Where, e is a mathematical constant and the approximate value is 2.7183. H_w is the values of Gaussian fuzzy membership, which are in the range between zero and one. H_i is the vector of voice. The symbols μ and σ are the mean and the standard deviation respectively. The average of each observation (H_w) of the voice vector is stored in another vector for the further step. The output of the Eq. 4 will be in the range between 0 to 1, so those values will multiply by 255 to make it in the range between 0 to 255 as grayscale intensities.

Step-4: Compute the square root of the vector length.

Step-5: Convert the vector into matrix based on the squared root value.

4 Speaker Recognition

As shown in Fig. 4, the voice signal is converting into the form of grayscale image by using the Voice Signal to Image (VSI) algorithm and fuzzy-based algorithm. in the speaker identification system, the overlapping fusion with the previous sample was utilized to generate a strong sample (signal-image). The grayscale image is filtered and pre-processed by applying the Local Binary Pattern (LBP) [11]. The Histogram of Oriented Gradient (HOG) [12] is a process to extract the features of signal-image. Different classification methodologies were applied for evaluating the proposed system such as Support Vector Machine (SVM) and Bagging ensemble algorithm.

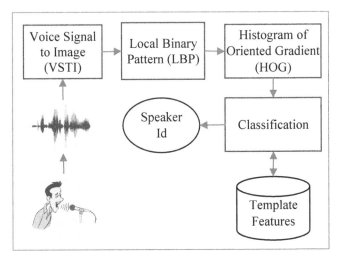

Fig. 4. The Proposed System for Speaker Identification

4.1 Local Binary Pattern Filter

This phase is to make the signal image ready for the further process. The Local Binary Pattern (LBP) was introduced 1990 by Ojala et al. [13] and it is one of the powerful techniques for understanding the texture futures.

The LBP, in this paper, used to filter the intensity of the signal image based on the neighbor intensities. Thus, the LBP needs to compare the eight-neighbor intensities against the central intensity and if the value of neighbor is greater than the intensity of centre, it is mentioned as one. If the value of neighbor is lesser or equal to the intensity of centre, it is mentioned as zero. In other way, this process is to threshold the eight neighbor pixels based on the centre pixels. The output of this process will be an 8-bit binary number which will be further converted into decimal number and it will be replace with the centre intensity [14] as shown in Fig. 3. Before applying the LBP on the signal image, those images are resized into 32 × 32 for obtaining a fixed length of feature vectors (as shown in Fig. 4). The following Fig. 5 is showing the output of LBP filter for the signal image.

4.2 Histogram of Oriented Gradient (HOG)

The HOG technique was introduced in 2005 for the human detection by N. Dalal and B. Triggs [12], also it is powerful technique in the pattern recognition system and it

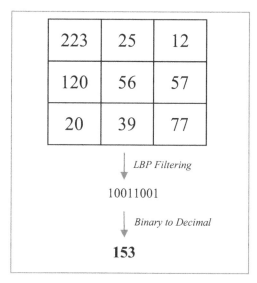

Fig. 5. Computing the LBP

has been used by many researchers as feature extraction technique. So, the HOG was proposed to extract features from the filtered image of the voice signal. The steps of the Histogram of Oriented Gradient as following:-

Step-1: The filtered image is divided into number of cells and blocks, where each cell is contained 8 × 8 pixels and the block are contained of 2 × 2 cells along with 50% overlapping from the block of next as revealed in Fig. 6, Fig. 7 and Fig. 8. In the signal image, the total number of blocks are 3 × 3 = 9 blocks.

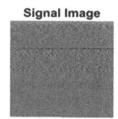
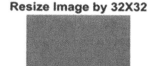

Fig. 6. Sample of signal image | **Fig. 7.** Sample of resizing image by 32 × 32 | **Fig. 8.** Sample of LBP Filtering for Signal Image

Step-2: In the range between 0 to 180°, 9 directions were selected. The orientation and gradient magnitude were calculated with the help of few equations as follows:-

$$dx = I(x + 1, y) - I(x - 1, y). \tag{5}$$

$$dy = I(x, y + 1) - I(x, y - 1). \tag{6}$$

$$m(x, y) = \sqrt{dx^2 + dy^2} \tag{7}$$

Step-3. Computing the histogram of each cells with the respect to the 9 bins direction

$$Total\ No.\ of\ Features = NB \times CB \times P \qquad (8)$$

Where, In the signal image, NB is the number of blocks, which are nine blocks. In the block, CB is the total number of cells as seen in Fig. 9. P is the 9 bins direction. So, the total numbers of HOG features are $9 \times 4 \times 9 = 324$.

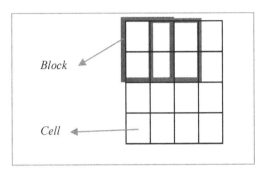

Fig. 9. The Blocks and Cells in the 32×32 Signal Image

4.3 Classifications

Different classifications were adopted for experimental evaluation of the proposed system for speaker identification such as SVM, KNN, Bagged Tree etc. The Classification Learner Tool in MATLAB used in this experiment. This tool is provided the accuracies and confusion matrix of the existing classification methods for the speaker identification system.

4.4 Database

The PDAs database [15] was developed by Y. Ochuchi in Mar 2003, so this dataset was used in this experiment for the speaker identification system. The PDAs database were recoded from 11 persons and from each individual 140 sentences. Thus, the dataset was divided into training and testing as 50% for each.

5 Voice Segmentation (Word-Level)

5.1 Pre-processing (Noise Reduction)

Noise reduction is a sub-branch of the speech technology. As seen in Fig. 10, Reducing noise region from noise-unwanted speech is necessity for the further processes. The next steps are proposed to remove the noise for the original signals:-

Step-1: Voice signal acquisition and analysis the properties of it as shown in Table 1.

Step-2: Basic processing of identification of each voice signals, which are related to the noisy signal model in the time domain as follows:

$$y(f) = x(f) + n(f) \tag{9}$$

Where, x(f) is the original signals, y(f) is the clean voice signals, n(f) is the noisy signals. In the domain of frequency, it may be denoted as:

$$Y(vf) = X(vf) + N(vf) \tag{10}$$

$$X(vf) = Y(vf) - N(vf) \tag{11}$$

Where Y(vf), X(vf) and N(vf) are Fourier transforms of y(f), x(f) and n(f) respectively. The vf is the frequency variable of the voice.

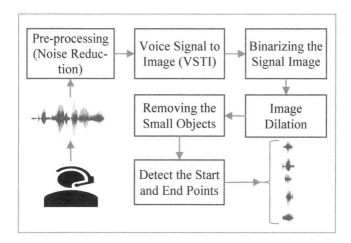

Fig. 10. The Proposed Methodology for Word Segmentation

Table 1. Acoustic Analysis of Each Sample of Voice

Criteria of voice	Depth
Frequency of Sampling	22050 Hz (22 kHz)
Parameter of Acoustic	FFT Spectrum (256 pts)
Size of Each Frame length	30 ms
Shifting of Each Frame	10 ms
Analysis window	Hamming Window
Noise Type	Acoustic Noise

Step-3: The noise signal and the signal of speech are replaced by their estimates:

$$\widehat{X}(vf) = Y(vf) - \widehat{N}(vf) \qquad (12)$$

The noise spectrum estimation $\widehat{N}(vf)$ is associated with the expected noise spectrum E $[|N(vf)|]$, which is computed using the time-averaged noise spectrum $\overline{N}(vf)$ taken from parts of the recording, where only noise is present.

Step-4: the noise estimation is given by:

$$\widehat{X}(vf) = E\big[|N(vf)|\big] \cong \big|\overline{N}(vf)\big| \qquad (13)$$

$$\widehat{X}(vf) = \frac{1}{K}\sum_{i=0}^{K-1}|N_i(vf)| \qquad (14)$$

Where,$|N_i(vf)|$ is the amplitude spectrum of the ith of the K frames of noise. In K^{th} frame, noise estimation may be obtained by filtering the noise as follows:

$$\widehat{N}_k(vf) = \big|\widetilde{N}_k(vf)\big| \qquad (15)$$

$$\widehat{N}_k(vf) = \lambda_n.\big|\widetilde{N}_{k-1}(vf)\big| + (1 - \lambda_n).|N_k(vf)| \qquad (16)$$

Where, $\widehat{N}_k(vf)$ is the smoothed noise estimate in i^{th} frame,λ_n is the filtering coefficient $(0.5 \le \lambda_n \le 0.9)$.

Step-5: Reducing of noise signals based on spectral subtraction (SS) method from an original voice file.

$$\widehat{X}_{vf}^{b} = Y_{vf}^{b} - \propto (vf)\overline{N}_{vf}^{b} \qquad (17)$$

Where,$\widehat{X}_{vf}^{b}$ is an estimation of the voice signal and magnitude spectrum to the power of b, $\overline{N}_{vf}^{b}$ is the time-averaged magnitude of noise spectrum to power the b, $\propto (vf)$ is controlled the amount of noise subtracted from the noise signals.

Step-6: Clean voice file: Reconstruct the voice signals.

5.2 Segmentation Algorithm

After the signal noises were removed, the voice signal was converted into a form of grayscale image by using the signal-range based conversation as shown in Fig. 11. The following algorithm is utilized to segment each word separately:-

Step-1: Read the voice file.

Step-2: Applied the Signal to Image algorithm for converting the signal into grayscale image.

Step-3: Binarizing the gray image into zero and one values.

Step-4: Applied the dilatation operator and label the objects in the image.

Step-5: Find the beginning and ending location of each object. The following equations were proposed to detect the starting-point and ending-point of word-object

$$Obj_i(x_j, y_j) \tag{18}$$

$$BL_i = ((x_{jb} - 1) \times NC) + 1 \tag{19}$$

$$EL_i = (x_{je} \times NC) \tag{20}$$

$$\sum_{i=1 \to n}^{n} Aud_i(BL_i : EL_i) \tag{21}$$

Where, Obj_i is the location of the interest objects in the signal image and i represents the number of objects. NC is the total number of columns. BL_i and EL_i are the beginning and ending location, which will be segmented from the original signal Aud_i.

5.3 Database

For the signal segmentation experiment, the local Marathi database was used. There were eight male speakers and seven female speakers. Manually, the phoneme acquisition has been collected the digits which were in Marathi language. 08 speakers of them uttered 0 to 10 and 07 speakers of those pronounced 1 to 10 in Marathi language. The size of phoneme acquisition is 158 phones. The frequency of all phones was 22 kHz.

6 Implementation and Results

The research experiment was carried out with the help of MATLAB software. The results of this paper are categorized into two phases: word segmentation and speaker identification.

6.1 Word Segmentation Results

The signal segmentation experiment was carried out on the word-level from the Marathi dataset. The proposed algorithm is aimed to segment the words which are presented in the voice file by using the computer vision techniques as shown in Fig. 11. From the Marathi dataset, 148 words were segmented correctly out of 159. 2 joined-words were segmented, and 13 repeated words were also segmented. 6 silent-parts were segmented.

The calculation of the accuracies and errors are presented in Table 2. The results of segmentation methods were computed manually by using the following equations:-

$$Segmentation\ Accuracy = \frac{Total\ number\ of\ Correct\ Segmented}{Total\ Number\ of\ Words} \times 100 \tag{22}$$

$$False\ Silent\ Segmentation = \frac{Total\ number\ of\ Silent\ Segmented}{Total\ Number\ of\ Words} \times 100 \tag{23}$$

$$False\ Repeated\ Segmentation = \frac{Total\ number\ of\ Repeated\ word\ Segmented}{Total\ Number\ of\ Words} \times 100$$

(24)

$$False\ Joined\ Word\ Segmentation = \frac{Total\ number\ of\ Joined\ word\ Segmented}{Total\ Number\ of\ Words} \times 100$$

(25)

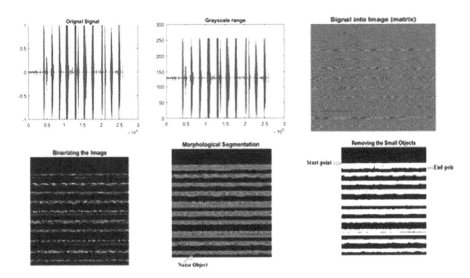

Fig. 11. The layout of the segmentation steps

Table2. Segmentation results

False Silent Segmentation	False Repeated Segmentation	False Joined Word Segmentation	Segmentation Accuracy
3.79%	8.22%	1.26%	93.67%

6.2 Speaker Identification Results

The HOG features were classified by using different kinds of existing techniques which are available in the classification learner tool in MATLAB from Table 3. The bagged tree classification method has shown the higher accuracy as compare to the other classification techniques. So, the confusion matrix of bagged tree only will be displayed in Fig. 12.

Table 3. Results of speaker recognition based on signal-range conversation

Classification method	Time in second	Recognition rate
Complex Tree	12.625	90%
Medium Tree	2.2916	83.5%
Simple Tree	2.1854	43.2%
Linear Discriminant	4.1004	95.8%
Linear SVM	14.353	95.2%
Quadratic SVM	10.806	96%
Cubic SVM	10.253	96%
Fine Gaussian SVM	11.672	29.1%
Medium Gaussian SVM	6.3808	94.9%
Coarse Gaussian SVM	4.6724	90.1%
Fine KNN	2.5063	89.7%
Medium KNN	1.3552	83.4%
Coarse KNN	1.7758	64.2%
Cosine SVM	1.7193	89.9%
Cubic KNN	14.535	80.1%
KNN of Weighted	1.2455	85.1%
Trees of Boosted	42.835	95.5%
Tree of Bagged	7.8311	96.9%
Subspace Discriminant	12.203	96.5%

Time Class	A	B	C	E	F	G	H	I	J	K	L
A	54	7		1	8						
B		66	3	1							
C			70								
E				67	3						
F	4	2		1	63						
G						68	2				
H						1	64	1	4		
I								67	3		
J								4	64	2	
K										66	4
L											70
	A	B	C	E	F	G	H	I	J	K	L
	Predicted Class										

Fig. 12. Confusion Matrix of Bagged Tree Classification for Speaker Identification (Fuzzy-based Conversation)

7 Conclusion

This paper presents the novelty of signal to image conversion and implement the computer vision techniques on the signal-image. Two methods were proposed for converting the signal into grayscale image; signal-range based and fuzzy-based. Where, two Digital Signal Processing (DSP) applications were developed for justifying the hypothesis, either be accepted or rejected. At the begging of this research work, the assumption was "since the image is transform into frequency domain in pattern recognition system, thus the signal can be transformed into spatial domain on a form of grayscale image". After that, the identification of speaker was developed on the basis of the proposed transformation and the system accuracy was found to be satisfactory. The second approach of automatic word segmentation was suggested, to make sure, either the proposed transforms can be work with the different challenges of DSP. However, the segmentation accuracy was reached up to 93.67% by using image processing techniques. The PDAs speech and Marathi databases were used in this experiment for speaker identification and signal segmentation, respectively. From the two proposed systems, the accuracies were higher than 90%, thus our hypothesis is accepted, and the signal can be implemented on a form of grayscale image. The main advantages of this research (voice to image conversion) are code optimization and simplified the implementation of signal processing system. The integration of computer vision techniques with signal processing will resolve some challenges of DSP.

8 Future Works and Scopes

Different features extracting techniques of image processing can be utilize on the signal image, where any signal can be convert into a grayscale image such as EGC signals.

References

1. Narasimha, M., Susheela Devi, V.: Pattern Recogniiton: An Algorithm Approach, pp. 1–6. Springer, London (2011). https://doi.org/10.1007/978-0-85729-495-1
2. Mukherjee, H., Obaidullah, S.M., Santosh, K.C., Phadikar, S., Roy, K.: Line spectral frequency-based features and extreme learning machine for voice activity detection from audio signal. Int. J. Speech Technol. 21(4), 753–760 (2018). https://doi.org/10.1007/s10772-018-9525-6
3. Mukherjee, H., et al.: Deep learning for spoken language identification: Can we visualize speech signal patterns? Neural Comput. Appl. 31(12), 8483–8501 (2019)
4. Zhang, Z., Wang, L., Kai, A.: Distant-talking speaker identification by generalized spectral subtraction-based dereverberation and its efficient computation. EURASIP J. Audio Speech Music Process. 2014(1), 1–12 (2014). https://doi.org/10.1186/1687-4722-2014-15
5. Ramos-Castro, D., Fierrez-Aguilar, J., Gonzalez-Rodriguez, J., Ortega-Garcia, J.: Speaker verification using speaker and text-dependent fast score normalization. Pattern Recognit. Lett. 28, 90–98 (2007)
6. Mathur,S., Choudhary, S.K., Vyas, J.M.: Speaker recognition system and its forensic implications 2(4), 1–6 (2013)
7. Damper, R.I., Higgins, J.E.: Improving speaker identification in noise by subband processing and decision fusion. Patter Recognition Lett. 24, 2167–2173 (2003)

8. Farrell, K.R., Mammone, R.M., Assaleh, K.T.: Speaker recognition using neural netoworks and conventional classfiers. IEEE Trans. Speech Audio Process. **2**(1), 194–205 (1994)

9. Reynolds, D.A.: Speaker identification and verifiation using Gaussian mixture speaker models. Speech Commun. **17**, 91–108 (1995)

10. Alsubari, A., Lonkhande, P., Ramteke, R.J.: Fuzzy-based classification for fusion of palmprint and iris biometric traits. In: Bhattacharyya, S., Pal, S.K., Pan, I., Das, A. (eds.) Recent Trends in Signal and Image Processing. AISC, vol. 922, pp. 113–123. Springer, Singapore (2019). https://doi.org/10.1007/978-981-13-6783-0_11

11. Ramteke, R.J., Alsubari, A.: Extraction of palmprint texture features using combined DWT-DCT and local binary pattern. In: 2nd International Conference on Next Generation Computing Technologies (NGCT), Dehradun, pp. 748–753 (2016)

12. Dalal, N., Triggs, B.: Histogram of oriented gradients for human detection. In: IEEE Computer Society Conference on Computer Vision and Pattern Recognition (CVPR 2005),1063–6919/05 (2005)

13. Ojala, T., Pietikinen, M., Harwood, D.: A comparative study of texture measures with classification based on featured distribution. Pattern Recognit. **29**(1), 51–59 (1996)

14. Alsubari, A., Satange, D.N., Ramteke, R.J.: Facial expression recognition using wavelet transform and local binary pattern. In: 2nd International Conference for Convergence in Technology (I2CT) (2017)

15. Obuchi, Y.: PDA speech database, carnegie mellon university. https://www.speech.cs.cmu.edu/databases/pda/index.html

Image and Signal Processing
in Agriculture

Automated Disease Identification in Chilli Leaves Using FCM and PSO Techniques

Sufola Das Chagas Silva Araujo[1]([✉]), V. S. Malemath[1],
and Meenakshi Sundaram Karuppaswamy[2]

[1] Department of Computer Science Engineering, KLE Dr. MSS College of Engineering
and Technology, Visvesvaraya Technology University, Belagavi, Karnataka, India
sufolachagas100@rediffmail.com, veeru_sm@yahoo.com
[2] Department of Engineering, Faculty of Engineering and Applied Sciences, Botho University,
Gaborone, Botswana
meenaksji@gmail.com

Abstract. Process of planting crop is the main source of revenue of rural India
as most of the population depends on farming for means of support. but due to
the annoying characteristics of diseases, and unforeseen changes in the climatic
conditions, the overall yield of the produce reduces. At large this will avoid losses
both in terms of value and volume of the agricultural harvest. The Regular methods
of investigating the infection in yield are not efficient and are not precise enough in
detecting the infection associated with plants. An expert system that can correctly
and effectively identify and detect plant infection will helps to enhance the yield
quality and quantity of plants

Keywords: Chilli disease · Chilli leaf image · Fuzzy C-means · Particle swarm
optimization · Pattern recognition

1 Introduction

Agriculture is the mainstay of India and about 3/4 of the people in India depends on
agriculture which plays a key role in economic development [14]. Dieses in plant crop
creates an undesirable effect on the countries whose financial prudence are principally
dependent on the farming.

Chilli is one of the main ingredient in Indian menu and have a good demand in
many parts of India [7]. It has been observed that Goa faced a major problem in chilly
crop cultivation as there where many infections observed at each stages of chilly crop
plantation for a variety called G-4 (Guntur-4) which lead to poor quality were grown.
So the identification and detection of infection in the primary stages could save from the
cumbersome losses.

So it is necessary to identify the disease accurately from the chilli leaf images,
classify them as healthy or unhealthy plant and then on the basis of level of infection,
prescribe suitable pesticides to prevent the loss in yield.

© Springer Nature Singapore Pte Ltd. 2021
K. C. Santosh and B. Gawali (Eds.): RTIP2R 2020, CCIS 1381, pp. 213–221, 2021.
https://doi.org/10.1007/978-981-16-0493-5_19

2 Literature Review

Shweta Joshi, Gayatri Jamadar, Sagar Nachan, R.R.Itkarkar, proposed an approach to detect chilli disease [1]. Images chilli plant leaves are are captured to check the well-being of each plant [7]. The technique Proposed, identifies the problems in the chilli plants [7]. The features computed are based on texture, colour, and area to be classified into disease chilli and normal chilli [27].

Everton Castelão Tetila, Bruno Brandoli Machado, paper presents the Simple Linear Iterative Clustering technique, to detect defect in the plant [3]. This employs the k-means clustering process for identifying regions, called super pixels [3]. G. Sandika Biswas, Bhushan Jagyasi, Bir Pal Singhy and Mehi Lalz, used Fuzzy-C means technique and classification was done with neural network to identify late blight disease of potato [20, 21]. This method achieved an accuracy of 93% for test images [21] captured in varying light condition, distances, changing orientations and background [21]. Jobin Francis, Anto Sahaya Dhas D, Anoop B K differentiated diseases in pepper [13] as Berry spot Disease and Quick wilt Disease [22]. Quick-wilt occurs with lack of Potassium, Magnesium and on some account Nitrogen. This analysis helped to improve the productivity of pepper [22, 33].

Angel Dacal-Nieto, Esteban Vázquez-Fernández, Arno Formella, Fernando Martin, Soledad Torres-Guijarro, Higinio González-Jorge extracted colour and projections features from HSV and RGB images of potato, using histogram and co-occurrence matrix [28, 15, 32]. 1-NN classifier was introduced and was optimized with an ad-hoc Genetic algorithm [9, 15]. Y. H. Sharath Kumar, G. Suhas extracted colour features for Identification and Classification of Fruit Diseases [34]. Muhammad Danish Gondal, Yasir Niaz Khan captured coloured leaf images to discriminate between three types of diseases [9, 17, 23]. The performance of the method showed 70 percent accuracy with a set of 50 images [9, 23].

3 System Architecture

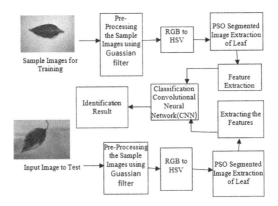

Fig. 1. Block diagram for proposed technique

The overall proposed technique for the system mentioned (see Fig. 1).

3.1 Image Acquisition

Leaf images are captured at different stages under laboratory condition using digital camera in a mini photo box of 30 cm × 45 cm × 35 cm. A total of 192 digital images were clicked. Some of the images were used for traing the system and some of which were used to test the system. The following Table 1 gives a summary of all the different types of leaves collected and the corresponding pesticides needed.

Table 1. Different infection stages

Sr.No	Month/Months	Stage	Pesticides
1	One month old	Healthy Leaf stage	Pesticides not required
2	Two months old	Stage one	Thiodicarp and Fipronil+
3	Three Months old	Stage two	Dimethoate and Acephate
4	Five months old	Stage three	Imidacloprid and Acephate
5	Five Months and above	Fully Grown	Cannot recover

3.2 Image Pre-processing

Pre-processing enhances the image wnich is passed through Guassian filter to decrease the overall noise. The filtered image is converted to HSV in order to make the object discrimination easier [26, 35] (see Fig. 2).

(a)   **(b)**   **(c)**

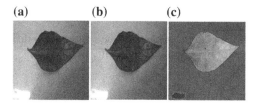

Fig. 2. (a) Resized Image (b) Gaussian Filtered Image (c) HSV Image

3.3 Image Extraction

Extracting meaningful information from the image is done using segmentation wherein multiple parts of interest of the image are clustered (see Fig. 3).

FCM is used to segment the image where each data object is member of multiple cluster with varying degrees of fuzzy membership between 0 and 1 in FCM [10].

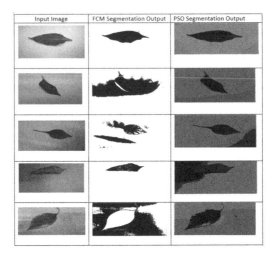

Fig. 3. FCM and PSO segmented output

Another method, Particle Swarm Optimization, segments the image by continuously trying to improve candidate solution, an stochastic optimization technique.

3.4 Feature Extraction

Features of interest are extracted with FCM and PSO segmentation. Second order statistical features are extracted using Gray-Level Co-Occurrence Matrix. Gray-Level Co-Occurrence Matrix is a method of examining texture features based on relationship of pixels gray levels co-occur in an image [9, 18]. Features of colour, shape and Texture are extracted.

3.5 Classification

CNN is trained for images classification based on features into different stages varying from healthy stage to partial infection stage to fully infected stage.

4 Proposed Methodology

4.1 FCM Technique

If the instructions have been followed closely, then only very minor alterations will be made to your paper. The FCM is clustering which allows one piece of data to belong to two or more clusters by taking hard decision on which cluster the pixel should be placed [1, 4, 29]. Sum of the membership value of a pixel to all clusters must be equal to 1 [29]. The description of the algorithm is as below

Step 1: Stopping condition and Sum of clusters, are set.
Step 2: Fuzzy partition matrix is initialized
Step 3: Cluster centroids and the value of J are calculated

$$J = \sum_{i=1}^{n} \sum_{k=1}^{n} u_{ik}^{m} |p_i - v_k|^2 \tag{1}$$

k-th cluster Centroid is calculated as below,

$$v_k = \left(\sum_{i=1}^{n} u_{ik}^{m} p_i \right) \div \left(\sum_{i=1}^{n} u_{ik}^{m} \right) \tag{2}$$

Step 4: When the value of J is less then the specified condition to stop and increment k. Go to step 4
Step 5: De-fuzzification.

4.2 Particle Swarm Optimization

PSO is a mathematical method used to optimizes a problem to a measure [6] of quantity by iteratively trying to improve a candidate solution based on the peculiar path taken by the swarms [4, 10, 19]. Kennedy and Eberhart proposed PSO which initializes random particles and then pursuits for optima by updating generation [11, 19]. The description of the algorithm is as below

Step 1: Cluster centres are initialised for each particle.
Step 2: Measure minimum distance from all pixels to its cluster center to form strong clusters.
Step 3: Fitness function computed and best solution is found as follows

$$v' = v + c1.r1.(pBest - x) + c2.r2.(gBest - x) \tag{3}$$

$$x' = x + v' \tag{4}$$

Where,
v is the present velocity,
v' the measured velocity,
x the present position,
x' the measured position,
r1 and r2 are random numbers in the interval [0, 1],
c1 and c2 are acceleration coefficients [2], Where c1 is a measure based on cognitive behaviour, and c2 is a measure based on social behaviour [19].
Step 5: Repeat the steps by calculating the cluster centres, until the cluster converges.

4.3 Convolutional Neural Network

Convolutional Neural Network is a Deep machine [6] Learning technique used by taking an image as input, and classifying by assigning weights [12, 30]. These weights are changed by a process of back propagation [12].

Input of the CCN is an images. Extracted important features are associated with subsequent middle layers in order to extract finer features [31]. Convolution Neural Network have neurons that have weights and biases, changed according to back propagation [24]. The network has one distinguishable outputs [25, 24].

5 Result

The proposed approach Specify what stage of infection the plant is in by automatically detecting the different infection stages. Depending on the plant infection stage, a suitable pesticide is suggested. The system was tested for various measures such as accuracy, sensitivity, specificity, precision, Recall and F-score (see Fig. 4).

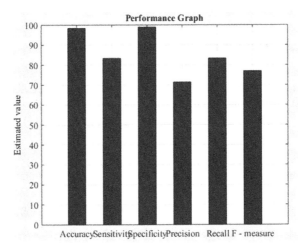

Fig. 4. Performance measure graph

6 Conclusion

Detecting the infection in the plant in the early stage is important as it will help increase the productivity of agricultural product. Several work has been proposed in similar lines. In this approach we extract features from both segmentation techniques i.e. Particle Swarm Optimization and Fuzzy C-Means. This helps in increasing the accuracy of detecting and identifying the respective image. The system aims to help the farmer to recognize the stage and take effective measures that will help him progress against economic losses. This system has an accuracy of 98.77%, sensitivity of 83.5% and with a precision of 72%.

7 Future Scope

The infections in the leaf, fruit and crop keeps growing due to environmental factors, viruses and pests so there is always a scope to tackle with this issue. Another work to be done is to implement new and better techniques for segmentation and feature extraction that can help identify diseases more efficiently and accurately.

References

1. Tari, Z., Fahad, A., Almalawi, A., Yi, X.: A taxonomy and empirical analysis of clustering algorithms for traffic classification. Institution of Engineering and Technology (IET) (2020)
2. Patil, D.P., Kurkute, S.R., Sonar, P.S., Antonov, S.T.: An advanced method for chilli plant disease setection using image processing. iCEST, 2017 Serbia Nis, 28–30 June 2017
3. Tetila, E.C., et al.: Identification of soybean foliar diseases using unmanned aerial vehicle images. IEEE Geosci. Remote Sens. Lett. **14**(12), 2190–2194 (2017)
4. Mohapatra, D.P., Patnaik, S. (eds.): Intelligent Computing, Networking, and Informatics. AISC, vol. 243. Springer, New Delhi (2014). https://doi.org/10.1007/978-81-322-1665-0
5. Santosh, K.C., Hegadi, R.S. (eds.): Recent Trends in Image Processing and Pattern Recognition. CCIS, vol. 1036. Springer, Singapore (2019). https://doi.org/10.1007/978-981-13-9184-2
6. Luhach, A.K., Kosa, J.A., Poonia, R.C., Gao, X.-Z., Singh, D. (eds.): First International Conference on Sustainable Technologies for Computational Intelligence. AISC, vol. 1045. Springer, Singapore (2020). https://doi.org/10.1007/978-981-15-0029-9
7. Husin, Z.B., Shakaff, A.Y.B.M., Aziz, A.H.B.A., Farook, R.B.S.M.: Feasibility study on plant chili disease detection using image processing techniques. In: 2012 Third International Conference on Intelligent Systems Modelling and Simulation (2012)
8. Disease identification in chilli leaves using machine learning techniques. Int. J. Eng. Adv. Technol. (2019)
9. Anthonys, G., Wickramarachchi, N.: An image recognition system for crop disease identification of paddy fields in Sri Lanka. In: 2009 International Conference on Industrial and Information Systems (ICIIS) (2009)
10. Anand, H., Dalal, V.: Comparative study of particle swarm optimization and fuzzy C-means to data clustering. Int. J. Comput. Appl. Technol. Res. **3**(1), 45–47 (2014)
11. Mohsen, F., Hadhoud, M., Mostafa, K., Amin, K.: A new image segmentation based on particle swarm optimization. Int. Arab J. Inf. Technol. **9**(5), 487–493 (2012)
12. A Comprehensive Guide to Convolutional Neural Network: https://towardsdatascience.com/a-comprehensive-guide-to-convolutional-neural-networks-the-eli5-way-3bd2b1164a53
13. Convolutional Neural Network for Visual Recognition: http://cs231n.github.io/convolutional-networks/
14. Gomathy, B., Nirmala, V.: Survey on plant diseases detection and classification techniques. In: 2019 International Conference on Advances in Computing and Communication Engineering (ICACCE) (2019)
15. Pujari, J.D., Yakkundimath, R., Byadgi, A.S.: SVM and ANN based classification of plant diseases using feature reduction technique. Int. J. Interact. Multimedia Artif. Intell. **3**(7), 6–14 (2016)
16. Patil, P., Yaligar, N., Meena, S.M.: Comparison of performance of classifiers - SVM, RF and ANN in potato blight disease detection using leaf images. In: 2017 IEEE International Conference on Computational Intelligence and Computing Research (ICCIC) (2017)

17. Ramapriya, B., Rajaselvi, D.: Disease segmentation in vegetables using K-means clustering. Int. J. Pure Appl. Math. **119**(10) (2018)
18. Akhloufi, M.A., Larbi, W.B., Maldague, X.: Framework for color-texture classification in machine vision inspection of industrial products. In: 2007 IEEE International Conference on Systems, Man and Cybernetics (2007)
19. Sun, G., Jia, X., Geng, T.: Plant diseases recognition based on image processing technology. J. Electr. Comput. Eng. **2018** (2018)
20. Kosamkar, P.K., Kulkarni, V.Y., Mantri, K., Rudrawar, S., Salmpuria, S., Gadekar, N.: Leaf disease detection and recommendation of pesticides using convolution neural network. In: 2018 Fourth International Conference on Computing Communication Control and Automation (ICCUBEA), 2018 Publication (2018)
21. Biswas, S., Jagyasi, B., Singh, B.P., Lal, M.: Severity identification of Potato Late Blight disease from crop images captured under uncontrolled environment. In: 2014 IEEE Canada International Humanitarian Technology Conference - (IHTC) (2014)
22. Francis, J., Dhas, D.A.S., Anoop, B.K.: Identification of leaf diseases in pepper plants using soft computing techniques. In: 2016 Conference on Emerging Devices and Smart Systems (ICEDSS) (2016)
23. Anthonys, G., Wickramarachchi, N.: An image recognition system for crop disease identification of paddy fields in Sri Lanka. In: 2009 International Conference on Industrial and Information Systems (ICIIS) (2009)
24. Godkhindi, A.M., Gowda, R.M.: Automated detection of polyps in CT colonography images using deep learning algorithms in colon cancer diagnosis. In: 2017 International Conference on Energy (2017)
25. Tan, Y.J., Sim, K.S., Ting, F.F.: Breast cancer detection using convolutional neural networks for mammogram imaging system. In: 2017 International Conference on Robotics, Automation and Sciences (ICORAS) (2017)
26. Abraham, A., Dutta, P., Mandal, J.K., Bhattacharya, A., Dutta, S. (eds.): Emerging Technologies in Data Mining and Information Security. AISC, vol. 755. Springer, Singapore (2019). https://doi.org/10.1007/978-981-13-1951-8
27. Dhingra, G., Kumar, V., Joshi, H.D.: Study of digital image processing techniques for leaf disease detection and classification. Multimedia Tools Appl. **77**(15), 19951–20000 (2017)
28. Dacal-Nieto, A., et al.: A genetic algorithm approach for feature selection in potatoes classification by computer vision. In: 2009 35th Annual Conference of IEEE Industrial Electronics (2009)
29. Fahad, A., et al.: A survey of clustering algorithms for big data: taxonomy and empirical analysis. IEEE Trans. Emerg. Top. Comput. **2**(3), 267–279 (2014)
30. Singh, P.K., Pawłowski, W., Tanwar, S., Kumar, N., Rodrigues, J.J.P.C., Obaidat, M.S. (eds.): Proceedings of First International Conference on Computing, Communications, and Cyber-Security (IC4S 2019). LNNS, vol. 121. Springer, Singapore (2020). https://doi.org/10.1007/978-981-15-3369-3
31. Garcia, C., Delakis, M.: Convolutional face finder: a neural architecture for fast and robust face detection. IEEE Trans. Pattern Anal. Mach. Intell. **26**(11), 1408–1423 (2004)
32. Randive, P.U., Deshmukh, Ratnadeep R., Janse, Pooja V., Gupta, Rohit S.: Discrimination between healthy and diseased cotton plant by using hyperspectral reflectance data. In: Santosh, K.C., Hegadi, R.S. (eds.) RTIP2R 2018. CCIS, vol. 1037, pp. 342–351. Springer, Singapore (2019). https://doi.org/10.1007/978-981-13-9187-3_30
33. Ghule, A., Deshmukh, R.R., Gaikwad, C.: MFDS-m red edge position detection algorithm for discrimination between healthy and unhealthy vegetable plants. In: Santosh, K.C., Hegadi, R.S. (eds.) RTIP2R 2018. CCIS, vol. 1037, pp. 372–379. Springer, Singapore (2019). https://doi.org/10.1007/978-981-13-9187-3_33

34. Sharath Kumar, Y.H., Suhas, G.: Identification and classification of fruit diseases. In: Santosh, K.C., Hangarge, M., Bevilacqua, V., Negi, A. (eds.) RTIP2R 2016. CCIS, vol. 709, pp. 382–390. Springer, Singapore (2017). https://doi.org/10.1007/978-981-10-4859-3_34

35. Mondonneix, G., Chabrier, S., Mari, J.M., Gabillon, A.: A machine learning algorithm for solving hidden object-ranking problems. In: Santosh, K.C., Hegadi, R.S. (eds.) RTIP2R 2018. CCIS, vol. 1035, pp. 640–652. Springer, Singapore (2019). https://doi.org/10.1007/978-981-13-9181-1_55

Deformation Behaviour of Soil with Geocell Using Image Analysis Techniques

Abhinav Mane[1](✉), Praful Gaikwad[2], and Shubham Shete[2]

[1] CSMSS, Chh. Shahu College of Engineering, Aurangabad 431008, Maharashtra, India
abhinav.vatsalya@gmail.com
[2] Vatsalya Consulting Services, Harsul, Aurangabad 431008, Maharashtra, India
praful.vatsalya@gmail.com, shubham.vatsalya@gmail.com

Abstract. Geocell is being used very conveniently for soil reinforcement and slope protection in the recent times. The main advantage of this synthetic material is provision of the confinement to the soil resulting in the higher resistance to shear failure of the soil. Geocell is a three-dimensional synthetic material having the varying wall thickness and opening sizes. The use of the confining properties of the geocell is needed to be investigated in the application of bearing capacity improvement of soil below the foundations. The present work demonstrates bearing capacity improvement of the soil using geocell as the confining reinforcement. A series of experimental model tests is performed in the present study with constant soil type (sand). Parameters considered for variation was the width (W), opening size (o_g) and height (h_g) of the geocell. The image analysis was performed mainly in ImageJ which is available on free license program with Digital Image Correlation and the Particle Image Velocimetry approach. The observations were made mainly in terms of maximum load carried by the subsoil and the depth of Rankine's zone obtained due to placement of geocell. Load bearing resistance of subsoil was in proportion with the width and height and in inverse proportion with the opening size of the geocell. The Rankine's zone also observed to shift downward due to geocell inclusion providing more shear surface and thus higher bearing capacity. Thus, the geocell may successfully be utilized below shallow foundations to obtain higher Safe Bearing Capacities.

Keywords: Geocell · ImageJ · PIV analysis · Image analysis · Template matching · Stack · Strip loading · Marker based image analysis · Digital camera

1 Introduction

Subsoil is major deriving part for stability of any structure. Different structure being used for various purpose such as Dams, Embankment, Roadways, Buildings and etc. The major part of subsoil is mainly consist of loose formation of soil where the construction of shallow foundation is highly impossible. At such location it is necessary to go with deep foundation to manage with the requirement of bearing capacity. However the construction of deep foundations in soil enforces large amount of expenses, highly precise equipment's and large amount of man work required. Despite doing all

© Springer Nature Singapore Pte Ltd. 2021
K. C. Santosh and B. Gawali (Eds.): RTIP2R 2020, CCIS 1381, pp. 222–237, 2021.
https://doi.org/10.1007/978-981-16-0493-5_20

this efforts it is highly unexpected that the project will be executed as planned. This may result in compromised safety of the structure. However construction of shallow foundation is mainly done in the extents that can be visualized exactly with the observer eye. So, providing reinforcement to the soil and enhancing its bearing capacity will provide a very good cheap alternative than the construction of deep foundation.

Geocell is a three dimensional honeycomb structure with multiple hollow cells connected together using ultrasonic method of welding. Geocell are made of high Density Polyethylene and are in folded forms in factory. This material has been evolved to use in soil as the reinforcement. Geocell is widely used over the last three decades, it contains interlocking cells which confines the soil in its pockets. Geocell are used as confinement in highways, retaining structures, slope protection and embankment of foundations. Figure 1 Theoretical dimensional view of geocell.

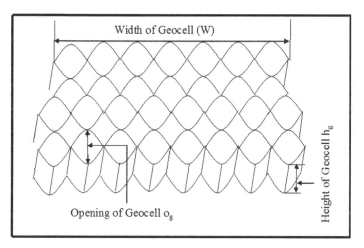

Fig. 1. Theoretical dimensional view for geocell

In soil reinforcement by using geocell, the compressive load are taken by the soil however the tensile load occurred due to applied load will be taken by geocell walls. This application is successfully in use in different infrastructure and other construction developments such as soil wall, soil slope, soil stabilization. However the utilization of such material for construction of shallow foundation for residential and public buildings has been limited. This is also expressed in our nest subsection literature review. Geocell acts generally in three dimensional manner providing the confined tensile resistance.

2   Literature Review

Recently geocell is being used as the reinforcing material providing the confining mechanism to soil. Different researchers have put their studies aiming towards various applications of the geocell its mechanism through small scale modelling. The concluding remarks and influencing research oriented outputs from different researchers are given in this section. Geocell provides three dimensional confinement which results in increase in overall stability of reinforcement foundation and reduction in lateral spreading of confined soil [1, 2]. [3] provided behaviour of geocell as wide slab which convey footing pressure below and distributes over wide width. Due to rigidity, the geocell touches the failure plane and forces deep into foundation soil, results in improvement of bearing capacity [3]. Past researchers exhibited advantages of geocell as confinement with the help of experimental and field execution Mane et al. (2016) [4, 5] have conducted strip loading with and without compressible inclusion was modelled through small-scale laboratory tests. In this research ImageJ software was used for analysis the result. [6] have conducted load tests on geocell supported model embankments over soft clay foundation The geometry effect of geocell reinforcement such as cell size, width, height pattern of formation and placement layer on improvement of performance studied thoroughly. [7, 8] researchers have put forward the variation in bearing capacity for the respective soil stratum and effect due to influence of soil reinforcement for the different stratum [9]. A series of plate load test on circular footing were performed over square shape paper grid cells to obtain modes of failure. The research concluded the optimum dimension of cell in which sand used as a confinement [10]. Studying the behaviour of soil by using geogrid as soil reinforcement. [11] reported case studies on geocell mattress supported road embankment. Several investigators reported beneficial use of geocell [12, 13], [14] defines the three aspect of main geocell layer which are membrane effect, lateral dispersion effect and vertical stress dispersion effect. The present study provided evaluation of bearing capacity due to inclusion of geocell below shallow foundation with varying parameters opening of geocell, with (10 mm, 20 mm) and (20, 30 & 40 mm) as height of geocell with width is equal to 1 & 2 times proportional to the width of footing (B & 2B) in sandy soil filled with 50% relative density.

3 Motivation

In the present work, strip loading applied over soil layer. Figure 2 shows the schematic cross section of the deformation behaviour of soil without geocell reinforcement and Fig. 3 shows with geocell reinforcement. When strip loading is gradually applied on the soil surface the pressure would transfer in the soil mass according to the Terzaghi's general bearing capacity theory. Forming a zone of elastic equilibrium and the plastic equilibrium the zones of plastic equilibrium extends the deformation of soil towards the surface of the soil resulting in formation of the heave. Geocell when positioned under strip footing at a depth of equal to width (B) of footing, geocell provides three-dimensional confinement for the capsulated soil which results in increase in load resistance due to increase in stiffness of soil and causes less deformation at higher load values. The rigidity of the soil mass placed in geocell increase due to confinement effect and so the loads from footing gets transferred at the bottom of geocell. It may again be experienced that the reduction in geocell opening size increases the rigidity and thus may enhanced bearing resistance.

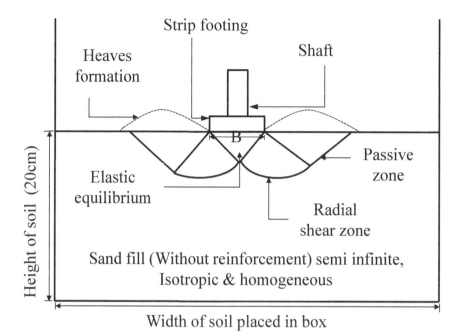

Fig. 2. Schematic cross section of model without geocell reinforcement

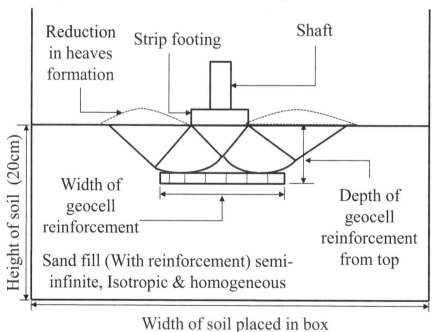

Fig. 3. Schematic cross section of model with geocell reinforcement

3.1 Materials

Sand

The sand used for this study was sandy soil used during casting of moulds in industry shown in Fig. 4. Which consist of fine rounded and sub rounded particles. Figure 5 shows particle size distribution (BIS: 2720-Part 4) of model sand. The sand categorised in as SM (silty sand) [Unified Soil Classification System].

Fig. 4. Front elevation for model sand

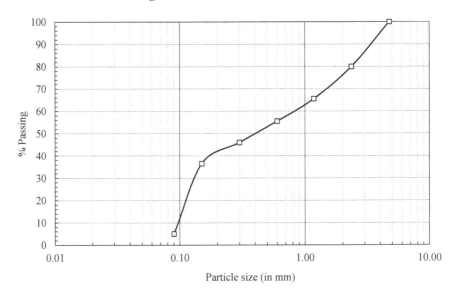

Fig. 5. Particle size distribution of model sand

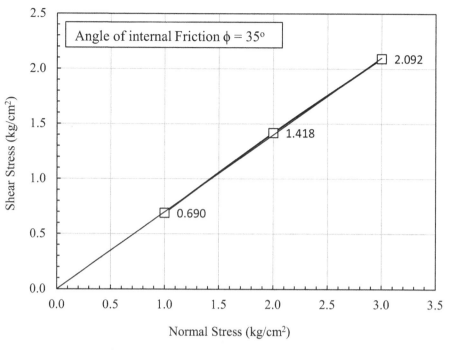

Fig. 6. Direct shear test parameters for model sand

The friction angle of sand was found to be 35° in direct shear (BIS: 2720-Part 13), at 50% RD with unit weight of 1.50 gm/cc shown in Fig. 6. The physical properties of sand summarizes in Table 1 used for test. Figure 4 provides photographic view of the sand used in the present study.

Geocell

Unique structure made from a flex material used as a three dimensional confinement to soil. Wide width tensile strength for the material shown in Table 1 and Fig. 9 (ASTM-D 4595 (2005)) material strips joined together forming a honeycomb cell structure. Opening of geocell made 20, 30 & 40 mm, height of geocell made as 10, 20 & 30 mm and width of geocell kept as 1 & 1.5 times proportional to width of footing B (B & 1.5B). Figures 7 and 8 shows the plane photographic view of geocell and shows photographic isometric view of geocell. Figure 1 Shows schematic diagram of geocell were 'o_g' termed as opening of geocell, W termed width of geocell, 'h_g' termed height of geocell.

Movable Markers

Figure 10 shows 'L' shaped movable marker inserted with soil layer for tracing displacement of soil particles within test.

Fig. 7. Isometric view of geocell

Fig. 8. Top view for geocell

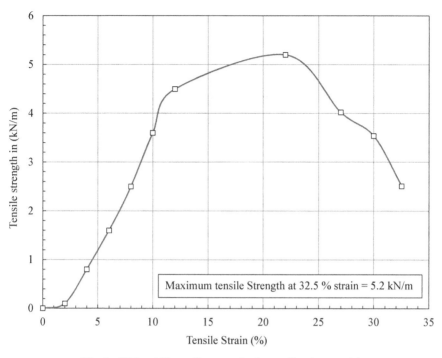

Fig. 9. Wide width tensile strength of geocell strip material

Table 1. Materials properties

Sand	Values
Void ratio maximum (e_{max})	0.849
Void ratio minimum (e_{min})	0.661
Angle of Internal friction (ϕ)	35°
Unit weight of sand at 50% relative density (gm/cc)	1.50
Flex	
Wide width tensile strength (kN/m)	5.2
Maximum tensile strain (%)	32.5

Fig. 10. Movable markers

4   Model Test Package

Figure 11 shows the filled-up soil in the strong MS metal box of dimensions 500 × 400 × 200 mm representing width, height & length of box. The soil filled in the model for test with 50% relative density i.e. with 1.50 gm/cm³ unit weight.

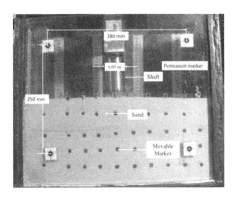

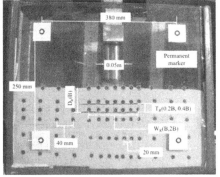

a) Without geocell reinforcement   b) With geocell reinforcement

Fig. 11. Front view of model test package

The Mild Steel case includes four side plate of thickness 10 mm. A 12 mm thick glass panel places in front side for observations of deformation & tracing soil particles through movable marker displacement during test & capturing set of images through digital camera at image per second. Before model testing the box was proof tested for different intensity of loading at different soil backfill. Grease shall apply on glass panel into inner side & thin polythene sheet placed on it in such a way that no friction occurs along boundary of box during the test, a bench mark was arranged to trace the displacement at the time of testing with the help of permanent marker glued at outer 4 side of front glass panel. Movable plastic marker 'L' shaped were glued into inner plane of glass panels and soil. At specific interval and instant of model filling. For the purpose of observation of movement of soil during test a digital camera installed on stand attached to model. MS square, plate, L-shaped attached to the box so that the camera can move along with the box which represent single point image capturing. Electric lamp mounted on stand used to maintained constant intensity of light & clarification throughout test. Universal Testing Machine with 2000 kN compression capacity used for all the experimental test model.at CSMSS Chh. Shahu collage of Engineering, Aurangabad. A load of 1 N/s applied gradually vertical and series of images captured using camera (cannon make 9 megapixel) at regular interval and the images were stored in connected computer which is located away from the setup.

4.1 Test Program

Table 2 provides details of model test performed in present work. Total 13 tests were performed in experiment includes one without geocell & 12 test with inclusion of geocell with (20, 30, & 40 mm) opening (B & 2B) width & (10 & 20 mm) height placed at a depth equal to (B) width of footing below top surface of footing.

Table 2. Details of test performed in present study

Test legend	Width of reinforcement (W)	Opening of geocell (o_g)	Height of geocell (h_g)
Without plastic	N.A.	N.A.	N.A.
M-01	B	20	10
M-02	2B	20	10
M-03	B	30	10
M-04	2B	30	10
M-05	B	40	10
M-06	2B	40	10
M-07	B	20	20
M-08	2B	20	20
M-09	B	30	20
M-10	2B	30	20
M-11	B	40	20
M-12	2B	40	20

5 Analysis and Interpretation

Table 3 shows the summary of test results obtained after compression loading on model without geocell reinforcement and with geocell reinforcement as per Table 2. The resulting graphs for maximum load vs. Normalised with of geocell drawn Fig. 12 for understanding effect of width of geocell reinforcement. In similar way depth of failure zone vs. Normalised width of geocell also plotted Fig. 13.

Table 3. Summary of test performed in present work

Test legend	Width of reinforcement (W)	Opening of geocell (og)	Height of geocell (hg)	Depth of Rankine's passive zone (cm)	Maximum load (N)	% increase in bearing load
Without plastic	N.A.	N.A.	N.A.	8.00	826.0	N.A.
M-01	B	20	10	11.25	1070.0	29.54
M-02	2B	20	10	12.16	2193.5	165.56
M-03	B	30	10	14.99	1124.5	36.14
M-04	2B	30	10	11.47	1600.0	93.70
M-05	B	40	10	9.04	945.0	14.41
M-06	2B	40	10	11.00	1693.5	105.02
M-07	B	20	20	9.00	1469.0	77.85
M-08	2B	20	20	11.38	2351.5	184.69
M-09	B	30	20	9.72	1327.0	60.65
M-10	2B	30	20	10.27	2141.5	159.26
M-11	B	40	20	9.85	1106.5	33.96
M-12	2B	40	20	10.70	1754.5	112.41

5.1 Image Analysis

Analysis were performed on series of images taken during the test. Analysis were done with the help of ImageJ which is free to access. The movement occurred above and around the reinforcement and footing were trapped with the help of plugins as advance template matching Fig. 14 and PIV (Particle Image Velocimetry) analysis. Figure 14 shows the deformed profile of experimental models with and without inclusion of geocell.

İmages captured from the beginning of the test up to failure obtained are arranged sequentially (Fig. 15) and further used in advanced template matching plugins. Template matching is the plugin mainly utilized to match all the sequential images concentrically over the bench marked region. The aligned images were then processed through PIV

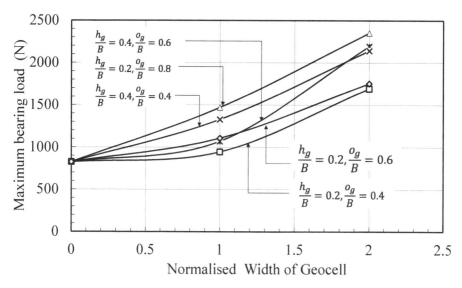

Fig. 12. Maximum bearing load vs. normalised width of geocell

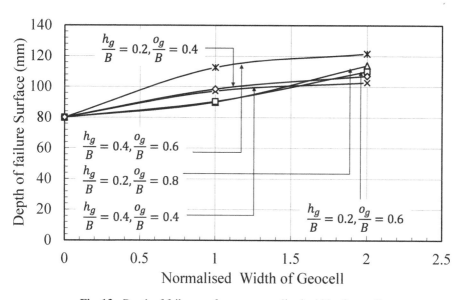

Fig. 13. Depth of failure surface vs. normalised width of geocell

analysis. PIV (Particle Image Velocimetry) is mainly used to track the movements of the particles in each sequential image with respect to previous image.

A marker based image analysis was performed over the images identifying the relative positions of the movable markers with respect to permanent markers. The displacement obtained through ImageJ computed as the position of markers in the first image and

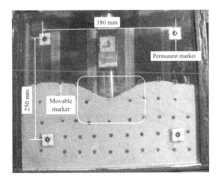 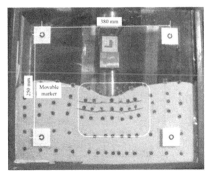

a) Without geocell reinforcement   b) With geocell reinforcement

Fig. 14. Deformed profile of geocell reinforcement

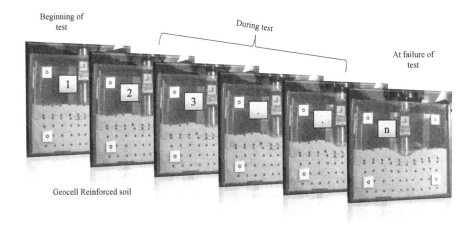

Fig. 15. Sequence of images considered for image analysis

displacement of the same marker in the second image. So to calculate the displacement at any instance of image addition of all the previous displacement should be done for the same particle in the image either in PIV analysis or marker based analysis. Vectors are scaled up for better visualisation.

A general shear failure was obtained when geocell placement was made. Prominent heaves were observed from both sides of foundation symmetrically. When depth of geocell was equal to width of foundation, the size of heaves were observed to decrease with increase in width of cell and decrease in cell height. Series of images were captured during the test displacements and movement of temporary markers. ImageJ is as open source application, which facilitates various analysis and interpretation technique on a sequences of images. The displacement vector diagram for with and without inclusion of geocell below footing is shown in Fig. 16 and 17.

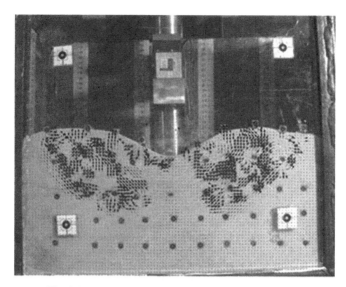

Fig. 16. Traced marker for without reinforcement case

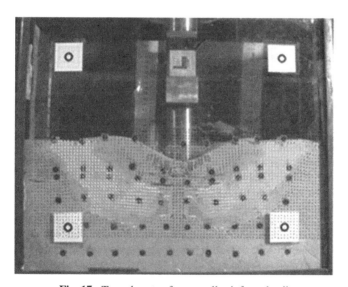

Fig. 17. Traced vector for geocell reinforced soil

6 Conclusions

Table 3 shows the summary of test performed on small scale model by using Geocell reinforcement. The test performed with varying in geocell opening, width of reinforcement and height of geocell. Based on the experimental investigation material assessment, image analysis & interpretation it is concluded that geocell may effectively be used below shallow isolated foundation in loose subsoil. Bearing resistance enhancement of a foundation due to placement of geocell is found to be proportionate with w & h_g, of geocell, however the enhancement was found to be inversely proportional to opening size of the geocell. The bearing resistance of the foundation enhances by placement of geocell below it. A maximum benefit of 184% was obtained due to width (w) 2B, height (h_g) 0.4, opening (o_g) 0.4B, of geocell placed at depth equals to width of footing (50 mm) below foundation, the graphical result as shown in Fig. 11. The depth of Rankine's passive zone found proportional to the normalized with of geocell shown in Fig. 12.

References

1. Madhavi Latha, G.: Investigation on the behavior of geocell supported embankments. Ph.D. thesis submitted to Indian Institute of Technology Madras, Chennai (2000)
2. Madhavi Latha, G., Rajagpaol, K.: Parametric finite element analyses of geocell supported embankment. Can. Geotech. J. **44**(8), 917–927 (2007)
3. Sitharam, T.G., Sireesh, S., Sk, D.: Model studies of a circular footing supported on geocell reinforced clay. Can. Geotech. J. **42**(2), 693–703 (2005)
4. Mane, A.S., Shete, S., Bhuse, A.: Earth pressure reduction on buried pipeline using geofoam. In: EUROGEO 2016, Europe (2016)
5. Mane, A.S., Shete, S., Bhuse, A.: Buried pipelines deformation behavior in different soils using geofoam. In: Pawar, P.M., Ronge, B.P., Balasubramaniam, R., Seshabhattar, S. (eds.) ICATSA 2016, pp. 947–955. Springer, Cham (2018). https://doi.org/10.1007/978-3-319-53556-2_95
6. Krishnaswamy, N.R., Rajagopal, K., Madhavi Latha, G.: Model studies on geocell supported embankment constructed over soft clay foundation. Geotech. Test. J. **23**, 45–54 (2000). ASTM
7. Mane, A.S.: Studies on deformation behavior of geogrid reinforced soil walls. In: EUROGEO 2016, Europe (2016)
8. Mane, A.S., Gawali, S.: Performance enhancement of geogrid reinforced soil wall with geofoam inclusion. In: ICSE 2016, Bangalore (2016)
9. Rea, C., Mitchell, J.K.: Sand reinforcement using paper grid cells. ASCE Spring Convention and Exhibit, Preprint 3130, Pittsburgh, PA, April 1978, pp. 24–28 (1978)
10. Mane, A.S., Viswanadham, B.V.S.: Centrifuge modeling of wrap-around geogrid reinforced soil walls. In: GEOtrendz Indian Geotechnical Conference 2010, IIT Mumbai (2010)
11. Cowland, J.W., Wong, S.C.K.: Performance of a road embankment on soft clay supported on a geocell mattress foundation. Geotext. Geomembr. **19**(4), 235–256 (1993)
12. Bathurst, R.J., Jarrett, P.M.: Large-scale model tests of geocomposite mattresses over peat subgrades. Transportation Research Record 1188. Transportation Research Board, Washington, DC, pp. 28–36 (1989)
13. Bathurst, R.J., Knight, M.A.: Analysis of geocell reinforced soil covers over large span conduits. Comput. Geotech. **22**, 205–219 (1998)
14. Tafreshi, S.N.M., Dawson, A.R.: Behavior of footing on reinforced sand subjected to repeated loading comparing use of 3D and planar geotextile. Geotext. Geomembr. **28**, 434–447 (2010)
15. Akinmusuru, J.O., Akibolade, J.A.: Stability of loaded footings on reinforced soil. J. Geotech. Eng. Div. **107**(6), 819–827 (1981). ASCE
16. American Society for Testing and Materials: Standard test method for tensile properties of geotextiles by wide-width strip method. ASTM D4595 (1986)
17. Binquet, J., Lee, K.L.: Bearing capacity tests on reinforced earth slabs. J. Geotech. Eng. Div. **101**(12), 1241–1255 (1975). ASCE

18. Butterfield, R.: Dimensional analysis for geotechnical engineers. Geotechniques **49**(3), 357–366 (1999)
19. Bush, D.I., Jenner, C.G., Bassett, R.H.: The design and construction of geocell foundation mattress supporting embankment over soft ground. Geotext. Geomembr. **9**, 83–98 (1990)
20. Chummar, A.V.: Bearing capacity theory from experimental results. J. Soil Mech. Found. Div. **98**(12), 1311–1324 (1972). ASCE
21. Fakher, A., Jones, C.J.F.P.: Discussion of bearing capacity of rectangular footings on geogrid reinforced sand, by Yetimoglu, T., Wu, J.T.H., Saglamer, A. J. Geotech. Eng. **122**(4), 326–327 (1996). ASCE
22. Farrag, K., Acar, Y.B., Juran, I.: Pull-out resistance of geogrid reinforcements. Geotext. Geomembr. **12**, 133–159 (1993)
23. Fragaszy, R.J., Lawton, E.: Bearing capacity of reinforced sand sub grades. J. Geotech. Eng. Div. **110**(10), 1500–1507 (1984). ASCE
24. Selig, E.T., Mckee, K.E.: Static and dynamic behavior of small footings. J. Soil Mech. Found. Div. **120**(12), 2083–2099 (1961). ASCE
25. Yang, X., Han, J., Parsons, R.L., Leshchinsky, D.: Three-dimensional numerical modeling of single geocell reinforced sand. Front. Archit. Civ. Eng. China **4**(2), 233–240 (2010). https://doi.org/10.1007/s11709-010-0020-7

Identification of Banana Disease Using Color and Texture Feature

Vandana V. Chaudhari[1(✉)] and Manoj P. Patil[2(✉)]

[1] Smt.G.G. Khadse College, Muktainagar, India
vandu2108@gmail.com
[2] School of Computer Sciences,
Kaviyatri Bahinabai Chaudhari, North Maharashtra University, Jalgaon, India
mpp145@gmail.com

Abstract. Plant diseases have grown-up in agriculture to be an impasse as it can cause dwindling in both quantity and quality of farming yield. This work explains a simple and adept method used to recognize leaf diseases by applying digital image processing and machine learning technology. In this study 24 color feature, 12 shape and 4 texture features were obtained from images of four kinds of diseases like, Sigatoka, Panama Wilt, Bunchy Top and Banana Streak Virus diseases. Principal component analysis (PCA) was achieved for reducing dimensions in features extracted and k-nearest neighbour classifiers used to identify banana diseases. The finest result was obtained when image identification was conducted based on PCA with K-nearest neighbour classifier.

Keywords: Banana leaf · K-nearest neighbour classifier · GLCM · Image processing

1 Introduction

Agriculture plays a vigorous part in the everyday life, it is one of the developing fields which attract thousands of researcher. Temperature, water, and light are the main aspects that control the growth and development of the disease. Environment changes affect many things like a plant, human being, Animal, etc., Change of climate effects on disease, yield, and quality of crops [13]. Diseases are one of the inevitable parts of living things; early recognition and diagnosis are required to avoid damages. The traditional method used to identify disease, by taking expert opinion or self-experience by the farmer. However, these methods need to be performed by a professional who requires extra cost and time [17]. Many of the diseases produce symptoms which are the main sign in diagnosis. Need for an automated system that provides early and accurate identification of disease. Computer technology has played an vital role in plant disease data processing. The computer helps to automatically recognize and detect the disease quickly and accurately for farmers and experts. By applying appropriate machine learning techniques it is easy to classify the diseases by in view of the

© Springer Nature Singapore Pte Ltd. 2021
K. C. Santosh and B. Gawali (Eds.): RTIP2R 2020, CCIS 1381, pp. 238–248, 2021.
https://doi.org/10.1007/978-981-16-0493-5_21

symptoms. The appearance of diseases is mostly viewed on the leaf. This paper focused on, to find the most arising diseases on the banana plants. The Principal Component Analysis technique is used to provide the best results to recognize the disease in the early stages. PCA is one of the trendiest techniques having the best statistical feature analysis capabilities. The main objective of PCA is to extract the key features, which are treated as a set of orthogonal variables which vary respect to diseases [5]. In order to find out the method for banana plant leaf diseases based on image processing techniques and machine learning, image recognition of four kinds of images of plant diseases. Including Sigatoka disease caused by Mycosphaerella fungus and Panama wilt caused by Fusarium Oxysporum fungus, Bunchy top caused through bunchy top virus and at last banana streak virus.

2   Banana Leaf Disease

Banana plant faces various diseases start from its initial stage. There are bacterial, fungal, viral diseases and diseases caused due to insects. Most of the symptoms are expressed on leaf, stem, fruits and roots [2]. Leaf spots are considered as key thing expressing the presence of the disease. Major banana diseases that put the symptoms on leaves are sigatoka disease, panama disease, moko disease, black spots, infectious chlorosis, bunchy tops, banana bract mosaic virus disease and banana streak virus [3]. Among these disease Sigatoka, Bunchy top, Banana Streak Virus and Panama Wilt diseases are discussed in following sections.

2.1   Sigatoka Disease

Sigatoka is a fungus disease which caused by the fungus Mycosphaerella gloeosporioides. Small yellow or brownish green narrow streaks on leaves and it spreads over the entire leaf. Brown spots are commonly found in the affected leaves. Leaves dry rapidly and cause to defoliate [3]. Following Fig. 1 shows the black sigatoka and yellow sigatoka disease images.

2.2   Banana Bunchy Top

Banana Bunchy Top is a virus disease which caused through the banana bunchy top virus. Main symptoms are size of the leafs is shortened and they become fragile [3]. Following Fig. 2 shows the Banana Bunchy top, Panama and Banana streak virus diseases.

2.3   Panama Disease

Panama is a fungus disease caused by a fungi Fusarium oxysporum f.sp. cubense. Symptoms are yellowing of the leaves lower most portions starting from margin to mid rib of the leaves. Extend yellowing region in upward direction leaving heart portion of the leaf alone in green. Leaves have cracks near the base and hangs down in the region of pseudo stem.

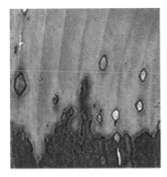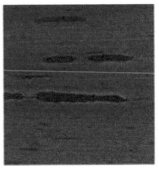

Fig. 1. Samples images of Black sigatoka and Yellow sigatoka (Color figure online)

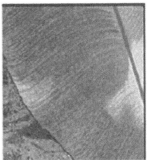

Fig. 2. Samples images of Buchy Top, Panama and Banana streak Virus disease

2.4   Banana Streak Virus

Banana streak virus is a common viral disease in banana leaves by having yellow vertical streaks in the leafs, which becomes gradually necrotic producing a black streaked arrival in older leaves [3]. Transmitted mostly through infected planting materials and more probably stops the growth of the plant.

3   Literature Review

Gulhane et al. explained an approach for cotton leaves disease diagnosis based on PCA with Nearest Neighborhood Classifier. Green channel is used into consideration for collection of feature in RGB image as disease or deficiencies on leaves are revealed well by green channel [5]. Wang et al. present an methodology for wheat and grapes disease recognition based image processing. Wheat and Grapes are selected for conducting experiments. Shape, texture and Color feature are preferred for feature extraction. Dimensions of features are reduced using PCA. Neural networks with backpropagation (BP), radial basis function (RBF) neural network, probabilistic neural networks and generalized regression networks (GRNNs) used as classifiers to recognize diseases of wheat and grape [16]. Vipindas et al. suggested an image processing based way for detection and identification of banana leaf diseases. The banana diseases considered are namely black sigatoka and panama wilt for conducting the experiment. Convert the RGB image into YCbCr color space and obtained the gray scale image by adopting the 'Y' component. Extract the color, shape and texture feature for classification. Classification was done using SVM and disease grading using ANFIS classifier. Confusion matrix used to compare SVM and ANFIS classifier [15]. Camargo et al. described the method to detect the visual indications of plant diseases. Proposed method development started by converting the RGB image into H, 13a, 13b. Image segmentation strategy described effectively by interpreting the distribution of intensities in the histogram, the threshold cut off value determine according to their position in the histogram. The proposed algorithm had the ability to detect the diseased region with a large range of intensities [2]. Surya Prabha et al. reviewed banana plant diseases as Panama and moko disease, Sigatoka and black spot disease, Bunchy top virus, Infectious chlorosis, and streak virus and the symptoms of the disease. Explain the techniques of image processing used in disease recognition systems like image formation, preprocessing, Analysis-Thresholding, Histogram, Region growing. Explain the importance of pattern classification for disease classification, Back propagations Neural Network, Support Vector Machine for classification, and Principal Component Analysis [3]. Rothe and Kshirsagar proposed a system to identify cotton leaf disease. Considered the Bacterial Blight, Myrothecium and Alternaria diseases and used the pattern recognition techniques to identify diseases. Images captured from the cotton research center, Nagpur. Used snake segmentation algorithm to isolate the diseases spot. Neural networks used as classifiers [9]. Maikar et al. proposed an approach to identify the disease of tomato and potato from the Pune district and cotton from the Marathwada region of Maharashtra. Color-based segmentation K-mean clustering applied to get diseased region. Green pixels are masked based on the threshold value set. The resulting image was converted from RGB to HIS color format. Color feature extraction was done and used neural networks for classification [10].

4   Proposed Framework

Figure 3 describe the step by step procedure of recommended system. Banana leaf image databases are essential to conduct the experiment for recognition of disease, Some of them used for training the system and remaining for testing purpose. During training, phase acquired the input image and image pre-processing is perform by resizing the images followed by feature extraction. Clustering is performed by implementing k-mean clustering and lastly, K-Nearest Neighbour is used for disease classification. For testing, the query image will be picked up from the user, and then image pre-processing and feature extraction is conducted with resulting the disease on leaves.

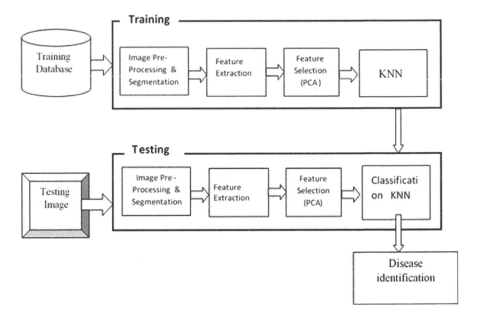

Fig. 3. Proposed architecture: automated plant disease identification

4.1   Image Acquisition, Pre-processing and K-MEANS Based Segmentation

The images are acquired by a commonly used method as a photograph by digital camera. The input images were resized into size 300×300 and used for experiment. Another method can also be used in this system; there will be a predefined dataset containing the images of banana Sigatoka and banana Wilt disease through which the images will be fetched to process. The captured images might be of various forms and dimensions. Hence the images are pre-processed to bring the same dimension images. Sample image of Bunchy Top and sigatoka disease is displayed in following Fig. 4. Segmentation is done to get the infected regions of the image. Segmentation is done by using a straightforward and fast method using the K-Means clustering algorithm. K-number of clusters is produced from the input images [11]. RGB space is transformed into $L*a*b$ and $a*b$ is the color space. The input image is splited into 3 clusters using the mentioned method. Next cluster with ROI (Region of Interest) chosen explicitly which used the input to the next process.

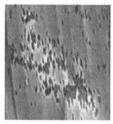

Fig. 4. Bunchy Top and Sigatoka disease affected images

4.2   Feature Extraction

Feature extraction acting a vital part in the area of computer vision and machine learning to analyse an object in the image. Each object has their specific shape, color, size and texture. Feature extraction is used to extract the object and classified into their significance class [18]. In proposed work the features constructed which describe the disease portion in the image depended on their color, shape and GLCM texture features. The extracted features have been used in feature selection process.

4.3   Color

Color is one of the mostly used visual features to evaluate the image because each disease identified by their symptoms and color is one of the best parameter which is used for recognition. Different color model used to represent the color of the image. RGB, HSV, YCbCr, HIS are the commonly used models which results in individual images analogous to each color space. Extract the 24 color feature of image to observe the disease area like, Mean, skewness, kurtosis, and standard deviation for RGB image and mean for YCbCr, HIS, HSV Lab images.

4.4   Shape

The shape features are significant in many pattern recognition problems. Shape feature define the geometric properties of the target region [8]. Canny edge used to detect the boundary of region. Features like, area, eccentricity, solidity, perimeter etc. 12 shape features are extracted from image.

4.5   Texture

The texture feature is used for images retrieval and the class determination. Many researcher focused plant leafs texture for the most important feature to identify the plants. Grayscale images used to calculate the GLCM features. Computes how recurrently a pixel and its contiguous pixel arises either vertically, horizontally and diagonally [14]. Four statistical features are presented in this paper as homogeneity, correlation, energy, contrast and which is derived by GLCM method as given in Table 1.

Table 1. Texture feature used in extraction

Sr.No.	GLCM Feature	Formula
1	Contrast	$\displaystyle\sum_{i,j=0}^{N-1} P(i,j)(i-j)2$
2	Correlation	$\displaystyle\sum_{i,j=0}^{N-1} p_{i,j}\left[\dfrac{(i-\mu_i)(j-\mu_j)}{\sqrt{(\sigma_i^2)(\sigma_j^2)}}\right]$
3	Energy	$\displaystyle\sum_{i,j=0}^{N-1} p(i,j)^2$
4	Homogeneity	$\displaystyle\sum_{i,j=0}^{N-1} \dfrac{p(i,j)}{1+(i-j)^2}$

4.6 Feature Selection

The significance of feature selection is to obtain the finest features from obtained features and served to a classifier, which growths the detection accuracy of the system. In the proposed system total 40 features are extracted during the feature extraction process. The extraneous and unnecessary features influence the system accuracy and performance time [18], therefore PCA used to increase the accuracy of the system. One of the most popular techniques having greatest statistical feature analysis capability is principal component analysis [5]. PCA is used for scrutinizing the image database defined by numerous sets of parameters to extract important features.

4.7 Training and Classification

K- Nearest Neighbour algorithm is used for training and classification of objects. Categorise the objects based on nearby training samples in the feature space. Store the feature vectors and class of the training images. Distance between the test image and training set is determined using the distance function $d_E(x, y)$ as shown in Eq. 1 [17], x and y are the scenarios composed of N features. Distance between the two scenarios is calculated using the Euclidean distance measure.

$$d_E(x, y) = \sum_{i=1}^{N} \sqrt{x_i^2 - y_i^2} \tag{1}$$

Calculate the minimum distance between the query image and training sets and get the k nearest neighbour. Query image is classified according to the highest number of class label from k neighbours.

5 Experimental Result

5.1 Dataset Preparation

To validate the projected system generated the dataset of banana leaf images containing 212 disease images with the help of field experts and, also used the predefined dataset available on [19] containing images for black Sigatoka and Panama wilt. Following Fig. 5 shows the sample input dataset.

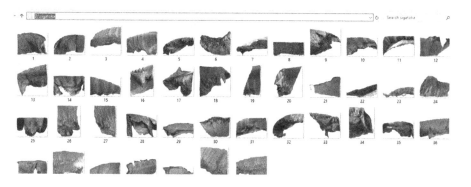

Fig. 5. Sample input dataset

5.2 Generating Clusters

Once the query image loaded, generate the clusters using K-mean segmentation techniques. In Fig. 6 shows the original image, Grayscale Image, and 3 different cluster form by original image. Grayscale Image is produced for indexing purpose to analyse the diseased area and leaf area pixels to further implement the segmentation techniques.

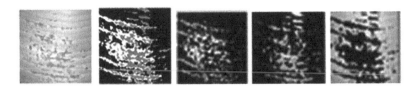

Fig. 6. Cluster of image

6   Result and Discussion

A total of 212 samples were considered for banana disease images (Sigatoka, Bunchy Top, Panama Wilt and Banana Streak Virus). To ensure the consistency of training model 10 fold method used. Total database is divided into 10 folds. First fold is taken for testing and the remaining samples were considered as the training. Same way upto 10 folds, each one at a time considered as testing and other for training purpose. Table 2 displays the accuracy based on each fold and Table 3 displays the detection rate of different diseases constructed using KNN classifier. As display in Table 2, the suggested method had good recognition rate. But, it contains different levels of inaccuracies in identifying the disease. Inaccuracies occur due to a number of causes as follows: Segmentation may be a challenging task as images are captured in natural condition; background encompasses leaves, plants, some part of soil and other unnecessary elements. A common postulation with specific disease recognition is the manifestation will always have the same appearances. However there is always some disparity in color, size and shape of symptoms. Brown or yellow narrow streaks and round spots on leaves were more obvious than yellow vertical streaks. Accordingly the Sigatoka disease had good recognition rate, and the recognition rate of banana streak virus was only 72%. Collected images were having different angles and distance which reduced the correspondence of characteristic parameters.

Table 2. Accuracy of classification using 10 fold

No. of fold	No. of image	Classified	Misclassified	Accuracy (%)
1	21	18	3	86
2	21	18	3	86
3	21	19	2	90
4	21	15	6	71
5	21	17	4	81
6	21	17	4	81
7	21	17	4	81
8	21	18	3	86
9	21	19	2	90
10	23	17	6	74

Table 3. Disease result

Disease category	Sample consider	Classified	Misclassified	Accuracy (%)
Sigatoka	55	47	8	85
Banana Bunchy Top	50	41	9	82
Panama Wilt	68	59	9	87
Streak virus	39	28	11	72

7 Conclusion

The automated system to detect banana leaf diseases based on image analysis with minimizing features using PCA and KNN was proposed. The disease regions were segmented by k-mean clustering methods and use the Euclidean distance technique to calculate the distance for the nearest element. The proposed model could efficiently detect and diagnose four banana leaf diseases; Sigatoka, Bunchy Top, Panama Wilt, and Banana Streak Virus. The result specified that the proposed detection method for banana leaf disease used to examine banana disease with good classification accuracy. This system is going to be very supportive for agriculture while it is effective than the manual system.

References

1. Bhange, M., Hingoliwala, H.: Smart farming: pomegranate disease detection using image processing. Procedia Comput. Sci. **58**, 280–288 (2015). Second International Symposium on Computer Vision and the Internet (VisionNet 2015). Elsevier
2. Camargo, A., Smith, J.S.: An image-processing based algorithm to automatically identify plant disease visual symptoms. Biosyst. Eng. **102**, 9–21 (2009)
3. Surya Prabha, D., SatheeshKumar, J.: Study on banana leaf disease identification using image processing methods. Int. J. Res. Comput. Sci. Inf. Technol. (IJRCSIT), 89–94 (2014)
4. Dev, A., Sharma, M., Meshram, M.: Image processing based leaf rot disease, detection of betel vine (Piper BetleL.). Procedia Comput. Sci. **85**, 748–754 (2016). International Conference On Computational Modeling And Security (CMS 2016). Sciencedirect, Elsevier
5. Gulhane, V., Kolekar, M.: Diagnosis of diseases on cotton leaves using principal component analysis classifier. Annual IEEE India Conference (INDICON) (2014)
6. Barbedo, J.G.A.: A review on the main challenges in automatic plant disease identification based on visible range images. Biosyst. Eng. **144**, 52–60 (2016)
7. Collins, J., Okada, K.: Improvement and comparison of weighted k nearest neighbors classifiers for model selection. J. Softw. Eng. **10**(1), 109–118 (2016). https://doi.org/10.3923/jse.2016.109.118
8. Zhu, J., Wu, A., Wang, X., Zhang, H.: Identification of grape diseases using image analysis and BP neural networks. Multimedia Tools Appl. **79**(21), 14539–14551 (2019). https://doi.org/10.1007/s11042-018-7092-0
9. Rothe, P.R., Kshirsagar, R.V.: Cotton leaf disease identification using pattern recognition techiques. In: International Conference On Pervasive Computing (ICPC). IEEE (2015)

10. Mainkar, P.M., Ghorpade, S., Adawadekar, M.: Plant leaf disease detection and classification using image processing techniques. Int. J. Innovative Emerg. Res. Eng. 2(4), 139–144 (2015)
11. Rastogi, A., Arora, R., Sharma, S.: Leaf disease detection and grading using computer vision technology and fuzzy logic. In: 2nd International Conference On Signal Processing and Integrated Networks (SPIN). IEEE (2015)
12. Dhivya, S., Shanmugavadivu, R.: Comparative study on classification algorithms for plant leaves disease detection. Int. J. Comput. Trends Technol. (IJCTT) 60(2), 115–119 (2018)
13. Ganie, S.A., Lone, A., Haq, S.: Impact of climate change on plant diseases. Int. J. Mod. Plant Anim. Sci. 1(3), 105–115 (2013). ISSN: 2327-3364
14. Sapkale, S.S., Patil, M.P.: Material classification using color and texture features. In: Santosh, K.C., Hegadi, R.S. (eds.) RTIP2R 2018. CCIS, vol. 1035, pp. 49–59. Springer, Singapore (2019). https://doi.org/10.1007/978-981-13-9181-1_5
15. Vipinadas, M.J., Thamizharasi, A.: Detection and grading of diseases in banana leaves using machine learning. Int. J. Sci. Eng. Res. 7(7), 916–924 (2016). ISSN 2229-5518
16. Wang, H., Li, G., Ma, Z., Li, X.: Image recognition of plant diseases based on principal component analysis and neural networks. In: 8th International Conference on Natural Computation (ICNC 2012). IEEE (2012)
17. Tian, Y., Zhao, C., Lu, S., Guo, X.: SVM-based multiple classifier system for recognition of wheat leaf diseases. In: Proceedings Of 2010 Conference On Dependable Computing (CDC 2010), 20–22 November 2010
18. Iqbal, Z., Khan, M.A., Sharif, M., Shah, J.H., ur Rehman, M.H., Javed, K.: An automated detection and classification of citrus plant diseases using image processing techniques: a review. Comput. Electron. Agric. 153, 12–32 (2018). www.elsevier.com/locate/compag
19. https://github.com/godliver/source-code-BBW-BBS

Enhanced HOG-LBP Feature Vector Based on Leaf Image for Plant Species Identification

Harsha Ashturkar(✉), A. S. Bhalchandra, and Mrudul Behare

Department of Electronics and Telecommunication Engineering, Government Engineering College, Aurangabad, India
harshaashturkar17@gmail.com, asbhalchandra@gmail.com, mbehare@gmail.com

Abstract. Plant species database has become essential as biodiversity is declining rapidly. With advance technology and technocrats, an attempt has been made of plant species identification depending on leaf image is implemented. Leaf characteristics are used to prepare feature vector. HOG and LBP are used as feature vectors, LDA and SVM as classifiers. When HOG and SVM are concatenated and used as feature vector, accuracy of classifiers is enhanced as compared to individual HOG and SVM as feature vector. LDA is proved to be better classifier than SVM for database mentioned.

Keywords: Histogram oriented gradient · Local binary pattern · Support vector machine · Linear discriminant analysis

1 Introduction

Environment refers to surroundings to living beings, it affects their lives too. Environment's main components are organisms, soil, water, air, solar energy etc. Plants are important component of environment, without which earth's ecology will have no existence. Globalization and urbanization, has effect on environment, like deforestation on large scale, pollution, ecology imbalance etc. Climate change and many other are leading to plants at the risk of extinction. Plant database is a move regarding safeguarding of earth's biosphere helping in protecting the plants and classify various kinds of flora varieties. Quick and precise plant identification is crucial for effective analysis and management of biodiversity. Botanist use different characters of plants to identify the plant species. Limited number of experts and rapid declining biodiversity leads to significant challenges to biological study and conservation.

These challenges are leading to concept of using advanced technology and approaches of computer vision.

With an availability of high-end portable devices like digital camera, scanners concept of plant database can be implemented. Computer vision algorithms can be used for identification and categorization of plant species. A detail analysis of plant database and comparison of plant species classification techniques is done [1]. With technology introduced in the field of plant species identification, a system for plant identification is

© Springer Nature Singapore Pte Ltd. 2021
K. C. Santosh and B. Gawali (Eds.): RTIP2R 2020, CCIS 1381, pp. 249–258, 2021.
https://doi.org/10.1007/978-981-16-0493-5_22

developed based on leaf image. Leaf images are acquired by digital camera or scanner. These images will be pre-processed which includes RGB to grey conversion, filtering, noise removal, image enhancement, segmentation etc. which makes them suitable for feature extraction. Extracted feature vector is passed on to classifier for classification.

1.1 Background

Major research work of plant identification is based on leaf analysis but flower, bark, fruit, full plant are also used [1]. Flowers are available in blooming seasons only; plant identification using flower with machine learning algorithms is a difficult task because it is a three-dimensional object. There is variation due to view point, scale and occlusion as compared to leaf images. Colour based analysis is easy. Shape based flower analysis is also possible. Individual shape of petals is considered but petals are usually soft and flexible which makes them to curl, twist makes it difficult to identify. Organs like, fruits, bark, full plant are also used for classification.

Feature reduces the dimension of the information by extracting the characteristics, patterns from image. Features of an image can be extracted by its content like shape, texture, colour, position, dominant edges of image items and regions.

Many researchers have worked upon various features mentioned above but there is no universal feature which can be implemented for all plant species. Different features or combination of features are also experimented and found to be successful. This is essential because leaf shape of some species may be same but its colour, texture may be different. Similarly, for flowers, it may have same colour but shape and texture may be different. Overall, general features can be categorized as shape, colour, texture. Shape feature can be categorized as Region based shape extractors and Contour based shape extractors [2, 3].

Main focus of this paper is Region based features. Simple and morphological shape features are described by diameter, major and minor axis, perimeter, centroid [4]. Based on these shape features morphological features are calculated like aspect ratio, rectangularity measure, circulatory measure, perimeter to area ratio, etc. [5–13]. Leaf specific features like LWF i.e. leaf width factor and AWF i.e. area width factor are also used [14, 15] respectively.

In Region based descriptors, moments are used for object classifier. They are steady to rotation, scale and translation. Hu has proposed six moments which is great contribution to this research [16]. These are combined with ZMI i.e. Zernike moment invariants and Legendre moment invariant i.e. LMI are used but their computational complexity is high [8, 17, 18].

Local Features SIFT, SURF, HOG are also used. Scale Invariant Feature Transform (SIFT) which is combination of feature extractor and feature detector. It is very robust against image scale, rotation, changes in illumination [19]. Speeded Up Robust Feature (SURF) is used for leaf classification [20]. Histogram of Orientated Gradient (HOG) is used on large scale [21]. HOG is calculated for cells which overlaps between neighbor blocks. HOG and Maximum Marginal Criteria (MMC) are combined to form feature vector [21]. Drawback of HOG is its responsiveness to leaf petiole direction. So pre-processing related to petiole orientation is essential. It is having lot of redundant information, leading to need of dimension reduction [22, 23].

Performance of the identification system is evaluated using classifier accuracy. Many researchers have used parameters mentioned above but when multi features are used, accuracy increases than that of using single parameter. Many data bases are available. Swedish Leaf dataset, Leafsnap dataset, Flavia dataset, ICL dataset, Oxfard Flower dataset are used for analysis of above-mentioned features.

Stephen Gang Wu, Forrest Sheng Bao et al. (2007) have used geometrical, digital morphological features, PCA is applied to reduce the dimensionality [24]. Chaki, Parekh et al. (2011) have used features like moment invariants, centroid-radii distances and fed as input to Neural Network for purpose of classification [25]. S. P. Raut, Dr. A. S. bhalchandra extracted shape based features for plant species identification [26] Abdul Kadir, Lukito Edi Nugroho et al. (2011) have used shape, texture features and neural net as classifier [8]. Minggang Du, Xianfeng Wang (2011) have used Histogram Oriented Gradient for depiction of shape and used PCA and LDA combinedly for dimensionality reduction [27]. Hang Zhang, paul yanne et al. (2012) have used geometric features, Local and Global features along with Support Vector Machine (SVM) as a classifier [28]. E. Aptoula, B.Yanikoglu (2013) have used two descriptors. One is morphological covariance on the leaflet outline and another is ring-shaped Covariance Histogram considering leaf venation attributes [29]. Tsolakidis, Kosmopoulos et al. (2014) have used Zernike moments and HOG feature and SVM as a classifier [30]. Trishen Munisami, Mahess Ramsurn et al. (2015) have used Geometric features along with KNN classifier [32]. Aimen Aakif, Muhammad Khan et al. (2018) have used various Morphological features along with SIFT and ANN is used as a Classifier to get the high rate of accuracy [5]. Muammer Turkoglu, Davut Hanbay used Local Binary Pattern (LBP) to acquire texture features from images without need of any kind Pre-processing and obtained remarkable results on three different datasets [31]. Slope of digital shape signature have also used for plant classification. Many combinations of features and classifiers are worked upon.

Finally, individually working with HOG as a feature is not giving satisfactory results so an attempt has been made to combine LBP with HOG as a feature. Two kinds of classifiers are used LDA and SVM for classification. Paper is arranged as follows: Sect. 2.1 gives Feature Vector description; Sect. 2.2 explains about Classifiers Sect. 3 describes Performance Evaluation and experimental results and paper concludes with Sect. 4.

2 Methodology

2.1 Feature Vector Description

HOG (Histogram of Gradient) [33]
Image is split into number of parts and each part into number of cells. Cell size is 8×8, 16×16, 32×32, 64×64 as per size of image. Gradient magnitude $G(x, y)$ and gradient direction $\theta(x, y)$ is calculated with given formula

$$G(x, y) = \sqrt{Gx^2(x, y) + Gy^2(x, y)} \tag{1}$$

$$\theta(x, y) = \arctan\left(\frac{Gy(x, y)}{Gx(x, y)}\right) + \frac{\pi}{2} \qquad (2)$$

Histogram of Oriented Gradient with 8×8 Cell size is shown below in Fig. 1

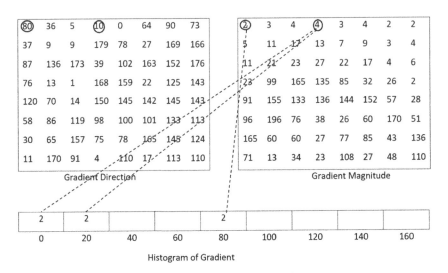

Fig. 1. Bin formation in HOG

Each pixel is taken into consideration while computing HOG so that even a minor change in images is enough to distinguish among two different images.

Image is split into number of blocks and every block is further split into number of cells for which the HOG is to be calculated. For 8×8 image patch, gradient of patch accommodates per pixel two value, leading to total number of values to be 128. These many numbers are depicted by using a 9-bin histogram further which is saved as a row of 9 values. Histogram vector consists 9 numbers showing the angles $0, 20, 40 \ldots 160$. Bin in the histogram vector is selected based upon direction and value based upon magnitude, contribution of all pixels in 8×8 cells are appended to generate 9 bin histogram.

LBP (Local Binary Pattern)

This operator gives information about texture better as compared to texture information provided by Gabor Co-occurrence and Wavelet approach [34].

It is used as local grey level structure representation. This operator considers adjacent pixels and thresholds the difference of grey level between them, generating binary values image patch as regional image descriptor. Originally it was decided for 3×3 neighborhood leading to 8-bit codes as shown in Fig. 2. Mathematically it is represented as

$$LBP(p_c + q_c) = \sum_{r=0}^{n} 2^r s(i_r - i_c) \qquad (3)$$

Where 'r' runs over 8 neighbors of central pixel c, i_r and i_c are grey level value at c and n.

$$s(u) = \begin{array}{l} 1, \, u > 1 \\ 0, \, otherwise \end{array} \qquad (4)$$

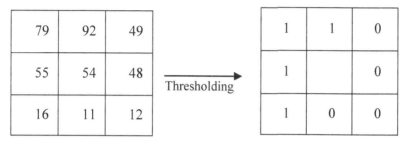

Fig. 2. Uniform LBP operator

There are many extensions or modifications in Local Binary Pattern. Uniform Local Binary Pattern is used because it is used for shape detection. Image is divided into grid of cells to generate Local Binary Pattern of that cell; finally, Local cell level LBP values are concatenated continuously to generate Global Local Binary Pattern vector. Usually LBP vector is high dimensional.

Final feature vector consists of HOG and LBP feature concatenated.

2.2 Image Classifiers

Linear Discriminant Analysis (LDA)

It is basically general form of Fisher's linear discriminant. It is accustomed machine learning or pattern recognition when data is to be classified in more than two classes. In general steps for LDA implementation for a given dataset is as follows [35].

- Mean vector for given number of classes m is calculated
- Within class and between class scatter is calculated
- Eigen vectors E1, E2, E3....Em and corresponding eigenvalues for scatter matrix are calculated for m classes
- Eigenvector is sorted and eigenvector with maximum eigen value is selected to form m × k dimensional matrix P, each column of P constitutes an eigenvector.
- Eigenvector array of size m × k is used to convert samples as a new subspace. It can be stated that, newly generated matrix Y = X × P is converted to dimension n × k in new subspace, where X is n × d dimensional array constituting n samples and Y are converted n × k dimensional samples in a new subspace.

Support Vector Machine (SVM)

Support Vector Machine is supervised type of classifier determined by separating hyperplane [36]. It creates a dividing hyperplane in the space, which establishes a border

between the two distinct datasets. It produces two planes on each side of dividing hyperplane between the two datasets to set up boundary.
The separating Hyperplane can be defined in equation form as

$$W^T X + b = 0 \qquad (5)$$

Where W is Weight, X is input and b is bias. It can be rewritten as

$$W^T X + b \geq 0 \ldots for d_i = +1 \qquad (6)$$

$$W^T X + b < 0 \ldots for d_i = -1 \qquad (7)$$

Where d_i is edge of division between hyperplane and nearest data location for weight W and bias b. With input X and Y it is represented as

$$F(x) = Y_i(W^T X + b) \qquad (8)$$

From above equation functional margin for classification is determined.

2.3 Dataset

Along with selection feature(s), Selection of database is another challenge. Images for database can be obtained from following categories: Scan Images, Pseudo-Scan Images and Photographs. While choosing the database, background of leaf plays important role. Most of the researchers use plain background and images are obtained by Scan and Pseudo-Scans to avoid overlaps. Swedish Leaf Database, Flavia Database, LeafSnap dataset, ICL Database are some of the publicly available Database on which large research has been carried out. Swedish Leaf Database contains scanned figure of isolated leaflet on plain framework. It includes 15 Swedish Plant Species having 75 Leaf images of each species. Flavia Database is sampled by using commonly available plants in China including 50–77 images per species of 32 different species. LeafSnap Database contains both, images taken by mobile camera with Controlled Light and other source is high quality lab images of pressed leaves. All together 185 tree species are considered. ICL Database is a huge database consisting of 220 plant species having individually 26 to 1078 images per species. This is a Chinese database Collected by Intelligent Computing Laboratory (ICL).

3 Performance Evaluation

Here a computer vision system is developed for plant identification based on LeafSnap image Database [37] and experimentation is carried out using MatLab 16a. Features used are HOG, LBP individually and concatenated with LDA and SVM as classifiers. Performance is evaluated with accuracy of classifiers.

Some samples of leaf images used from database are displayed in Fig. 3. All images of these categories are sequentially processed. Cell size of HOG and LBP is fixed to 16 in order to capture small scale spatial information. There is no optimal cell size

Fig. 3. Types of leaf

Table 1. Classifier accuracy

Sr. No	Feature	Classifier	Accuracy in %
(1)	HOG	SVM	83
(2)	LBP	SVM	85
(3)	HOG	LDA	86
(4)	LBP	LDA	88
(5)	HOG+LBP (concatenated)	SVM	91
(6)	HOG+LBP (concatenated)	LDA	93

and hence the best size depends on our data. HOG and LBP are calculated and passed on to Classifiers Learners Application, a facility available in, Statistics and Machine Learning Toolbox of MATLAB for classification. The accuracies are achieved by the k-fold cross-validation technique where k = 5.

In training the HOG and LBP with Classifiers like SVM and LDA, time complexity is based on the size of the Feature vector. Normally Classification Learner Application requires time of 1–2 min to train all parameters on well performing computer system.

4 Conclusions

Plant species identification system using leaf images is implemented with MatLab 16a software with LeafSnap database.

Two features, HOG (Histogram of Gradient) and LBP (Local Binary Pattern) are used. HOG is calculated for whole image. It is calculated for cell with overlap in neighbor block. It has redundant information as well so use of LBP is suggested to improve the classification accuracy. HOG and LBP in concatenated form gives better results as compared to individual use of them.

Linear Discriminant Analysis (LDA) and Support Vector Machine (SVM) are used as classifiers. LDA is analytical solution. It focusses on data points to estimate the covariance matrix. It tries to minimize the within class disperse and maximizes between class disperse, whereas SVM is an optimization problem. It is optimized over data points which lie on separating margin. It performs better on two class problem whereas LDA handles multi class problem.

LDA as a classifier performs better as compared to SVM with same input data.

References

1. Wäldchen, J., Mäder, P.: Plant species identification using computer vision techniques: a systematic literature review. Arch. Comput. Methods Eng. **25**, 507–543 (2018). https://doi.org/10.1007/s11831-016-9206-z
2. Santosh, K.C., Lamiroy, B., Wendling, L.: DTW-radon-based shape descriptor for pattern recognition. Int. J. Pattern Recogn. Artif. Intell. **27**(3), 1350008 (2013)
3. Santosh, K.C., Lamiroy, B., Wendling, L.: DTW for matching radon features: a pattern recognition and retrieval method. In: Blanc-Talon, J., Kleihorst, R., Philips, W., Popescu, D., Scheunders, P. (eds.) Advanced Concepts for Intelligent Vision Systems, pp. 249–260. Springer, Heidelberg (2011). https://doi.org/10.1007/978-3-642-23687-7_23
4. Wu, S.G., Bao, F.S., Xu, E.Y., Wang, Y., Chang, Y., Xiang, Q.: A leaf recognition algorithm for plant classification using probabilistic neural network. In: IEEE International Symposium on Signal Processing and Information Technology, Giza, pp. 11–16 (2007). https://doi.org/10.1109/ISSPIT.2007.4458016
5. Aakif, A., Khan, M.: Automatic classification of plants based on their leaves. Biosyst. Eng. (2015). https://doi.org/10.1016/j.biosystemseng.2015.08.003
6. Caballero, C., Carmen Aranda, M.: Plant species identification using leaf image retrieval. In: CIVR 2010 - ACM International Conference on Image and Video Retrieval (2010). https://doi.org/10.1145/1816041.1816089
7. Du, J.X., Wang, X.F., Zhang, G.J.: Leaf shape-based plant species recognition. Appl. Math. Comput. (2007). https://doi.org/10.1016/j.amc.2006.07.072
8. Kadir, A., Nugroho, L., Susanto, A., Santosa, P.: A comparative experiment of several shape methods in recognizing plants. Int. J. Comput. Sci. Inf. Technol. (2011). https://doi.org/10.5121/ijcsit.2011.3318
9. Kalyoncu, C., Toygar, O.: Geometric leaf classification. Comput. Vis. Image Underst. (2014). https://doi.org/10.1016/j.cviu.2014.11.001
10. Novotný, P., Suk, T.: Leaf recognition of woody species in Central Europe. Biosyst. Eng. **115**, 444–452 (2013). https://doi.org/10.1016/j.biosystemseng.2013.04.007
11. Pauwels, E.J., de Zeeuw, P.M., Ranguelova, E.B.: Computer-assisted tree taxonomy by automated image recognition. Eng. Appl. Artif. Intell. **22**(1), 26–31 (2009). ISSN 0952-1976
12. Wang, X.-F., Du, J.-X., Zhang, G.-J.: Recognition of leaf images based on shape features using a hypersphere classifier. In: Huang, D.-S., Zhang, X.-P., Huang, G.-B. (eds.) ICIC 2005. LNCS, vol. 3644, pp. 87–96. Springer, Heidelberg (2005). https://doi.org/10.1007/11538059_10
13. Yahiaoui, I., Mzoughi, O., Boujemaa, N.: Leaf shape descriptor for tree species identification, pp. 254–259 (2012). https://doi.org/10.1109/ICME.2012.130

14. Hossain, J., Amin, M.A.: Leaf shape identification-based plant biometrics. In: 13th International Conference on Computer and Information Technology (ICCIT), Dhaka, pp. 458–463 (2010)
15. Yanikoglu, B., Aptoula, E., Tirkaz, C.: Automatic plant identification from photographs. Mach. Vis. Appl. **25**, 1369–1383 (2014). https://doi.org/10.1007/s00138-014-0612-7
16. Ming-Kuei, H.: Visual pattern recognition by moment invariants. IRE Trans. Inf. Theory **8**(2), 179–187 (1962)
17. Wang, X.-F., Huang, D.-S., Du, X., Xu, H., Heutte, L.: Classification of plant leaf images with complicated background. Appl. Math. Comput. **205**, 916–926 (2008)
18. Zulkifli, Z., Saad, P., Mohtar, I.: Plant leaf identification using moment invariants general regression neural network. In: Proceedings of the 2011 11th International Conference on Hybrid Intelligent Systems, HIS, pp. 430–435 (2011)
19. Che Hussin, N.A., Jamil, N., Nordin, S., Awang, K.: Plant species identification by using scale invariant feature transform (SIFT) and grid based colour moment (GBCM). In: IEEE Conference on Open Systems (2013), ICOS 2013, pp. 226–230 (2013)
20. Le, T., Nguyen, Q.-K., Pham, N.-H.: Leaf based plant identification system for Android using SURF features in combination with bag of words model and supervised learning. In: International Conference on Advanced Technologies for Communications (2013)
21. Xiao, X.-Y., Hu, R., Zhang, S.-W., Wang, X.-F.: HOG-based approach for leaf classification. In: Huang, D.-S., Zhang, X., Reyes García, C.A., Zhang, L. (eds.) ICIC 2010. LNCS (LNAI), vol. 6216, pp. 149–155. Springer, Heidelberg (2010). https://doi.org/10.1007/978-3-642-14932-0_19
22. Du, M., Wang, X.: Linear discriminant analysis and its application in plant classification. In: 2011 Fourth International Conference on Information and Computing (ICIC), pp. 548–551 (2011)
23. Zhang, S., Feng, Y.Q.: Plant leaf classification using plant leaves based on rough set. In: International Conference on Computer Application and System Modeling (ICCASM 2010), Taiyuan, pp. V15-521–V15-525 (2010)
24. Wu, S.G., Bao, F.S., Xu, E.Y., Wang, Y., Chang, Y., Xiang, Q.: A leaf recognition algorithm for plant classification using probabilistic neural network. In: 2007 IEEE International Symposium on Signal Processing and Information Technology, Giza, pp. 11–16 (2007)
25. Jyotismita, C., Parekh, R.: Plant leaf recognition using shape based features and neural network classifiers. Int. J. Adv. Comput. Sci. Appl. (IJACSA) **2**(10), 41–47. https://doi.org/10.14569/IJACSA.2011.021007
26. Raut, S.P., Bhalchandra, A.S.: Plant recognition system based on leaf image. In: Second International Conference on Intelligent Computing and Control Systems (ICICCS), Madurai, India, pp. 1579–1581 (2018). https://doi.org/10.1109/ICCONS.2018.8663028
27. Du, M., Wang, X.: Linear discriminant analysis and its application in plant classification. In: 2011 Fourth International Conference on Information and Computing (ICIC), Phuket Island, pp. 548–551 (2011). https://doi.org/10.1109/ICIC.2011.147
28. Zhang, H., Yanne, P., Liang, S.: Plant species classification using leaf shape and texture. In: International Conference on Industrial Control and Electronics Engineering, Xi'an, pp. 2025–2028 (2012)
29. Aptoula, E., Yanikoglu, B.: Morphological features for leaf based plant recognition. In: IEEE International Conference on Image Processing, Melbourne, VIC, pp. 1496–1499 (2013)
30. Tsolakidis, D.G., Kosmopoulos, D.I., Papadourakis, G.: Plant leaf recognition using zernike moments and histogram of oriented gradients. In: Likas, A., Blekas, K., Kalles, D. (eds.) SETN 2014. LNCS (LNAI), vol. 8445, pp. 406–417. Springer, Cham (2014). https://doi.org/10.1007/978-3-319-07064-3_33
31. Türkoğlu, M., Hanbay, D.: Leaf-based plant species recognition based on improved local binary pattern and extreme learning machine. Phys. A Stat. Mech. Appl. (2019). https://doi.org/10.1016/j.physa.2019.121297

32. Munisami, T., Ramsurn, M., Kishnah, S., Pudaruth, S.: Plant leaf recognition using shape features and colour histogram with K-nearest neighbour classifiers. Procedia Comput. Sci. **58**, 740–747 (2015)

33. Pang, Y., Yuan, Y., Li, X., Pan, J.: Efficient HOG human detection. Sig. Process. **91**, 773–781 (2011). https://doi.org/10.1016/j.sigpro.2010.08.010

34. Savelonas, M., Iakovidis, D., Maroulis, D.: LBP-guided active contours. Pattern Recogn. Lett. 1404–1415 (2008). https://doi.org/10.1016/j.patrec.2008.02.013

35. https://www.apsl.net/blog/2017/07/18/using-linear-discriminant-analysis-lda-data-explore-step-step/. Accessed 12 Sept 2019

36. Ng, A.: CS229 Lecture notes. Part V, Support vector Machine. https://cs229.stanford.edu/notes/cs229-notes3.pdf

37. Kumar, N., Belhumeur, P.N., Biswas, A., Jacobs, D.W., Kress, W.J., Lopez, I.C., Soares, J.V.B.: Leafsnap: a computer vision system for automatic plant species identification. In: Fitzgibbon, A., Lazebnik, S., Perona, P., Sato, Y., Schmid, C. (eds.) ECCV 2012. LNCS, pp. 502–516. Springer, Heidelberg (2012). https://doi.org/10.1007/978-3-642-33709-3_36

Intelligent Irrigation System Using Machine Learning Technologies and Internet of Things (IoT)

Sarika Patil$^{(\boxtimes)}$ and Radhakrishana Naik

Dr. BAM University, Aurangabad, India
sarikapatil181@gmail.com, naikradhakrishna@gmail.com

Abstract. Scare water resources necessitates technological involvement in irrigation scheduling, that can help to manage water according to the weather condition of different seasons, crop growth stage and landscape information. The proposed method calculates actual water required using machine learning model and Evapotranspiration. The model is trained using real time weather data to predict actual water requirement. Reference Evapotranspiration is calculated with the help of Penman-Monteith Method. Before starting with this real time system proposed model is implemented with the help of past 10 years web scraped weather data. Proposed algorithms of water requirement and Irrigation scheduling is executed on scrapped data. After successful results system is implemented for real time use. Furthermore, system consists of eight Arduino nodes that acting as a slave to read weather, soil landscape and rain data. In addition, three Raspberry-pi equipped with the Wi-Fi module, acting as server to send collected data to a remote web server. These databases are used as input for machine learning algorithm. As per the observations of proposed system water usage is getting reduced in large quantity as compared to the traditional irrigation system used for irrigation.

Keywords: Weather data · Irrigation · Machine learning · Temperature · Agriculture · Water · Soil · Algorithm · Crop growth · Scheduling · Sensors · Communication · Evapotranspiration · Automation · Crop

1 Introduction

Agricultural productions consume more than 85% of freshwater over the planet than any other water uses. Freshwater consumption for agricultural uses will continue to increase because of increasing population and food demand [1]. Day by day freshwater resources getting reduced so it is mandatory to optimize water use in agriculture.

In India, most of agricultural part is irrigated through the manual system. The manual system gives irrigation only based on observation. The microcontroller and sensor-based automated system that will provide irrigation based on sensor data has many advantages [2]. Proposed system has been considering all aspects of water requirement and irrigation system like water requirement based on crop growth stage, canopy of crop, crop age, landscape information like slope of farm, sea-level elevation, soil type, total sunshine

© Springer Nature Singapore Pte Ltd. 2021
K. C. Santosh and B. Gawali (Eds.): RTIP2R 2020, CCIS 1381, pp. 259–267, 2021.
https://doi.org/10.1007/978-981-16-0493-5_23

hours, dew point temperature, relative humidity, actual and saturation vapor pressure etc. Some static information is fed at the time of installation, and remaining information updated periodically to execute the algorithm and estimate water requirement, if any, also adjust any aperiodic task with water requirement calculations.

This paper focus on the different methodology used previously for automated irrigation system followed by the proposed methodology. A proposed system is fully automatic, as compared to other method that send alerts to farmer for further processing. The system uses Machine learning approach to fulfill irrigation requirements in modern days by making full automation and smart water utilization. Once configured this system calculates irrigation needs and provides an irrigation schedule that includes time, duration and interval of irrigation that is handled automatically by sending signal to the solenoid valve. Solenoid valve opens automatically as per irrigation schedule and also taking extra care of maintaining the water level in the tank using ultrasonic sensors.

The water requirement is different in every stage of the crop. Proposed system considering three crop stages i.e. Initial, Mid, Harvesting stage. Based on that crop coefficient gets changed in the machine learning algorithm.

This paper discusses related work followed by the proposed framework research methodology, assumptions, mathematical work, algorithm, evaluation, optimization, and a conclusion.

2   Related Work

Evapotranspiration is an important parameter for most of the agricultural evaluation process and water resource management studies. Water evaporated through crop is estimated with the help of evapotranspiration and crop coefficient, which depends on many things like crop growth stage, canopy and age of crop [5]. Different methods exist to estimate reference evapotranspiration, among them Ravazzanietal method [8] is used when only temperature data is available. In this method, the Hargreaves coefficient is adjusted based on local elevation.

$$ET0 = (0.817 + 0.00022z)0.0023Ra(Tmean + 17.8)(Tmax - Tmin)^{0.5} \quad (1)$$

Another method is Hargreaves and Samani (1985) method [4] that is radiation based method used when some weather data is missing [8]. It is expressed as:

$$ET0 = 0.0023Ra(Tmean + 17.8)(Tmax - Tmin)^{0.5} \quad (2)$$

Here Ra is extraterrestrial radiation, daily Temperature is represented by Tmin, daily minimum Temperature is represented by $T_{(min)}$ and daily max Temperature by $T_{(max)}$. Another method is solar radiation-based method (Irmak, 2003). It is expressed as

$$ET0 = 0.489 + 0.289R_s + 0.023T_{mean} \quad (3)$$

Here R_s is solar shortwave Radiation and Teman is average of Min and max temperature [10].

Net Radiation based method (Irmak, 2003). It is expressed as,

$$ET0 = 0.489 + 0.289R_n + 0.023T_{mean} \quad (4)$$

Here R_n is Net radiation and Tmean is average of Tmax and Tmin [10].

By using the FAO Penman-Monteith equation, crop coefficients can be calculated by using this equation.

Crop Coefficient = Crop Evapotranspiration/Reference.

Evapotranspiration Penman method giving best results to calculate water requirements.

Artificial neural network and Fuzzy logic are used to optimize green roof irrigation [11]. The model is trained using Artificial neural network and Fuzzy logic to calculate soil moisture for irrigation scheduling. Aperiodic conditions are not considered in this system. This system uses weather data from nearby weather station.

CommonKADS expert system [12] provided solution to optimize water usage for crops. The main objective of the CommonKADS expert system is to determine optimal water quantity and time. Parameters considered are the crop, climate, water, soil, and farm data. Fababean crop is taken for case study. Knowledge base of Fababean crop is extracted from the existing knowledge base of fababean crop built based on the Common-KADS methodology. The system is considering all required parameters. While considering any crop for case studies, its knowledge base should be exist to use commonKADS methodology. Not any a-periodic tasks are not considered in the system.

Gene Expression Programming (GEP) [13] is used to develop predictive model for furrow irrigation infiltration. Genetic Programming is the successor of Genetic Algorithm. GEP encodes information in linear chromosomes, which then expressed in expression trees. GEP model is developed using ready dataset of available literature. Z under furrow irrigation is estimated using this model. In this method data set is randomly distributed as training, testing and validation set. In training stage model is trained to get the best performance. Testing phase measures statistical fit of the model. Furrow irrigation is a complex phenomenon. The model is giving good results for furrow irrigation.

IRRINET is developed using Artificial Intelligence Expert – Decision support system (DSS) based on daily water balance methodology [14] of soil-plant-atmosphere. It determines daily irrigation schedule including water quantity and duration of watering. Web based interface is used to gather all data on Central sever for further analysis. Parameter used are climate and soil data. Real time data are used for irrigation scheduling. Web based server is used to collect data on a central server. Aperiodic conditions are not considered in this system.

A feed forward neural network and fuzzy logic-based hybrid smart decision support system (DSS) [15] is implemented to determine soil moisture for irrigation. The model is trained using real time data to predict soil moisture contents and Evapotranspiration is calculated by Blaney-Criddle methodology. Advantages of this system are real time monitoring of data and quick SMS alerts provided to farmer. Soil moisture is measured with the help of training data.

To summarize, Most of the methods are based on Artificial Intelligence and fuzzy logic. The actual soil data collection is challenging task, to overcome that most of researcher train their models to predict soil moisture [11, 15]. Some methods use ready dataset from previous literature instead of real time data. Proposed method using real time weather data and trying to optimize the use of water and making irrigation fully automatic task for farmers. In proposing method Modern Machine Learning algorithm with multiple regression analysis is used to predict water requirement using real time data. IOT architecture is designed and implemented to read real time weather data. This method tries to reduce the error deviation of actual and predicted water requirement with a small error value among all available methods in literature.

3 Details of Proposed Framework

3.1 Mathematical Model

Over the last 50 years, several methods developed by various scientists all over the world with various climatic parameters. However, accuracy testing of the method is time-consuming and costly and not possible to calculate it in short duration, Evapotranspiration data are repeatedly required at a short interval to schedule the irrigation. Among the method given in the paper [6], the modified Penman-Monteith method was offered better results with a tiniest possible error about a living grass reference crop.

$$ETo = \frac{0.408\Delta(R_n - G) + \gamma\frac{900}{T+273}u_{2)}(e_s - e_a)}{\Delta + \gamma(1 + 0.34u_{2)}} \tag{5}$$

Where ET0 is reference Evapotranspiration [mm day-1],
Rn is net radiation at the crop surface [MJ m-2 day-1],
G is soil heat flux density [MJ m-2 day-1],
T is mean daily air temperature at 2 m height [$°C$],
u_2 is Wind speed at 2 m height [m s $-$ 1],
e_s is Saturation vapor pressure [kPa],
e_a is Actual vapor pressure [kPa],
$e_s - e_a$ is Saturation vapor pressure deficit [kPa],
Δ is Slope vapor pressure curve [kPa $°C - 1$],
Γ is psychrometric constant [kPa $°C - 1$].
Rn is net radiation expressed as the difference of Net shortwave (solar) radiation (Rns) and Net Longwave radiation (Rnl).

The Soil Heat flux is presented here follows the idea of soil temperature follows air temperature.

$$G = C_s\frac{T_i + T_{i-1}}{\Delta t}\Delta z \tag{6}$$

Where G soil heat flux [MJ m-2 day-1], Cs soil heat capacity [MJ m-3 $°C - 1$], Ti air temperature at the time i [$°C$], Ti $- 1$ air temperature at the time i $- 1$ [$°C$], Δt length of time interval [day], Δ effective soil depth. For one day to ten days, period soil heat flux considered approximate to zero.

Wind speed instrument placed at the site may be at a different height, so wind speed data is adjusted at 2-m height using following logarithmic wind speed profile calculations.

$$u_2 = u_z\frac{4.87}{ln(67.8z - 5.42)} \tag{7}$$

Where u_2 wind speed at 2 m above ground surface [m s − 1], u_z measured wind speed at z m above ground surface [m s − 1], z height of measurement above ground surface [m]. Saturation vapor pressure is related to air temperature

$$e^0(T) = 0.0618exp\left[\frac{17.27T}{T + 237.3}\right] \tag{8}$$

Where e^0 (T) saturation vapor pressure at the air temperature T [kPa], T air temperature [°C].

Because of non-linearity of Eq. 8, the mean saturation vapor pressure for any period must be computed as the mean between the saturation vapor pressure at the mean daily maximum and minimum air temperatures for that period:

$$e_s = \frac{e^0(T_{max}) + e^0(T_{min})}{2} \tag{9}$$

Slope vapor pressure can be calculated using the following formula,

$$\Delta = \frac{4098e^0(T)}{(T + 237.3)^2} \tag{10}$$

Where: Δ slope vapor pressure curve [kPa°C − 1] T air temperature [°C] e^0(T) saturation vapor pressure at temperature T [kPa].

Figure 1 shows a scatter plot of ET0 (Dependent variable) and Max Temperature (Independent variable). As per plot Temp_max data appearing closely with a regression line. It shows strong linear relationship with ET0.

$$Y = 0.3296x - 5.1516 \tag{11}$$

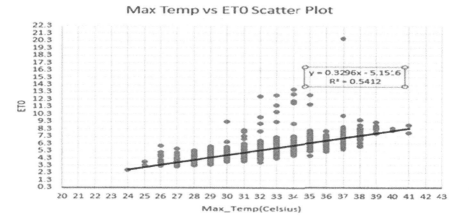

Fig. 1. Scatter plot of ET0/Max_Temp

Regression summery shown in Fig. 2. Having multiple r value is 0.7 and p value is 0. As per p and r value. So both of them showing significant relationship with ET0. Above summery giving double validation of high linearity between Max Temp and ET0.

SUMMARY OUTPUT

Regression Statistics	
Multiple R	0.735657832
R Square	0.541192446
Adjusted R Square	0.541006091
Standard Error	0.740233326
Observations	2464

ANOVA

	df	SS	MS	F	Significance F
Regression	1	1591.279552	1591.2796	2904.0843	0
Residual	2462	1349.041517	0.5479454		
Total	2463	2940.321068			

	Coefficients	Standard Error	t Stat	P-value	Lower 95%	Upper 95%	Lower 95.0%	Upper 95.0%
Intercept	-5.151613156	0.19927531	-25.851738	1.41E-130	-5.542377693	-4.76084862	-5.54237769	-4.76084862
Max_Temp	0.329553404	0.006115348	53.889556	0	0.317561647	0.34154516	0.31756165	0.34154516

Fig. 2. Regression output

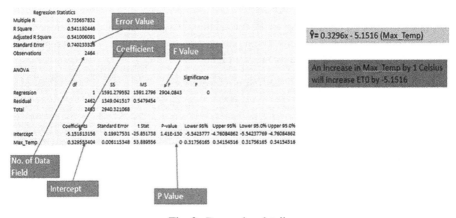

Fig. 3. Regression details

3.2 Algorithm

Configuration of the proposed system with static parameters required for machine learning algorithm.

Configuration of system with crop plantation date, harvesting date and growth status data.

Calculation of the crop coefficient that is mandatory information while executing machine learning algorithms.

Calculation of Reference Evapotranspiration.

Calculation of crop Evapotranspiration using reference Evapotranspiration and crop coefficient.

Calculation of Adjusted crop Evapotranspiration for different stages of crop.
Calculation of total water requirement by considering all the above information.
Calculate Irrigation interval and duration.
Irrigation scheduling.
Fine-tuning of aperiodic activity.

4 Experimentation

For a case study Vineyard is considered. 8 Arduino nodes are used to measure weather, soil and landscape data by applying different orientation. Raspberry-pi is used and equipped with WIFI module acting as gateway to send data over the internet. Soil, Temperature, Rain, Humidity and ultrasonic sensor are placed using different orientation. Vineyard main season started from mid - October and ends at March. October to November is the initial stage, December–January is mid stage and February- March is the end stage of vineyard. Type of soil in selected region is loam/clay. The Site is located in Nashik district at 19.9975° N, 73.7898° E. Elevation level of Nashik is 980m. Hardware and site design details are as given in below.

As shown in Fig. 3 the system uses 8 Arduino Uno R3 controller that acting as slaves. Each Arduino is connected to Bluetooth HC-05, Temperature and Humidity sensor (DHT11) and soil moisture sensor. Raspberry – Pi 4 model is connected to Wi-Fi (ESP8266) model to send data to a central server. Ultrasonic sensors are used to maintain adequate water levels in a water tank. The test bed of 20 feet * 20 feet is used for plantation of 40 crops. 2 emitters of speed ½ gph for each plant are considered, thus for 40 plant 80 emitters of speed ½ gph are required. The total water requirement is calculated with the help of proposed irrigation water requirement algorithm. As per calculations 40 plants we require 40 gallons water per hour per plant (Fig. 4).

Calculate total no of hour of irrigation per month

$$= \frac{\text{Total water required}}{\text{Irrigation Rate per hour}} = \frac{240 \text{ gallon per month}}{40 \text{ gallon per hour per plant}} = 6 \text{ hours per month}$$

Irrigation schedule Method 1: Suppose we decided to give irrigation after every 6th day means 5 times in a month.

$360/5 = 71$ min $=$ watering cycle time.
$240/5 = 48$ gallons water is required for each cycle.

Irrigation Schedule method 2: Suppose we decided to give irrigation after every 4th day $(30/4 = 7)$ means 7 times in a month.

$360/7 = 51$ min watering time.
$240/7 = 34$ gallons water is required for each cycle.

The time complexity of the system is O (N). Three loops are used in the implementation, but only one of them is executed based on crop stage. If the day is between 1st

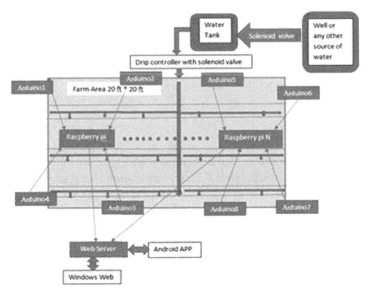

Fig. 4. Test bed design with IOT framework

of October and 30th November then 1st for loop get executed because this period is the initial stage of crop. So N is the number of days and each statement in the loop is executed in O (1) time.

5 Conclusion

Irrigation methods used in literature are based on fuzzy logic, Knowledge base expert system, Genetic programming. Fuzzy logic based methods train the model to predict soil moisture values and trying to overcome soil moisture reading challenge. Knowledge base methods require a ready knowledge base of required crop. IRRINET system using web based server to store tank scheduling. Main objective of proposed method is not only irrigation but also optimization of water used in irrigation. Modern Machine Learning approach used in the proposed system to make it more intelligent and perform all task in irrigation automatically based on real time data and proposed algorithms. As compared to other method proposed system does not require any manual interruption once installed on the farm. The proposed system is totally automatic collecting real time data, executing algorithms, generating irrigation schedule and executing irrigation schedule by automatically opening solenoid valve.

References

1. Jury, W., Vaux, H.: The emerging global water crisis: managing scarcity and conflict between water users. Adv. Agron. **95**, 1–76 (2007)
2. Yang, G., Liu, Y., Zhao, L., Cui, S., Meng, Q., Chen, H.: Automatic irrigation system based on the wireless network. In: Proceedings of 8th IEEE International Conference on Control and Automation, Xiamen, pp. 2120–2125 (2010)

3. Ravazzani, G., Corbari, C., Morella, S., Gianoli, P., Mancini, M.: Modified Hargreaves-Samani equation for the assessment of reference evapotranspiration in Alpine river basins. J. Irrig. Drain. Eng. **138**, 592–599 (2012). ASCE
4. Hargreaves, G.H., Samani, Z.A.: Reference crop evapotranspiration from temperature. Appl. Eng. Agric. **1**, 96–99 (1985)
5. Djaman, K., Irmak, S.: Actual crop evapotranspiration and alfalfa-and grass-reference crop coefficients of maize under full and limited irrigation and rainfed conditions. Irrig. drainage Eng. **139**, 433–446 (2013)
6. Doorenbos, J., Pruitt, W.O.: Guidelines for predicting crop water requirements, FAO irrigation and drainage. Paper no. 24 (1977)
7. Allen, G., Walter, I.A., Elliot, R.L., et al. (eds.): The ASCE standardized reference evapotranspiration equation. Standardization of Reference Evapotranspiration Task Committee Final Report. American Society of Civil Engineers (ASCE), Reston (2005)
8. Djaman, K., Balde, A.B., Sow, A., Muller, B., et al.: Evaluation of sixteen reference evapotranspiration methods under Sahelian conditions in the Senegal River Valley. J. Hydrol. Reg. Stud. **3**, 139–159 (2015)
9. Maina, M.M., Amin, M.S.M., Aimrun, W., Asha, T.S.: Evaluation of different ET0 calculation methods: a case study in Kano State, Nigeria. Philippine Agric. Sci. **95**, 378–382 (2012)
10. Alkaeed, O., Flores, C., Jinno, K., Tsutsumi, A.: Comparison of several reference evapotranspiration methods for Itoshima Peninsula area, Fukuoka, Japan, vol. 66. Memoirs of the Faculty of Engineering, Kyushu University (2006)
11. Tsang, S.W., Jim, C.Y.: Applying artificial intelligence modeling to optimize green roof irrigation. Energy Build. https://doi.org/10.1016/j.enbuild.2016.06.005
12. Hazman, M.: Crop irrigation schedule expert system. In: 13th International Conference on ICT and Knowledge Engineering, pp. 78–83. IEEE (2015)
13. Yassin, M.A., Alazba, A.A., Mattar, M.A.: A new predictive model for furrow irrigation infiltration using gene expression programming. Comput. Electron. Agric. **122**, 168–175 (2016)
14. Manini, P., Genovesi, R., Letterio, T.: IRRINET: large scale DSS application for on-farm irrigation scheduling. Procedia Environ. Sci. **19**, 823–829 (2013)
15. Mohapatra, A., Lenka, S.: Neuro-fuzzy-based smart DSS for crop specific irrigation control and SMS notification generation for precision agriculture. Int. J. Converg. Comput. **2**, 3–22 (2016)

Evaluation of Oh Model for Estimating Surface Parameter of Soil Using L-Band and C-Band SAR Data

Ajit Yadav$^{(\boxtimes)}$, Momin Raisoddin, B. Sayyad Shafiyoddin, and R. Mohammed Zeeshan

Department of Computer Science, Milliya Arts, Science and Management Science College, Beed 431122, Maharashtra, India

Ajitkumaryadav1986@gmail.com, syedsb@rediffmail.com, zeeeshan.shaikh@gmail.com

Abstract. In this proposed work we have focuses on the estimation of soil moisture using L-band and C-band polirimetric (HH, HV, VH, VV) and dual band (VV, VH) SAR (synthetic aperture radar) data set. The empirical model derived by Oh for finding scattering parameter from bare soil surface was implemented. Polarimetric radar backscattering were conducted for bare soil surface for a different values of surface roughness and surface soil moisture at L-band and C-band at different incidence angle from 20° to 60°. A series of field data were collected from different areas as per the standard method given in literature. The data in this paper were taken from two days near Bardoli and Ahmedabad city during the field experiment conducted in 2017. Currently we analyze the collected data to understand the relation between field parameter (surface soil moisture, dielectric constant, surface roughness) and SAR data (radar backscattering). In addition radiative transfer model and radar backscattering model are used to simulate the L-band, C-band data observations. In this work we have simulated empirical Oh model for estimating surface soil moisture using L-band, C-band SAR data set. We have applied all the models given by Oh in different research paper between (1992–2002). Further this results were used to find dielectric constant and use Topp's model to derived surface soil moisture. A good agreement was observed between the estimated and simulated values. Performance of Oh model is promising for L-band and quite good for C-band SAR data for surface soil moisture estimation.

Keywords: Poliremetric · Oh model · SAR data · Surface soil moisture · Surface roughness · Backscattering coefficient

1 Introduction

Surface soil moisture is an important parameters agriculture and plays important role in different crop production, water and energy balance, water requirement, and atmosphere change. Soil surface parameters are very important but retrieval of these parameters are

© Springer Nature Singapore Pte Ltd. 2021
K. C. Santosh and B. Gawali (Eds.): RTIP2R 2020, CCIS 1381, pp. 268–277, 2021.
https://doi.org/10.1007/978-981-16-0493-5_24

very difficult from microwave SAR data. SAR data is extensively investigated in the past four decades. Surface soil moisture are the key values for agricultural and hydrological applications [1], SAR data is widely used for estimating surface soil parameters over the large area [2–7]. SAR sensitivity to surface soil parameter are use full for water management, drought monitoring, flood forecasting, and sustainable agriculture [8]. The SAR back scattering is depending upon physical and electrical parameter of the target material [9–12] and also on SAR configuration (polarization, incidence angle, frequency, etc.). As per the literature the most dependent parameter on which SAR scattering parameters depend is geophysical parameter of soil surface (surface soil moisture, surface roughness, dielectric constant, etc.) [13]. Most of the researcher have been explain the effect of soil surface roughness is more as compared to the surface soil moisture on SAR back scattering coefficient [14–16]. Till date many researchers have studied the bare soil surface parameter and also have given many empirical, semi-empirical and physical models to estimate surface soil moisture using SAR data. One can divide all these method into two group's physical and empirical model, unlike empirical model, physical model does not depends on the site and also does not required field calibration. The most popularly used physical model is the IEM (integral equation model) model developed in 1992 by Fung and Chen [17]. In this model SAR backscatter ($\sigma°$) considered as a function of SAR configuration and surface soil parameters. Another estimating models are empirical or semi- empirical model, which is based on the field parameters and required field calibration. Oh model is an empirical model and give significant results on bare soil surfaces for a large range of microwave signals (L-, C-, X-band) [18]. Theoretical scattering models [19, 20], and radar measurement show that the σ^0 is more sensitive to surface roughness than soil moisture. Oh model can be prefer over the other due to its simplicity and it also gives the direct relation between the physical parameter (surface soil moisture, surface roughness and dielectric constant, etc.) and backscattering coefficient ($\sigma°$).

In SAR images ($\sigma°$) which is backscattering coefficient returning from the target to SAR antenna system that is influenced by the soil surface parameters soil moisture (dielectric constant) [21]. SAR data can be used for retrieving the soil surface parameter using backscattering and physical and theoretical models that gives the relation between the target parameter (soil moisture and surface roughness) and SAR configurations (incidence angle, polarization and frequency) [22]. As we have two parameters on which backscatter coefficient depends are soil moisture and surface roughness, so multiple equations (two or more than two) are required which are multi- configuration SAR data solution [11].

Oh et al. in 1992 have evaluated these inversion techniques with extensive data set in situ measurements. Oh have given the first empirical model [18] in 1992 for inversion technique for radar scattering from bare soil surface. In this he was taken bare soil surface under variety of soil moisture and soil surface roughness at different frequency bands (L-, C-, X-bands). This model further modified by himself in 2004 [23], in this model three modified equations are explained and also give inversion method from SAR data. In this study we are focusing on Oh model to retrieve the soil surface physical parameter, for which we have used two data set one is from the extensive study of Oh et al. in [18, 22]. Other dataset is composed of SAR images and in situ measurement (real data). For each data set we have simulated Oh empirical model and also compare with the real composite dataset.

The aim of this study is to estimate the soil surface parameter using Oh model and also evaluate the potential of this model. A real data set is collected to perform the evaluation. As usual real dataset is divided into two parts, the first part is used to validate the simulation of $\sigma°$ of Oh model and second part is used to measure the accuracy of M_v (volumetric soil moisture) from the model for L-band, C-band. Finally estimated M_v based on the real SAR observation obtained from L-band SAR and C-band SAR images are compared with in situ measurements.

The Oh model and study area and method is explained in Sect. 2. The work start from the implementation of the different back scattering relations are given by Oh et al. [18, 23]. It may be noted that this model can be applied on different frequency bands. Simulations and development of Oh model is explained in Sect. 3. Results of experimental data accusation and comparisons with simulated model are given in Sect. 4. Conclusions of this study are given in Sect. 5.

2 Study Area and Data Collection

2.1 Study Site

The study area is located in Bardoli which located at *21.12°N 73.12°E*, in Gujarat and covers an area of 46 sq. km. The location of these area is shown in (Fig. 1) which is an airborne SAR images collected on 17-jun-2017. Figure shows a combination of all four bands (HH, HV, VH, and VV) as the red green and blue channel. The Bardoli district which we have taken in to consideration for study is lying between (**Lat**: *21.1475 – 21.2101*, **Long**: *73.14688 – 73.15707*). More than 50 samples were collected from the different area of various land cover. After that these samples were used for analysis of soil surface parameter (soil moisture).

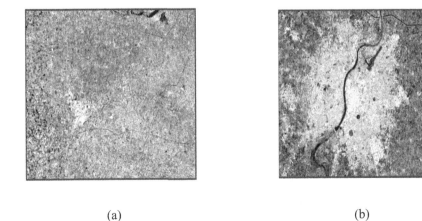

(a)   (b)

Fig. 1. Sentinel 1 data dated 25/06/2018 (a) Bardoli district (b) Ahmedabad city

2.2 SAR Data

Sentinel 1. Is a microwave remote sensing satellite having SAR payload used on C-band (5.35 GHz). Sentinel 1 is a European radar imaging satellite launched in 2014. This radar has a capability of imaging in dual pole VV, and VH. Sentinel images is freely available any one can down load and use the data, the detail specification of sentinel 1 is given in Table 1. The collected images is covering the required studying area portion. Another data is taken from the NISAR mission, this mission is a joint project between NASA (national aeronautics and space administration) and ISRO (Indian space research organization). NISAR mission will consist of dual frequency (L-, and S-band), fully polarimetric radar. The detail specification of NISAR mission is given in Table 2.

Table 1: Specifications of sentinel 1 SAR.

Parameter	Interferometric wide-swath mode (IW)
Swath width	250 km
Incidence angle range	29.1°–46.0°
Sub-swaths	3
Azimuth steering angle	±0.6°
Azimuth and range looks	Single
Polarisation options	Dual VV+VH
Maximum noise equivalent sigma zero (NESZ)	−22 dB
Radiometric stability	0.5 dB (3σ)
Pixel size (meter)	10

Table 2: Specification for NISAR, L & S Band SAR (given in NISAR handbook)

Sr. No	Parameter	Specification		
1	Aircraft type	Beech craft		
2	Aircraft altitude	8.0 km		
3	Ground trace velocity	120 m/sec		
4	Operating frequency	1.25 GHz (L-band) & 3.20 GHz (S-band)		
5	Chirp bandwidth	25 MHz	50 MHz	75 MHz
6	Resolution (m) (Azimuth × Slant range)	2 × 6	2 × 3	2 × 2
7	Antenna configuration	Co-Located Antenna		
8	Antenna polarization	Linear Dual		

(*continued*)

Table 2: (*continued*)

Sr. No	Parameter	Specification
9	Effective antenna dimensions (m)	1.0 (Azimuth) × 0.35 (Elevation)
10	Antenna roll bias	37°, 51° and 64° (Nominal - 37°)
11	Look angle with antenna roll bias	24° to 77°
12	Incidence angle range	24° to 77° (corresponding to antenna roll angle of 37° & 64°)
13	SAR mode	Strip map
14	Polarization modes	Single (HH/VV); Dual (HH+HV/VV+VH); Compact (RH+RV/LH+LV); Full-pol (HH+HV+VH+VV)
15	Interferometric (InSAR) imaging	Not Available
16	Imaging swath (Nominal)	Overlapped Swath (S+L) 5.9 km @ 37°; 10 km @ 51°; 19 km @ 64°
17	Integrated ambiguities	< -20 dB
18	Noise equivalent sigma-0 (NEσ°)	< -20 dB (Threshold); < -25 dB (Desirable)
19	Radiometric resolution	3 dB-Single Look, 1 dB-Multi Look
20	SAR data format/Products	HD5/GeoTiff Format, Level-1 SLC; Level-2 Multi looked Geo-coded

3 Methodology for Estimating of Surface Soil Parameter

The method is explained by the flow chart as shown below in (Fig. 2). First geo-referencing operation is performed on the SAR row images with collected ground control point, after geo-referencing backscattering coefficient has been calculated as per the different algorithm and formula provided by the different SAR image provider agencies ie. (ISRO, ESA). After that speckle noise is removed by Lee-Sigma filtering method using SNAP and ENVI software tools.

After that vegetation correction has been done, for this purpose a semi empirical model called Radiative Transfer Model were applied for minimizing the impact of vegetation on SAR backscattering coefficient from soil surface. As we know that the backscattering coefficient is due to the collective of surface and vegetation canopy. Here we have applied water-cloud model [24] as a Radiative Transfer Model, which is given by the following equations.

$$\sigma^{\circ}_{canopy} = \sigma^{\circ}_{veg} + \gamma^2 \sigma^{\circ}_{soil} \tag{1}$$

$$\sigma^{\circ}_{veg} = AV_1 \cos \vartheta \left(1 - \gamma^2\right) \tag{2}$$

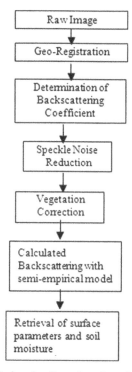

Fig. 2. Estimation flow chart for radar image.

$$\gamma^2 = \exp\left(-2V_2B\big/\cos\vartheta\right) \tag{3}$$

Where γ^2 is the canopy transmitting factor. V_1 and V_2 is canopy descriptors, A and B are coefficient and obtained by the regression analysis [25]. We considered canopy descriptor as LAI for both V1 and V_2. Using the above Eq. (1) we can easily find σ° _soil_ for different crop cover, can be used further calculations. After doing the vegetation correction we have to calculate backscattering coefficient with Oh empirical model which is explained in detail [23]. The basic idea of implementation of this model is explained in the subsequent section. After that we are now a position to retrieve the surface parameter of soil moisture. In literature many researchers have compared the different physical and empirical models, and also introduce some methods ie. Genetic algorithm (GA) [26], neural network (NN) [8] in addition to empirical model. We have observed that Oh model is one of the simplest model which gives good results in different frequency bands as well as he has explain the retrieval model in details [18, 23].

4 Model Evaluation and Analysis

In this work we have used Oh empirical model for finding soil surface parameter. In this section we will discuss the implementation of Oh model and also do the analysis of the model. Oh has done an extensive experiment with polarimetric scatterometer which is given in [18] we are here using that extensive data to simulate the back scattering parameter. Implementation of the model has been done by implementing the following equations given in [23], which are as follow.

$$\sigma_{vh}^{\circ} = 10\log 10\left(0.11(M_V)^{0.7}\left(\cos(!pi/180\vartheta)\right)^{\wedge 2.2}\left[1 - \exp\left(-0.32(ks)^{1.8}\right)\right]\right) \quad (4)$$

$$p = 10\log 10\left(1 - (2!pi/180\vartheta/!pi)^{(0.35(M_V)\wedge(-0.65))}\exp\left(-0.4(ks)^{1.4}\right)\right) \quad (5)$$

$$q = 10\log 10\left(0.95(0.13 + \sin(1.5!pi/180\vartheta))^{1.4}\left(1 - \exp\left(-1.3(ks)^{0.9}\right)\right)\right) \quad (6)$$

where, σ_{vh}° is backscattering coefficient of cross polarization (VH) band, M_v is soil moisture, θ is incident angle, ks roughness parameter, p and q are co-polarization ratio $(\sigma_{hh}^{\circ}/\sigma_{vv}^{\circ})$ and cross-polarization ratio $(\sigma_{vh}^{\circ}/\sigma_{vv}^{\circ})$ respectively.

We have simulated the backscattering coefficient by taking the half of in situ measured data and rest of the data is used to validate the implemented model. The simulations were conducted by taking some initial values with some step size.

M_v is taken from 0.03–0.3 with step size 0.09, ks is initially taken 0.13–6.98 with step size of 0.52 and θ is varied from 10° to 70° with step of 10°. After simulating the result we have analyze the result on some measured value from ground we have found that the result gives the near value as computed from the model. Some of the result as shown in Fig. 3 below.

For the model, M_v is soil moisture, θ is incident angle, ks roughness parameter, p and q are co-polarization ratio $(\sigma_{hh}^{\circ}/\sigma_{vv}^{\circ})$ and cross-polarization ratio $(\sigma_{vh}^{\circ}/\sigma_{vv}^{\circ})$ respectively are obtained from the image and ground measurement are used as known parameter simulate p and q parameters and then the simulated values are compared with the same values from sentinel SAR and AIR born L-band SAR values. The accuracy is estimated by calculating RMSE (root mean square error) and the correlation coefficient (R^2). For the said study we have calculated (RMSE = 1.0235609418 dB for ratio p in L-band and 2.12091834 dB in C-band) we can observed that L band estimation is better than C-band.

The Oh retrieval algorithm is simple to implement, one can interpret that the backscattering coefficient is the function of the soil surface roughness, soil moisture, incidence angle and can be expressed as.

$$\sigma^{\circ} = f(\vartheta, ks, M_v) \quad (7)$$

And the permittivity of soil can be correlate with surface soil moisture is given in [28].

Co-polarized ratio Vs surface roughness

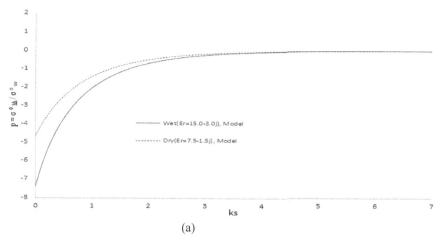

(a)

Cross-Polarizaed ratio Vs Surface roughness

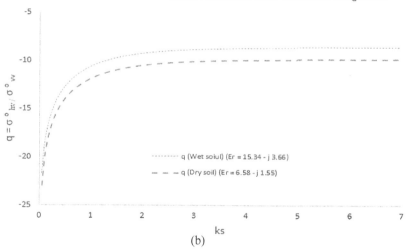

(b)

Fig. 3. Shows the simulated result a) co-polarization verses surface roughness b) cross-polarization verses surface roughness.

$$\varepsilon = (a_0 + a_1 S + a_2 S_i + a_3 C) + (b_0 + b_1 S + b_2 S_i + b_3 C) * (c_0 + c_1 S + c_2 S_i + c_3 C) * n \quad (8)$$

Where a, b, c, are empirical coefficient m is soil surface moisture and S is the percentage soil texture.

Proposed method is applied on L-band and C-band data set. Samples were collected from different field cover (sugarcane, barren land, okra (flowering), lamba village).for June 2017. Observed and estimated value shows good agreement with minimum RMSE value.

5 Conclusions

L-band and C-band SAR images of different area are analyzed for the estimation of surface soil parameter (soil moisture, surface roughness and incidence angle) using Oh empirical model. It is very complex to estimate by other theoretical and physical model, empirical model are simple and give direct relation between backscattering coefficient and other important soil surface parameters. A semi- empirical model based on Oh et al. given in subsequent year of their research (1992, 1994, 2002, and 2004) were used to estimate the surface parameter. Estimated values are in good agreement with the observed values. This estimation can be improved by taking the surface roughness and soil texture with high accuracy. This type of analysis are very use full for landscape, vegetation and other planning management people. This article is to evaluate the empirical model and to show the potential and technique to estimate surface parameters. The result shows that Oh model nearly estimate the parameters, in the different models equations given by Oh in (2004) is more suitable for the estimation in both the band of data.

References

1. Lecomte, V., King, C., Cerdan, O., Baghdadi, N., Bourguignon, A.: Use of remote sensing data as alternative inputs in the stream Runoff model. In: Proceedings of the International Symposium on Physical Measurements & Signatures in Remote Sensing, Aussois, France, 8–12 January, pp. 699–704 (2001)
2. Aubert, M., et al.: Analysis of TerraSAR-X data sensitivity to bare soil moisture, roughness, composition and soil crust. Remote Sens. Environ. **115**, 1801–1810 (2011)
3. El Hajj, M., Baghdadi, N., Zribi, M., Bazzi, H.: Synergic use of Sentinel-1 and Sentinel-2 images for operational soil moisture mapping at high spatial resolution over agricultural areas. Remote Sens. **9**, 1292 (2017)
4. Gao, Q., Zribi, M., Escorihuela, M., Baghdadi, N.: Synergetic use of Sentinel-1 and Sentinel-2 data for soil moisture mapping at 100 m resolution. Sensors **17**, 1966 (2017)
5. Mattia, F., et al.: High resolution soil moisture content from sentinel-1 data. In: Proceedings of the 2017 IEEE International Geoscience and Remote Sensing Symposium (IGARSS), Fort Worth, TX, USA, 23–28 July 2017
6. Paloscia, S., Pettinato, S., Santi, E., Notarnicola, C., Pasolli, L., Reppucci, A.: Soil moisture mapping using Sentinel-1 images: algorithm and preliminary validation. Remote Sens. Environ. **134**, 234–248 (2013)
7. King, C., Baghdadi, N., Lecomte, V., Cerdan, O.: The application of remote-sensing data to monitoring and modelling of soil erosion. CATENA **62**, 79–93 (2005)
8. Mirsoleimani Reza, H., Sahebi, R., Baghdadi, N., Hajj, M.I.: Bare soil surface moisture retrieval from sentinel-1 SAR data based on the calibrated iem and dubois models using neural networks. Sensors **19**, 3209 (2019)
9. Dobson, M.C., Ulaby, F.T.: Active microwave soil moisture research. IEEE Trans. Geosci. Remote Sens. **1**, 23–36 (1986)
10. Sahebi, M., Angles, J.: An inversion method based on multi-angular approaches for estimating bare soil surface parameters from RADARSAT-1. Hydrol. Earth Syst. Sci. **14**, 2355–2366 (2010)
11. Ulaby, F.T., Batlivala, P.P., Dobson, M.C.: Microwave backscatter dependence on surface roughness, soil moisture, and soil texture: Part I-bare soil. IEEE Trans. Geosci. Electron. **16**, 286–295 (1978)

12. Panciera, R., Tanase, M.A., Lowell, K., Walker, J.P.: Evaluation of IEM, Dubois, and Oh radar backscatter models using airborne L-band SAR. IEEE Trans. Geosci. Remote Sens. **52**, 4966–4979 (2013)

13. Altese, E., Bolognani, O., Mancini, M., Troch, P.A.: Retrieving soil moisture over bare soil from ERS 1 synthetic aperture radar data: Sensitivity analysis based on a theoretical surface scattering model and field data. Water Resour. Res. **32**, 653–661 (1996)

14. Rahman,M., et al.: Mapping surface roughness and soil moisture using multi-angle radar imagery without ancillary data. Remote Sens. Environ. **112**, 391–402 (2008)

15. Satalino, G., Mattia, F., Davidson, M.W., Le Toan, T., Pasquariello, G., Borgeaud, M.: On current limits of soil moisture retrieval from ERS-SAR data. IEEE Trans. Geosci. Remote Sens. **40**, 2438–2447 (2002)

16. Fung, A.K., Li, Z., Chen, K.-S.: Backscattering from a randomly rough dielectric surface. IEEE Trans. Geosci. Remote Sens. **30**, 356–369 (1992)

17. Baghdadi, N., Holah, N., Zribi, M.: Calibration of the integral equation model for SAR data in C-band and HH and VV polarizations. Int. J. Remote Sens. **27**, 805–816 (2006)

18. Oh,Y., Sarabandi, K., Ulaby, F.T.: ,An empirical model and an inversion technique for radar scattering from bare soil surfaces. IEEE Trans. Geosci. Remote Sens. **30**, 370–380 (1992)

19. Baghdadi, N., Saba, E., Aubert, M., Zribi, M., Baup, F.: Comparison between backscattered TerraSAR signals and simulations from the radar backscattering models IEM, Oh, and Dubois. IEEE Geosci. Remote Sens. Lett. **6**, 1160–1164 (2011)

20. Baghdadi, N., et al.: Semi-empirical calibration of the integral equation model for co-polarized L-band backscattering. Remote Sens. **7**, 13626–13640 (2015)

21. Bertoldi, G., Chiesa, S.D., Notarnicola, C., Pasolli, L., Niedrist, G., Tappeiner, U.: Estimation of soil moisture patterns in mountain grasslands by means of SAR RADARSAT2 images and hydrological modeling. J. Hydrol. **516**, 245–257 (2014)

22. Sahebi, M.R., Angles, J., Bonn, F.A.: Comparison of multi-polarization and multi-angular approaches for estimating bare soil surface roughness from spaceborne radar data. Can. J. Remote Sens. **28**, 641–652 (2002)s

23. Oh, Y.: Quantitative retrieval of soil moisture content and surface roughness from multipolarized radar observations of bare soil surfaces. IEEE Trans. Geosci. Remote Sens. **42**, 596–601 (2004)

24. Attema, E., Ulaby, F.T.: Vegetation model as water cloud. Radio Sci. **13**, 357–364 (1978)

25. Susan, M., Vidal, A., Troufleau, D., Ionue, Y., Mitchell, T.A.: Ku- and C-band SAR for discriminating agricultural crop and soil conditions. IEEE Trans. Geosci. Remote Sens. **36**, 265–272 (1998)

26. Singh, D., Katapaliya, A.: An efficient modeling with GA approach to retrieve soil texture, moisture and roughness from ERS-2 SAR data. PIER **77**, 121–136 (2007)

Greenhouse Microclimate Study for Humidity, Temperature and Soil Moisture Using Agricultural Wireless Sensor Network System

Mangesh M. Kolapkar[1][(✉)] and Shafiyoddin B. Sayyad[2]

[1] Vidya Pratishthan's Arts Science and Commerce College, Baramati, Maharashtra, India
mmkolapkar1999@gmail.com
[2] Milliya Arts, Science and Management College, Beed, Maharashtra, India

Abstract. In present work, there are three wireless sensor nodes developed for the measurement of microclimate at three different locations inside greenhouse by designing a wireless sensing node that can measure various atmospheric parametric conditions like carbon dioxide, Oxygen, Humidity, Temperature and light intensity inside greenhouse as well as Soil moisture contents. The designed and developed nodes were placed in a star topology inside greenhouse at Sangvi village at Baramati tehsil and district Pune in Maharashtra state of India. During experimentation, data was collected at different times and on various dates for proper testing and evaluation.

Keywords: Agricultural Wireless Sensor Network (AG-WSN) node ·
Greenhouse · Humidity · Temperature · Soil moisture · Microcontroller · Sensor

1 Introduction

In order to have an accurate climate control inside greenhouse it is essentially need to measure local area climate at different locations inside the greenhouse. Present system practices large length of wired sensors which not only require more power but also suffers from delayed response, fixed area of location, maintenance, less accuracy. In addition to this such system needs power source continuously. In comparison to this Agricultural Wireless Sensor Network (AG-WSN) node based system have multiple advantages such as wireless communication with very less power consumption, in built dc power supply to provide continuous power to AG-WSN node which consists of a tiny high speed microcontroller with wireless communication facility and highly accurate analog or digital output sensors that are interfaced, compact size so easy to locate and can carry anywhere inside and outside of greenhouse because it can communicate well within the radius of greenhouse and it also has cost advantage because it has very less maintenance.

Inside greenhouse, wherein the crops are grown in controlled environment where physical parameters like temperature and humidity of environment, soil moisture percentage, Oxygen and Carbon dioxide concentrations, light intensity etc. physical parameters are measured continuously and also controlled, which can result into good agricultural produce yield [1]. Greenhouse is practiced to monitor and maintain constant

© Springer Nature Singapore Pte Ltd. 2021
K. C. Santosh and B. Gawali (Eds.): RTIP2R 2020, CCIS 1381, pp. 278–289, 2021.
https://doi.org/10.1007/978-981-16-0493-5_25

required air temperature, humidity, soil moisture for the complete growth of crop grown inside. The sunlight increases air temperature in greenhouse and also of the crop inside greenhouse during day time. During the absence of sunlight plant leaves that are growing in greenhouse affects strongly on greenhouse air temperature. If this temperature difference is not compensated immediately there are more chances in development of harmful fungus and weed on the crop and strongly leads to other types of diseases (Korner and Challa, 2003b: Komer and olst, 2005).

Many researchers have developed their own models of greenhouse designs using mathematical calculations and techniques (Kempkes and Van de Braak, 2000: Wang and Deltour, Zhang et al., 1997, 2002). With development of new technology there are many advance techniques have been implemented for the estimation of required range of microclimatic parameters inside greenhouse (Boonen et al., 2000; Van Pee et al., 1998) [2].The protection from excessive heat and solar radiations can be provided by a film, which can absorb long wave infrared radiations, can be used for the greenhouse plants. For measurement and to fix optimum values of temperature and relative humidity, CO2 concentration, air circulation, ventilation etc. inside greenhouse, an advanced electronic system is highly expected [3]. A wireless embedded system using advanced microcontroller can provide more accurate and reprogrammable as per need type instrument which can measure the physical and Oxygen and Carbon dioxide gas percentage in air and can share these parameters using wireless communication. Therefore, a wireless sensing node (WSN) network system it is designed using pre calibrated analog output sensors and by using sixteen bit advance microcontroller MSP EZ430 RF 2500T. This sophisticated instrumentation system can display wirelessly measured parameters of greenhouse like humidity, temperature as well as oxygen and Carbon dioxide levels in PPM (parts per million) in particular.

2   Origin of the Problem

In the field of environment controlled agricultural engineering, while dealing with plant growth, most sophisticated technology is essential [4]. In case of greenhouse, the various parameters such as intensity of light, temperature, humidity etc. are controlled. On survey, it is found that the temperature and humidity of the greenhouse are controlled manually. In some cases electromechanical systems are also used. Normally, the thermostats and humidistat are used to turn the fans and the pumps ON and OFF to maintain the indoor climate. However, the use of thermostats is not so accurate and reliable to maintain the temperature inside greenhouse. The humidistat can also be used to ON/OFF the foggers and ventilators. However, humidistat is less reliable and exhibits hysteresis in the measured values. In short, the sophisticated instrumentation is rarely found in agriculture based field. To overcome present day problems of environment control, one can go with the electronic sensors and instrumentation. The sensors such as temperature sensors, humidity sensors (capacitive or resistive) etc. having promising characteristics are available. Moreover, the highly precise and reliable measurement instruments such as highly reliable and accurate sensors interfaced wireless embedded microcontroller based system can be used by considering the needs of sophisticated electronic instrumentations to measure the parameters, such as humidity, temperature and other greenhouse related greenhouse physical parameters [5]. Plant growth mainly depends on temperature, humidity, ultra violet filtered light, soil moisture as well as the levels of the carbon dioxide and oxygen in air as microclimate inside greenhouse [6]. It is highly essential to continuously monitor and control these microclimatic parameters same at each and every areas of greenhouse for better plant growth [7]. In greenhouse required microclimatic continuous adjustment can help in increase of agricultural produce [8]. In Old generation greenhouse systems the parameters under measurement had one or two long cabled humidistat and thermostat based measurement system that was located in the central part of greenhouse and at a distance of ten to eleven foots above the ground level to provide the present humidity and temperature information to greenhouse system. This system was not able to provide accurate information about measured parameters due to its limitations and was inefficient to control other areas except local climate effectively. So to measure the soil moisture, humidity and temperature using wireless communication and to increase accuracy of readings an Agricultural Wireless Sensor Network (AG-WSN) multimodal system is necessary to design (Fig. 1).

3 Agricultural Wireless Sensor Network Multimodal System Architecture and Working

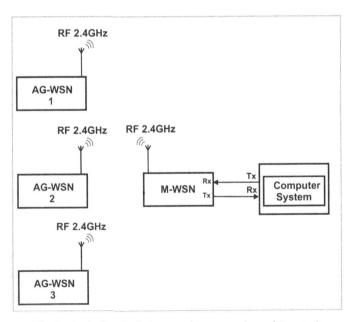

Fig. 1. Agricultural wireless sensing system (complete setup).

The Wireless Sensing Node (WSN) developed here is a smart and advanced microcontroller based embedded system which is used to measure greenhouse related microclimatic parameters such as temperature in degree Celsius (°C), relative humidity in percent

RH and soil moisture in percentage [9]. The said system consists of three numbers of independent local WSN microclimatic data transmitting nodes that communicates to a master receiver single hosting node for collecting the data from these three independent WSN nodes and display it on host computer system (Fig. 2).

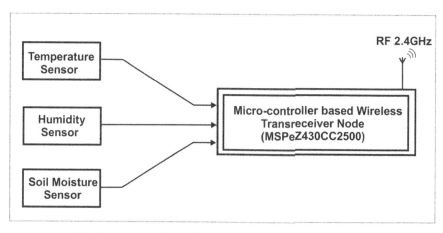

Fig. 2. Agricultural wireless sensing network (AG-WSN) node

All three numbers of independent AG-WSN nodes are placed at three different locations inside the same greenhouse. The greenhouse is built on nearly 1.2 acers of agricultural land. Each node is placed in middle areas of greenhouse but at a distance of 150 m away from each other at different locations. Each node is kept at the distance of 1 foot from the land with soil moisture sensor leads are dipped inside the land at a distance of 5 cm. The first node was placed at starting area while second node was place in middle area and third node was place at end area of same greenhouse. These nodes are powered by single alternating current supply which is converted in to required direct supply voltage for working if individual node and connected sensors to it.

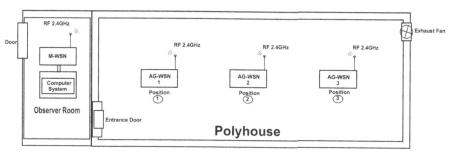

Fig. 3. Placement of complete node setup in polyhouse or greenhouse

Each node senses the local weather parameters around it and transmits these parameters to a master receiver node which collects information for these three nodes placed at different locations inside same greenhouse. Sensor gathers information in analog output form which is converted into it's digital equivalent output with 10 bit analog to digital converter present inside microcontroller and after this data is transmitted to master node 2-FSK (Frequency Shift Keying) technique.

The data collected is used for understanding exact difference in above said microclimatic parameters measured by each independent Wireless Sensing Nodes (WSN) located in different areas inside greenhouse [10]. The collected data provides present local microclimate inside greenhouse at different locations and after comparison with expected values a decision making control system can get activated to adjust required parameter values by providing ventilation or by switching on and off process of mist foggers in required area of greenhouse and in addition to this, the system continuously records in order to create database. It is also important to design and built a user friendly and reliable WSN system for this application with innovation. The system developed here is highly sophisticated, advanced and can measure six microclimatic parameters related to greenhouse in three different sections inside the greenhouse.

Inside the greenhouse, microclimatic parameters data is collected by independently located wireless sensing node and is transmitted to the hosting node which is dedicatedly used for data collection from these independent wireless sensing nodes. The received data is displayed on specially programmed main window of the host computer system. These microclimatic parameters can be used as reference values for data collection can be used for achieving more accurate values for the different crops that can be grown inside greenhouse this attempt will make the wireless embedded system smart and intelligent. The recorded data is collected from different locations inside greenhouse using three independent wireless sensing nodes and displayed, stored and is plotted in simple graphical format. The distance between wireless sensing data transmitter node and the host wireless receiver node has limitations. It is essential to use an internet equipped computer system for sharing this collected information anywhere using internet TCP-IP protocol to overcome the mentioned limitations. As per the proposed work, wireless system is designed and developed. It is an intelligent, robust and user friendly wireless sensor nodal communication system for the measurement of greenhouse related microclimatic parameters.

4 Analog Output Sensor Used for Interface

The analog output based sensors along with their signal conditioning module are used for interfacing with the microcontroller MSPez430 are tabulated as follows (Table 1).

Table 1. Sensor characteristic details

Sensing parameter	Soil moisture	Temperature	Humidity
Sensor and characteristic	YL - 69 (Silver coated)	LM35, Internal calibrated	SYHS220, Internal calibrated
Response type	Uses volumetric method	10 mv/ °C change in output	Provides analog output Voltage
Measurement detail	Surface contact type	−55 °C to +150 °C range of operation	±5% accuracy
Measurement characteristic	Change in resistance to change in moisture	Minimum self-heating effect of 0.08 °C in still air	0° to 60°celsius range of operation
Repeatability	Dependent on soil water resistivity	Approved 0.5 °C accuracy at 25 °C	Highly reliable in in range of 30%–90% RH
Output type	analog and digital output	analog output only	analog and digital output
Supply Voltage (DC type)	+3.3 / +5.0 V	+3.3 / +5.0 V	+5.0 V

5 Material and Methods

To establish wireless communication between master and slave wireless communication nodes it was decided use of MSPez430RF2500T is a 16 bit data processing compact pen drive size complete development tool which is USB based target board embedded system which is similar to Universal Serial Bus (USB) memory device for single sensor node, that includes 16-bit microcontroller along with eight (8) analog- input channels (AN0-AN7) interface with 10-bit or 12-bit Successive Approximation (S/A) type analog to digital converter (ADC). CC2500 microcontroller is used for wireless communication as radio frequency (RF) trans-receiver [11]. At wireless data receiving end the same master trans-receiver and Spy Bi-Wire and MSP 430 application universal asynchronous receiver transmitter (UART) is employed along with the host computer (base station). For establishing wireless communication, special communication protocol software called SensitiviTI [TM] establishes radio frequency (RF) communication between Agricultural Wireless Sensor Network (AG-WSN) nodes and base site. Besides this data is processed and displayed in customize graphical and tabular format by the user computer using front end Java programming language and data received from Agricultural Wireless Sensor Network (AG-WSN) node is saved in notepad file format on daily basis, the data collection is done in real time but data averaging is done and displayed and on computer and saved in file by selection of 1 min, 5 min, 30 min and in an hourly mode. This information is useful for ON/OFF cooling fans and foggers manual or automatic. This information can also be relay via internet (Fig. 4 and Fig. 5).

There are some of the additional features of MSPez430F2274 microcontroller such as 16 bit Ultra Low Power RISC type CPU with 16 MIPS, two 16 bit timers, a watchdog

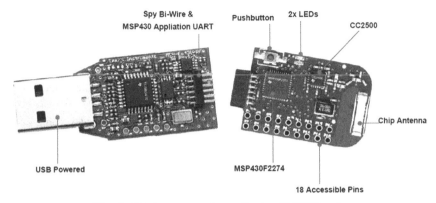

Fig. 4. Wireless master data collection M-WSN node

Fig. 5. Wireless acquisition slave module (MSPeZ430-RF2500)

timer, reset facility (brownout), eleven (11) General Purpose Input and Output pins. MSPez430F2274 microcontroller is interfaced with peripherals using different types of interface such as Universal Asynchronous Receiver and Transmitter (UART) modes, Universal Serial Communication Interface (USCI), Serial Peripheral interface (SPI) and Inter-Integrated Circuit Interference (I2C). MSPez430F2274 microcontroller has built in Flash memory support (32 KB + 256 KB), 1 KB RAM. It can accept eight (8) analog sensor input interface.

The trans-receiver is also associated with radio frequency communication link using CC2500RF. It produces the RF performance. CC2500RF has Low current consumption about 13.3 mA in RX with high sensitivity. During sleep mode it consumes only 0.4 mA current with 240 μs fast startup time during receiving (Rx) and transmitting (Tx) data mode. It has programmable power output up to +1 dBm (decibels relative to one milliwatt). Its data communication frequency range is from 2.4 GHz to 2.4835 GHz, Industrial, Scientific and Medical (ISM) frequency band and Short Range Device (SRD) frequency band [12]. It supports the analog features such as OOK, 2-FSK, GFSK, and

MSK. These features are suitable for frequency hopping and multichannel systems due to a fast settling frequency synthesizer with 90 μs settling time and inbuilt analog temperature sensor. The digital features of CC2500RF are efficient SPI interface that is all registers can be programmed with one "burst" transfer. It has programmable channel filter bandwidth. There are few important consumable applications of CC2500RF such as wireless keyboard and mouse, wireless audio systems, wireless game controllers as well as radio frequency (RF) enabled remote controls in toys.

6 Programing of Access Point and Nodes

The programing of wireless sensing node and access point programming is done by using C language each node communicates and calculates the data input from each sensor connected. The data averaging is done as per user need for decided time form thirty seconds to five minutes and after this time readings from each wireless sensing node collected by master node is displayed in separate window of screen in terms of numeric values. The display of readings on screen is done using Java language. The wireless sensing node programming is done using MSPeZ430-RF2500CC development tool known as Code Compose Studio using sample code and data sheet. The sample program is well understood and new programming is done for each sensor interfaced. All nodes are programmed using C language. In order to establish communication between master node with computer universal serial port a Spy Bi-Wire and MSP 430 application universal asynchronous receiver transmitter (UART) that allows to display all data collection from each independent agricultural wireless sensing network (AG-WSN) node, a specially designed separate window is programmed on screen of personal computer is written in Java language for displaying all the three agricultural wireless sensing network (AG-WSN) node values in detail at instant time. The sensor data backup files on daily basis are stored separately. These values can be used for plotting the graph on daily basis.

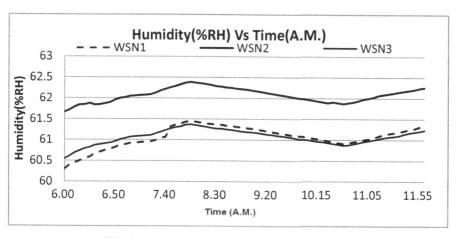

Fig. 6. Humidity (%RH) Versus Time (A.M.) graph.

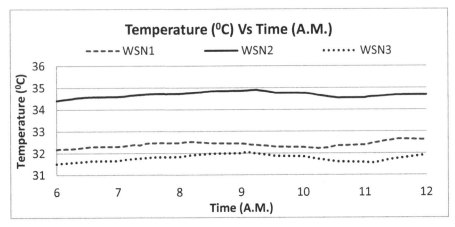

Fig. 7. Temperature (°C) Versus Time (A.M.) graph.

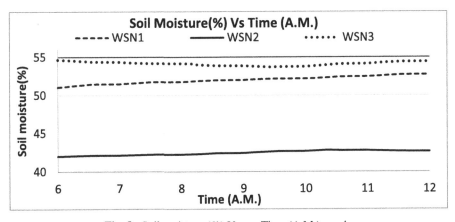

Fig. 8. Soil moisture (%) Versus Time (A.M.) graph.

Figures 6, 7 and 8 represents graphs of humidity, temperature and soil moisture values captured by three independent AG-WSN nodes respectively at three different locations inside same polyhouse respectively.

7 Plant Development Related Environmental Factors

The minimum and maximum values of microclimate inside greenhouse can be set in fully automated greenhouse [13, 14]. In case of manually controlled greenhouse by the close and continuous monitoring of parameters one can on off inlet and exhaust fan for controlling inside temperature. Humidity and soil moisture control can be achieved by turning on and off foggers or drip water system. Required values of temperature and humidity keeps away developments of pathogens and pests and fungus which is dangerous to plant development. The required temperature range is between 22 °C to

32 °C as well as humidity range is between 55% RH to 70% RH also required levels of carbon dioxide is 1000 to 3000 ppm as well as oxygen levels varies in between 20.95% to 21% [15].

8   Comparison with Other Systems

The system used for the measurement of air temperature and humidity values for comparison with leaf temperature and leaf humidity mentioned to find out close relation in between [16]. The system discussed [16] here is similar in terms of microcontroller series of MSP 430 and transceiver wireless communication device CC2420, Chipcon make and for measurement air humidity and air temperature sensor SHT71 and IRtec Rayomatic 10 noncontact infra-red (IR) temperature sensor for measurement of leaf temperature and also Model 237 a plant leaf wetness measurement sensor from Cambell Scientific, Inc. Which is changes it's resistance in accordance with leaf wetness. The system discussed here not only shows large variations between air temperature and humidity values between plant leaf temperature and plant leaf humidity but also confirms large variations in air temperature and humidity inside the same greenhouse at the same time it also confirms that there is no variation inside entire greenhouse recorded by other wireless sensing nodes which seems to be unrealistic and can create ambiguous conclusions.

Whereas present system discussed here does not shows large variations in the measurement of air temperature and air humidity inside same greenhouse and can clearly differentiate between areas of microclimate inside same greenhouse which confirms that system is more accurate.

9   Result and Discussion

It has been observed that there are variations in all the three parameters that are measured at different three locations inside greenhouse as shown in Fig. 3. The central position of AG-WSN 2 shows humidity and temperature values are slightly on higher side than AG-WSN 1 and 3, because it was placed at the center place of greenhouse as well as soil moisture values are completely depends on values of humidity and temperature. When these two values are on their optimum levels soil moisture percentage is always low due to the soil type and soil temperature. Soil type in polyhouse is very important and it is highly porous also incapable of retaining moisture for longer period. Soil moisture is nearly same in all three locations inside greenhouse [17]. The middle or center area of greenhouse is warmer than starting and end areas of greenhouse. Soil moisture is less in middle area of greenhouse. It is wearisome to establish exact relationship between all three parameters under measurement [15]. The continuous increase in temperature and decrease in humidity can increase evapotranspiration rate of plant leaves of crop growing inside and decrease plant moisture [18].

10 Conclusion and Future Scope

It has been concluded that the presented system has proved its importance for the measurement and showed differentiation in greenhouse microclimate in terms of humidity, temperature and soil moisture parameter measured at different three locations inside same greenhouse or polyhouse. Our future work will mainly focus on more data collection and analysis of collected data using sensors array. Economical solar back up power supply will be designed to locate AG-WSN nodes at any point in greenhouse and more nodes will be designed. An automated control will be practiced for making on or off fans and mist foggers located inside greenhouse.

References

1. Saad, R., Mohsin, R.P., Salman, H.K.: The design and analysis of automated climatic control for greenhouse, technology forces. J. Eng. Sci. (2010)
2. Liai, G., Man, C., Juan, T.: A wireless greenhouse monitoring system based on solar energy. Telkomnika 11(9), 5448–5454 (2013)
3. Timmerman, G.J., Kamp, P.G.H.: Computerized environmental control in greenhouses. PTC, The Netherlands, pp. 15–124 (2003)
4. Greenhouse guide. https://www.littlegreenhouse.com/guide.shtml (2008)
5. Mancuso, M., Bustaffa, F.: A wireless sensors network for monitoring environmental variables in a tomato greenhouse. In: 6th IEEE International Workshop on Factory Communication Systems in Torino, Italy (2006)
6. Tarran, J., Torphy, F., Burchett, M.: Use of living pot plants to cleanse indoor air–research review. In: Proceedings of 6th International Conference on indoor Air Quality, Ventilation and Energy Conservation in Buildings, vol. 3, pp. 249–256. Sendai, Japan (2007)
7. Yongxian, S., Chenglong, G., Yuan, F., Juanli, M., Xianjin, Z.Z.: Design of greenhouse control system based on wireless sensor networks and AVR microcontroller. J. Netwk. 6(12) (2011)
8. Liu, H., Meng, Z., Cui, S.: A wireless sensor network prototype for environmental monitoring in greenhouses. In: International Conference on Wireless Communications, Networking and Mobile Computing (WiCom 2007), pp. 2344–2347 (2007)
9. Jeng-Nan, J., Radharamanan, R.: Low cost soil moisture system: a capstone design project. In: International Conference on Intelligent Computation Technology and Automation, pp. 1012–1014 (2010)
10. Dae-Heon, P., Beom-Jin, K., Kyung, R., Chang-Sun, S.: A study on greenhouse automatic control system based on wireless sensor network. In: Wireless Personal Communications, Springer US (2009). https://doi.org/https://doi.org/10.1007/s11277-009-9881-2 1572-834X, 1572-834X
11. Radu, G.B., Vlad, C., Robert G. L.: New Assitive Technology for Communicating with and Telemonitoring Disabled People. In: 20th International Symposium for Desing and Technology in Electronic Packaging (SIITME), IEEE (2014)
12. Gaurav, S., Siddharth, M., Shivprasad, S., Romit, P.: Implementation of a wireless sensor node using programmable SoC and CC2500 RF module. In: 2014 International Conference on Advances in Communication and Computing Technologies (ICACACT 2014) (2014)
13. Malan, D., Thaddeus, R.F., et al.: Codeblue An adhoc sensor network infrastructure for emergency medical care. In: Proceedings of the MobiSys, Workshop on Applications of Mobile Embedded System (2004)

14. Wenbin, H.W., Jianglei, L., Fengqi, G., Jianhui, C.: Research of wireless sensor networks for an intelligent measurement system based in ARM. In: International Conference on Mechatronics and Automation (2011). https://doi.org/10.1109/ICMA.2011.5985809

15. Nikhilesh, M., Dharmadhikari, Y.B.: Designing and applications of PIC microcontroller based greenhouse monitoring and controlling system. Int. J. Electron. Commun. Eng. **8**(2), 107–121 (2015) ISSN 0974–2166

16. Dae-Heon, P., Beom-Jin, K., Kyung-Ryong, C.: A study on greenhouse automatic control system based on wireless sensor network. Wirel. Person. Communi. **56**(1), 117-130 (2011) https://doi.org/https://doi.org/10.1007/s11277-009-9881-2, Print ISSN: 0929-6212, Online ISSN: 1572-834X

17. Baldwin, K.R.: Soil Quality Considerations for Organic Farmers, pp. 1–14. Organic Production, Center for environmental farming systems (2006)

18. Drake, B.G., Salisbury, F.B.: Aftereffects of low and high temperature pretreatment on leaf resistance, transpiration and leaf temperature in Xanthium. Plant Physiol. **50**, 572–575 (1972). https://doi.org/10.1104/pp.50.5.572

Vulnerability Assessment of Climate-Smart Agriculture

Ramdas D. Gore[✉] and Bharti W. Gawali[✉]

Department of Computer Science and Information Technology, Dr. Babasaheb Ambedkar
Marathwada University, Aurangabad 431004, Maharashtra, India
ramdasgore1888@gmail.com, drbhartirokade@gmail.com

Abstract. The aim of this research is to highlight the advantages of Climate-Smart Agriculture and the progress accomplished by implementing information technology to make agriculture intelligent. The research article also includes the circumstances for policy and investment to succeed under climate change in sustainable agricultural growth for food security. It also involves a cropping calendar that differs for males, females, and kids to classify the gender division of labor and access and resource control. Climate information is taken from various tools such as the Department of Meteorology, Satellite Images, WSN, and IoT Tool Kit. Accordingly, tree cultivating, preservation agriculture, minimum tillage, and natural resource management are accumulated under the single umbrella of Climate-Smart Agriculture (CSA). The aim of this paper is to have a glance of relationships between CSA and its application.

Keywords: Remote sensing · GIS · Climate-Smart Agriculture · Climate calendar · Crop calendar · Adaptation · Mitigation · FieldSpec

1 Introduction

The agricultural field is playing a vital role in both the global and national economy. 70% of the Indian population is dependent on agriculture and it also contributes to the GDP of the Indian budget. It offers fundamental livelihood requirements for millions of individuals and helps both developing and advanced nations achieve food safety. The relationship between agribusiness and change in the environment is a subject of growing interest [1]. Under climate change predictions, worldwide agricultural production is anticipated to decline, pretending to pose a threat to worldwide food safety. It is also essential for agricultural contribution for a substantial quantity of worldwide emissions annually, which increases to satisfy greater demand with intensification or extension of manufacturing. Global deforestation attributes agriculture to an estimated 80% [2].

Climate-Smart Agriculture (CSA) may have the ability to deliver triple-win advantages from enhanced adaptation, efficiency, mitigation, and a prospective approach to tackle climate variation and food security issues. CSA includes the use of different atmospheric brilliant farming technologies to produce returns or domesticated animals, which is reduced weight on forests for agricultural use as it is possible to maintain or

© Springer Nature Singapore Pte Ltd. 2021
K. C. Santosh and B. Gawali (Eds.): RTIP2R 2020, CCIS 1381, pp. 290–301, 2021.
https://doi.org/10.1007/978-981-16-0493-5_26

improve profitability, build flexibility to alter the environment and moderate the elevated emanations of the division [3]. Many global agreements and declarations acknowledge agriculture as a significant concern for climate change independence, such as sustainability and biodiversity conservation, natural resources, wildlife habitat and financial development [4]. The global tasks are scheduled to assist national-level operations in enhancing sustainability and reducing the natural impact of horticultural development. It is essential to understand that the effects of climate change on agricultural manufacturing could have an adverse effect, and agricultural production is also a significant cause of emissions. Policy strategies are therefore needed to balance the need for both food security and mitigation of climate change from the agricultural sector [5–7].

The global population increases continually to over 9 billion by 2050, interest in nutritional products will increase. Worldwide rural generation may need to be 60–70% created by 2050 to take care of future demand at a comparable age. It's an adverse impact on the atmosphere's horticultural development. The change in the environment could cause crops to decline by as much as half in some extremely vulnerable territories [8].

In its continuing assessment study, the Intergovernmental Panel on Climate Change is observed remarkable improvements in global temperature, examples of precipitation, and frequency of exceptional climate opportunities. Possibly the most complex and natural issue examined by the present situation is environmental change which is increasingly seen as an extreme challenge to agriculture in general and to food health in particular [9]. IPCC describes environmental change as an alteration to the state of the climate which can be recognized (e.g. by means of observable tests) through improvements in say, changeability of its properties, and which persists for an all-inclusive duration, usually decades or longer [10]. Despite the reality that in its case and results environmental change is global, it is creating countries like India that are gradually confronting unfavorable results. Universally, environmental change is seen as deception of market elements where polluters do not have to pay for negative externalities [11].

The following is the organization of this document. The fundamental concept about CSA in Sect. 2. Methods and techniques described in Sect. 3 and Sect. 4 are concluded.

2 Climate-Smart Agriculture (CSA)

CSA is improving infrastructure, policies, institutions and development of investment. It involves on-cultural mediations such as soil fecundation, mulching, cross-cropping, the promotion of improved livestock, combination of dry spell-tolerant crops, and protection of the environment against opportunity, as well as past intercessions such as carbon finance, the development of efficient markets and the improvement of climate estimation. The problem is also all that the mediation process considers [12, 13].

The CSA has three primary pillars enhancing productivity, resilience (adaptation), reducing/removing GHGs, enhancing national food safety and growth objectives. The challenge in agribusiness is to create the homestead harvests and to transport the item to the end client at the most optimal price. Over the last few decades, monitoring the ecological source plays a crucial role. Agriculture has played a crucial role in most nation's economy [14]. These factors are physically estimated by an individual by a customary methodology and these manual estimates are checked every single day. Remote sensing

Table 1. Development of agriculture tools/instruments

Year	Agriculture tools/instuments
Before 1900	
1901-1919	
1920-1939	
1940-1959	
1960-1979	
1980-1999	
2000-2020	

and GIS, IoT and Cloud techniques are used in CSA to monitor soil fertility, soil quality, temperature, humidity to improve horticultural yield [15]. The combination of IoT and cloud techniques has advanced farming improvement and influenced them to recognize

Smart Agriculture and a successful approach to addressing the issue of horticulture (Fruit crops, Vegetable crops, Tuber crops, Plantation and Spices crops, etc.).

The tractor is the first technology in agriculture. Today we are using very sophisticated agricultural technology such as Remote Sensing, GIS, IoT, Artificial Intelligent (AI) and many more techniques. It is used to save time, improve productivity, improve crop quality, assist decision-making, and use the alternative solution in a critical situation. It shows in Table 1 [16, 17].

Cultivating is the politically fragile region for a few reasons, including the general people's prerequisite for the country district to provide an appropriate supply resisting increasing interest. Procedures aimed at alleviating discharges from rural areas but declining subsistence safety is not politically popular in open doors either nationally or globally. Since the livelihoods and employment of a large number of rural poor people rely on agriculture.

Meanwhile, as urban populations depend on the expected supply of rural goods on neighborhood food markets to meet their nutritional requirements [18–20].

3   Methods and Techniques for CSA

The continuing advances in the field of Topographic Data Framework and Remote Detection have opened up countless real applications in distinct areas. Using these advances, processes can be linked to different areas [21]. National farming creation, considered in a practical structure, relies on the effective utilization of normal assets such as soil, water, creature assets, hereditary assets editing/planting, and so on. So as to achieve a financially stable society with an ecologically friendly development and competent use of distinctive resources, It is vital that a far-reaching data framework should be produced that provides methodological and occasional data to organizers, leaders and training offices [22, 23].

The soil is a very important part of the environment, wildlife, human well-being, and so on. Appropriate land use in agriculture needs a strong perception of soil (soil properties) characteristics. An assessment of the soil quality used by different types of hyperspectral tools shows in Table 2 [24]. Table 3 and Table 4 shows the International and National level CSA methods, and smartness.

Table 2. Hyperspectral tools.

Sr. no.	Tools	Range
1	Vis/NIR spectrophotometer	200–1050 nm
2	FieldSpec Pro FR	325–1075 nm
3	FieldSpec Pro, FieldSpec TM FR Spectroradiometer, Field Spectroradiometer, FieldSpec Pro FR spectrometer, Spectroradiometer, ISCO S.R. Spectroradiometer, IRIS sensor and Lansat-TM	350–2500 nm
4	FT-NIR Spectrometer	400–2500 nm
5	FieldSpec 3 Spectroradiometer	700–2500 nm
6	MIR spectrum	400–4000 cm^{-1}
7	AvaSpec spectrometer	24 channels

Table 3. International level CSA

Methods	Smartness	Sites
Cropping calendar, Historical calendar, Climate calendar, Organizational mapping, Farmer interviews	Carbon, Water, Nitrogen, Energy, Knowledge, Weather	Acholi sub-region of Uganda, Tanzania's Southern Corridor
Soil quality	Carbon, Nitrogen	United States
Biodiversity, Questionnaire, Crop response, Cost structure	Carbon, water, soil	Gautemala, America
Climate data, Census data, Qualitativedata	Crop (Rice, Tea, Peanut, Cassava, Acacia)	Ky Anh dist, Hanoi, Vietnam
Multi-Criteria Decision Making (MCDM) CSA Practices, Questionnaires, ArcGIS, Biophysical, Social, Economic	Satellite	Kenya
ArcGIS, WSN, IoT, Zigbee, Robot	Satellite, Drone	Malaysia
Questionnaire, Census data, PRA, RRA, Biophysical, Social, Economic	Composting, Mulching, Intercropping, AnimalFeeding, Drought Tolerant Crop	Kenya, Sub-Saharan Africa
CLIMSAVE IA	Climate scenarios, Scio-economic scenarios (literacy, gender)	Europe
Trade-off Analysis for Multi- Dimensional Impact Assessment (TOA-MD) model	Local grass, Napier grass, Maize, sunflower, cotton (local cows), farm size	Lushoto distric of Tanzania

3.1 FieldSpec Spectroradiometer

The spectroradiometer scanned the soil samples and collected a speed of 0.1 s per spectral scan with a wavelength of 350–2500 nm using FieldSpec, FieldSpec Pro and FieldSpec FR Spectroradiometer are collected a total of 2151 data points per reading [25]. It has software from third parties like RS3, View-Spec Pro, Indico Pro, and software. FieldSpec Spectroradiometer can export the.asd file format into the ASCII code and display the location of the field latitude and longitude [26].

Table 4. National level CSA

Methods	Smartness	Sites
Climate data and scenarios, Adaptation strategies	Nitrogen, Crop (rice, cotton)	Telangana, India
Empirical model,	Rabi crop and Kharif crop	Rajasthan, India
AGRIBOT robot, IoT, PIC microcontroller	Temperature, Humidity	KIT, Krishnan Koit, Tamil Nadu, India
Wireless Sensor Network (WSN), IoT, Zigbee	Temperature, Humidity, Soil Moisture, Rainfall, Water flow	Dr. AIT, Bangalore, India
WSN, IoT, Zigbee	Temperature, Soil Moisture, Rain Detector, Water level	SKN Sinhgad, Pune, India
ArcGIS, NDVI	Water Resources	Latur and CoEP college, India
WSN, IoT,Zigbee, microcontroller, Robot	Temperature, Soil Moisture, Rain Detector, Water level	Pune, India
IoT, Cloud Server	Soil Fertility, WaterContent Temperature, Humidity	REC Thandalam, Chennai, India
Remote Sensing and GIS(Spatial data and Soil survey and Land Use Planning)	Rainfall, Temperature, Soil Texture, Drainage, Flood Hazard	Bihar, India
Remote Sensing and GIS (LISS III image)	Water bodies, Dense Forest, Vegetation, Fallow and Barren Land	India
Remote Sensing and GIS	Agroforestry	India

3.2 Database Collection

Multiple techniques and methods are used for database collection. The database is gathered from Satellite Images, Sensors, IoT Kits, Census Data, Metrological Department, Toposheet, Questionnaires, Farmers' Interviews, Smart work such as Water, Energy, Nutrient, Carbon, Weather, and Knowledge Techniques. Weather data for the development of the Cropping and Climate Calendar are taken from the Metrological Department and Remote Sensing and GIS.

3.3 Remote Sensing and GIS Techniques

Remote sensing and GIS tools and techniques are no longer limited to the areas of computer science, geography, cartography or environmental science, but are now used and implemented successfully in many other sectors such as marketing, accounting, media, property management, business, forest, food services, government agencies, health services, education, hospitals, retail, transportation [27]. Weather events. Nilam Cyclone of 2012, Uttrakhand Landslides and Floods of 2013, Hudhud Cyclone of 2014, these are the latest occurrences in India. Satellite data are taken from the LISS III, LANDSAT 7, EO-1, SPOT, MODIS, AVHRR, C-OPS, NASA's Coastal Zone Color Scanner, SeaWiFS, NASA's Terra, NASA's Aqua, ISRO's Hyperspectral Sensors, Unmanned Aerial Vehicle (UAV) and many more for agriculture area.

3.4 Ground Base Technology

From the field, the data is collected and analyzed. It is collected from FieldSpec Spectroradiometer, Drone, Robots, IoT, WSN, Sensors (Temperature, Humidity, Soil Moisture and Rainfall), GPS, Census data, Questionnaires and Farmers Interview [28].

3.5 Cropping Calendar for Agriculture

Cropping Calendar for Agriculture has two main objectives: 1) Cropping Calendar is widely used to crop classifications, define primary crops, and when related production activities occur. 2) It is used to classify the gender division of labor and how to access and regulate resources that differ for men, women and children. Cropping Calendar for Agriculture is essential to assess (e.g. how labor supplies from different organizations are observed and how gender-based use of time can affect their value) [29, 30].

3.6 Climate Calendar for Agriculture

It is used to classify the features and irregular weather patterns such as Heavy rainfall, flooding, extreme wet and dry years, times of specific strains, liability, and discusses the general impact of weather fluctuations on the agricultural building. The first member discusses a typical year in tiny organizations prepared by gender and classifies the rainy season and summer season [31]. The rainy season is identified with blue paper stripes and red paper summer season. Then participants are asked to determine an abnormally summer season [32]. That year is written down on paper and the classification process is repeated in which months it has rained and which it has not. Finally, for an abnormally dry year, respondents repeat this process. This provides a chance to investigate and comprehend the effects of periodic and dangerous circumstances and how farmers cope with such rainfall variability [33].

3.7 Cropping Season

The period of agriculture is usually begin from June to July. Cropping season is split into two primary seasons depending on monsoons such as the season of Kharif (monsoon) and the season of Rabi (winter). There's a brief season in April-July between the Kharif and Rabi season. It's called a crop called Zaid. These plants are cultivated on irrigated soil and are not waiting for monsoons, crop kinds are pumpkin, cucumber, bitter land, etc. The Kharif season begins in July through October. Cotton, Rice, Corn, Soybeans, Sorghum, Groundnut, Pearl Millet (Bajri), Finger Millet/Ragi, Arhar, etc. [34]. Rabi season is started from October to March. The rabi season crops are included Wheat, Barley, Oats, Chickpea/Gram, Linseed, Mustard, etc.

3.8 Climate Smartness

İt includes:

- **Water-Smart**

Rainwater Harvesting: A source of rainwater that does not allow run-off and is used in rained/dry areas for agricultural purposes [35].

Drip Irrigation: Water is directly applied to the plant's root zone which eliminates the loss of water.

Laser Land Leveling: Ground leveling ensures consistent field water distribution and eliminates water loss.

Furrow Irrigated Bed Planting: This system provides extra productive irrigation and evacuation control and improves fertilizer usage efficiency throughout the Monsoon.

Drainage Management: Disposal by a water control system of excess water (flood).

Cover Crop Method: Reduces the loss of evaporation of soil water and adds nutrients to the soil [36].

• **Energy-Smart**

Zero Tillage/Minimum Tillage: Reduces property scheduling energy usage. It is also improved water infiltration and the preservation of organic matter in the soil over the long term.

• **Nutrient-Smart**

Site-Specific Integrated Nutrient Management: Optimum supply of soil nutrients after a while and region matching crop demands with the correct product, frequency, time and place. Bean cultivation increases nitrogen and nutrient quality in a crop method [29].

Leaf Color Chart: Quantify the quantity of nitrogen used to support the green crop. Mostly used in paddy for split portion execution, but jointly used to detect nitrogen deficiencies in maize and wheat crops. Intercropping pulses: vegetable cultivation in replacement lines or mixed with other fundamental crops. This method increases the consistency of soil and nitrogen [32].

• **Carbon-Smart for Soil**

Agro-forestry: Promote sequestration of carbon including management of land use.

Concentrate Feeding for Livestock: Reduces nutritional shortages and allows a minimum volume of food for domesticated animals.

Fodder Management: Promote sequestration of carbon and the management of land use.

Integrated Pest Management: Reduces the use of chemicals [37].

• **Weather-Smart**

Climate Livestock Housing: Protection of livestock against unforeseen climatic occurrences (heat/cold stresses).

Agro-advisory Crop Based on Climate: Agro-advisories were applied to farmers primarily based on climate data.

Crops Insurance: Crop insurance to offset income loss due to unforeseeable weather conditions [38].

- **Climate and Agriculture Knowledge-Smart**

Conditional Crop Planning: Climate risk management plan during the harvest season for significant weather- related events such as drought, flooding, heat/cold stress.

Enhanced Crop Varieties: Drought-tolerant crop varieties, flooding, and heat/cold stress.

Seed and Fodder Banks: Conservation of plants and seeds for climate risk management [39].

- **Challenges**

CSA faces a number of difficulties linked to conceptual comprehension, exercise, policy environment, and approach funding such as drought, hailstorm, heavy rain, flood, frost, cyclone, and other abiotic stresses that are explained as an impact of change in the climate. Specific challenges considered to be the following:

- Lack of funding, technical expertise or capacity, risk management data, information and suitable analytical instruments at the national and international level [40].
- Discouragement arising from potential short-term yield cuts.
- Insecurity of tenure.
- Customs of society [41].
- Lack of access to new data dissemination techniques or difficulties with them [42].

4   Conclusion

For both the national and global economy, the agricultural field is essential. CSA's core benefits are increased productivity through sophisticated techniques like Remote Sensing, and GIS, IoT, and ICT. There are a number of methods and techniques that affect the extent to which farmers adopt CSA technologies at a specific place. These methods and techniques could be altered and used by stakeholders to increase practices emission growth, and to help stakeholders make choices about the reform of the agriculture system, methods, and techniques. The population continues to grow and is anticipated to double by 2050, the land will be challenged to satisfy its people's food security and nutritional needs while ensuring ongoing economic growth and sustainable livelihoods on the soil where agriculture is the backbone of the economies of many nations. CSA will provide an assertive alternative to agricultural vulnerability issues such as food security, climate risk, soil quality, soil organic matter, soil erosion, productivity, decision-making, greenhouse gas (GHG) reduction and farming methods. Further evaluations are required to investigate techniques that address the constraints of huge information gap and ambiguity about the effect of different Climate Smart Agriculture procedures in particular manufacturing systems.

References

1. Reeta, R., Pushpavathi, V., Sanchana, R., Shanmugapriya, V.: A deterministic approach for smart agriculture using IoT and cloud. Int. J. Pure Appl. Math. **118**(18), 2413–2424 (2018)
2. Gondchawar, N., Kawitkar, S.: Smart agriculture using IoT and WSN based modern technologies. IJIRCCE **4**(6), 12070–12076 (2016)
3. Aruna, G., Lawanya, G., AnbuNivetha, V., Rajalakshmi, R.: Internet Of Things based innovative agriculture automation using AGRIBOT. SSRG Int. J. Electron. Commun. Eng. 163–166 (2017). Special Issue, ISSN 2348-8549
4. Savale, O., Managave, A., Ambekar, D., Sathe, S.: Internet of Things in precision agriculture using wireless sensor networks. Int. J. Adv. Eng. Innov. Technol. (IJAEIT) **2**(3), 1–4 (2015). ISSN 2348-7208
5. Manjunatha chari, S., Sivakumar, B.: Development of smart network using WSN and IoT for precision agriculture monitoring system on cloud. Int. Res. J. Eng. Technol. (IRJET) **04**(05), 1502–1505 (2017)
6. Sethi, P., Sarangi, S.R.: Internet of Things: architectures, protocols, and applications. Hindawi J. Electr. Comput. Eng. **2017**, 1–25 (2017)
7. Li, S., Ji, W., Chen, S., Peng, J., Zhou, Y., Shi, Z.: Potential of VIS-NIR-SWIR spectroscopy from the chines soil spectral library for assessment of nitrogen fertilization rates in the Paddy-Rise Region, China. Remote Sensing **7**, 7029–7043 (2015). Open Access
8. Gore, R.D., Nimbhore, S.S., Gawali, B.W.: Understanding soil spectral signature though RS and GIS Techniques. Int. J. Eng. Res. Gen. Sci. **3**(6), 866–872 (2015)
9. Gore, R.D., Chaudhari, R.H., Gawali, B.W.: Creation of soil spectral library for Marathwada region. Int. J. Adv, Remote Sens. GIS **5**(6), 1787–1794 (2016)
10. Gholizadeh, A., Amin, M.S.M., Boruvka, L., Saberioon, M.M.: Models for estimating the physical properties of paddy soil using visible and near infrared reflectance spectroscopy. J. Appl. Spectrosc. **81**(3), 534–540 (2014). https://doi.org/10.1007/s10812-014-9966-x
11. Todorova, M., Mouazen, A.M., Lange, H., Astanassova, S.: Potential of near-infrared spectroscopy for measurement of heavy metals in soil as affected by calibration set size. Water Air Soil Pollut. **225**(8), 19 (2014). 2036. https://doi.org/10.1007/s11270-014-2036-4
12. Saleh, M., Belal, A.B., Arafat, S.M.: Identification and mapping of some soil types using field spectrometry and spectral mixture analyses: a case study of North Sinai. Egypt. Arab. J. Geosci. **6**(6), 1799–1806 (2011). https://doi.org/10.1007/s12517-011-0501-6
13. Garfagnoli, F., Martelloni, G., Ciampalini, A., Innocenti, L., Moretti, S.: Two GUIs-based analysis tool for spectroradiometer data pre-processing. Earth Sci. Inf. **6**(4), 227–240 (2013). https://doi.org/10.1007/s12145-013-0124-4
14. Khadse, K.: Spectral reflectance characteristics for the soils on baseltic terrain of central Indian plateau. J. Indian Soc. Reomte Sens. **40**(4), 717–724 (2011). https://doi.org/10.1007/s12524-011-0187-y
15. Lei, Z., Yao, M., Liu, M., Li, Q., Mao, H.: Comparison between fertilization N, P, K and No fertilization N, P, K in paddy soil by laser induced breakdown spectroscopy. IEEE Intell. Comput. Technol. Autom. **1**, 363–366 (2011)
16. Yang, H., Kuang, B., Mouazen, A.M.: Affect of different preprocessing methods on principal component analysis for soil classification. In: IEEE, ICMTMA, vol. 1, pp. 355–358, January 2011

17. Kusumo, B.H., Hedley, M.J., Hedley, C.B., Tuohy, M.P.: Measuring carbon dynamics in field soils using soil spectral reflectance: prediction of maize root density, soil organic carbon and nitrogen content. Plant Soil **338**(1), 233–245 (2011). https://doi.org/10.1007/s11104-010-0501-4

18. Dematte, J.A.M., Fiorio, P.R., Araujo, S.R.: Variation of routine soil analysis when compared with hyperspectral narrow band sensing method. Remote Sens. **2**(8), 1998–2016 (2010). Open Access

19. Bilgili, V., van Es, H.M., Akbas, F., Durak, A., Hively, W.D.: Visible/near infrared reflectance spectroscopy for assessment of soil properties in a semi-arid area of Turkey. J. Arid Environ. **74**(2), 229–238 (2010)

20. Bellinaso, H., Dematte, J.A.M., Romeiro, S.A.: Soil spectral library and its use in soil classification. Scielo, Revista Brasileira de Ciencia do Solo **34**(3) (2010)

21. Du, C., Zhou, J.: Evaluation of soil fertility using infrared spectroscopy: a review. Environ. Chem. Lett. **7**(2), 97–113 (2008). https://doi.org/10.1007/s10311-008-0166-x

22. Li, Z., Yu, J., He, Y.: Use of NIR spectroscopy and LS-SVM model for the discrimination of varieties of soil. In: Li, D., Zhao, C. (eds.) CCTA 2008. IAICT, vol. 293, pp. 97–105. Springer, Boston, MA (2009). https://doi.org/10.1007/978-1-4419-0209-2_11

23. Wu, J., Liu, Y., Chen, D., Wang, J., Chai, X.: Quantitative mapping of soil nitrogen content using field spectrometer and hyperspectral remote sensing. In: IEEE, ESIAT, vol. 2, pp. 379–382, July 2009

24. Yunusa, I., Whitley, R., Zeppel, M., Eamus, D.: Simulation of evapotranspiration and vadose zone hydrology using limited soil data: a comparison of four computer models. In: IEEE, ICECS, pp. 152–155, December 2009

25. Xue, L., Li, D., Huang, S., Wu, C.: Spatial variability analysis on soil nitrogen and phosphorus experiment based on geostatisics. In: IEEE, ETTANDGRS, vol. 2, pp. 237–240, December 2008

26. Linker, R.: Soil classification via mid-infrared spectroscopy. In: Li, D. (ed.) CCTA 2007. TIFIP, vol. 259, pp. 1137–1146. Springer, Boston, MA (2008). https://doi.org/10.1007/978-0-387-77253-0_48

27. Rossel, R.A.V., Jeon, Y.S., Odeh, I.O.A., McBratney, A.B.: Using a legacy soil sample to develop a mid-IR spectral library. Soil Res. **46**(1), 1–16 (2008)

28. Brown, D.J.: Using a global VNIR soil-spectral library for local soil characterization and landscape modeling in a 2nd-order Uganda watershed. Siecnce Direct Geoderma **140**(4), 444–453 (2007)

29. Sain, G., et al.: Costs and benefits of climate-smart agriculture: the case of the Dry Corridor in Guatemala. Agric. Syst. **151**, 163–173 (2017)

30. Hochman, Z., et al.: Smallholder farmers managing climate risk in India: 2. Is it climate-smart?. Agric. Syst. **151**, 61–72 (2017)

31. Simelton, E., Dao, T.T., Ngo, A.T., Le, T.T.: Scaling climate-smart agriculture in North-central Vietnam. World J. Agric. Res. **4**(4), 200–211 (2017)

32. Khatri-Chhetri, A., Aggarwal, P.K., Joshi, P.K., Vyas, S.: Farmers' prioritization of climate-smart agriculture (CSA) technologies. Agric. Syst. **151**, 184–191 (2017)

33. Rioux, J., et al.: Planning, implementing and evaluating Climate-Smart Agriculture in Smallholder Farming Systems. In: 11 Mitigation of Climate Change in Agriculture Series, The Experience of the MICCA Pilot Projects in Kenya and the United Republic of Tanzania, Food and Agriculture Organization of the United Nations (FAO) Rome (2016)

34. Rama Rao, C.A., et al.: District level assessment of vulnerability of Indian agriculture to climate change. Curr. Sci. **110**(10), 1939–1946 (2016)

35. Holman, I.P., Brown, C., Janes, V., Sandars, D.: Can we be certain about future land use change in Europe? A multi-scenario, integrated-assessment analysis. Agric. Syst. **151**, 126–135 (2017)

36. Newaj, R., Chavan, S.B., Prasad, R.: Climate-smart agriculture with special reference to agroforestry. Indian J. Agrofor. **17**(1), 96–108 (2015)
37. Malhotra, S.K.: Horticultural crops and climate change: a review. Indian J. Agric. Sci. **87**(1), 12–22 (2017)
38. Brandt, P., Kvakic, M., Butterbach-Bahl, K., Rufino, M.C.: How to target climate-smart agriculture? Concept and application of the consensus-driven decision support framework "target CSA." Agric. Syst. **151**, 234–245 (2017)
39. Shirsath, P.B., Aggarwal, P.K., Thornton, P.K., Dunnett, A.: Prioritizing climate-smart agricultural land use options at a regional scale. Agric. Syst. **151**, 174–183 (2017)
40. Shikuku, K.M., et al.: Prioritizing climate-smart livestock technologies in rural Tanzania: a minimum data approach. Agric. Syst. **151**, 204–216 (2017)
41. Notenbaert, A., Pfeifer, C., Silvestri, S., Herrero, M.: Targeting, out-scaling and prioritising climate-smart interventions in agricultural systems: lessons from applying a generic framework to the livestock sector in sub-Saharan Africa. Agric. Syst. **151**, 153–162 (2017)
42. Mwongera, C., et al.: Climate smart agriculture rapid appraisal (CSA-RA): a tool for prioritizing context-specific climate smart agriculture technologies. Agric. Syst. **151**, 192–203 (2017)

Machine Learning Model Based Expert System for Pig Disease Diagnosis

Khumukcham Robindro$^{(\boxtimes)}$, Ksh. Nilakanta Singh,
and Leishangthem Sashikumar Singh

Department of Computer Science, Manipur University, Imphal 795003, India
rbkh@manipuruniv.ac.in, nilakanta.kakching@gmail.com,
saskulei@manipuruniv.ac.in

Abstract. This paper describes the importance of machine learning (ML) algorithms in the design and development of diagnostic expert systems. The designed system is intended on how ML techniques can be used in the prediction of pig diseases through visible symptoms and their past behaviours. These types of systems are very useful in situations where the domain experts in the field are not readily available. It is important to make an accurate diagnosis of pig diseases to provide control strategies for any symptom shows in a pig to prevent related health issues. The main concerned in the study begins with the collection of disease symptoms and related information of pigs during their life span from the domain experts and literatures in the domain. The acquired knowledge and information are then pre-processed to be used for being trained different ML algorithms. The system development involves the training and analysis of different ML classifier models and their performances. It has thus been found in the study that SVM classifier can produce more accurate results in disease prediction. The paper finally presents the design and development of an expert system for the diagnosis of Pig diseases using SVM classifier which is trained with pig disease dataset.

Keywords: Expert system · ML algorithms · Pig diseases · SVM

1 Introduction

In recent years, there has been an increased recognition that more attention needs to be paid in piggery sector as it could play a supportive role to meet the challenges of economic development and livelihood of farmers and nutritional requirements of ever growing population of the country. It has also been observed that piggery provides a rich source for meat production in India. Pig farming generally offers a tremendous advantage to weaker sections of the society due to its high fecundity, better-feed conversion efficiency, early maturity and short generation interval. It also provides an ideal opportunity for the pig farmers in the rural area to get supplementary incomes and improving their living standards.

© Springer Nature Singapore Pte Ltd. 2021
K. C. Santosh and B. Gawali (Eds.): RTIP2R 2020, CCIS 1381, pp. 302–312, 2021.
https://doi.org/10.1007/978-981-16-0493-5_27

The people in the north east region are widely consuming pork and Pig farming is a traditional and important livelihood resource for the poor and tribal people in this region [1]. At the same time, the farmers are facing many problems in pig rearing. It is thus required to build up the capacity of pig farmers in this region by interventions on knowledge build up, shelter management and veterinary services for health related issues for pigs in this region. Majority of pigs in this region are reared in traditional small scale production systems and providing for the conversion of their waste into fertilizer for agricultural crops [2]. It is thus important to understand the impact of pig farming in improving the economic status of the population of the region.

The main practical problem here in the study is the lesser availability of quality pigs which are free from diseases leading to quality pork. It is very important to detect the infectious diseases in the early stage in Pig for prevention and possible treatment to ensure the production of quality pork to the consumers. The method used in this proposed system is the development of a Disease Prediction Expert System to support in solving these problems. Such types of approaches have successfully been used for the identification of diseases in pig when the Vet-experts are not readily available. It has also been proven the usefulness of these systems effectively with reliable results in far corner areas where the reachability of Vet-experts is limited.

The major limitations in the development procedures of these disease prediction expert systems are the methods used which are primarily based on the techniques developed in rule based knowledge representation techniques. The methods presented here in these kinds of systems exploit prior knowledge about the disease diagnosis from the experts. Moreover, it is very time consuming to deal with the exploited knowledge which has to be represented in the form of rules. In this study, an efficient method for the prediction of probable disease from the past observation of diseases is to be evaluated by observing the use of different machine learning algorithms in predictive systems to overcome these limitations. A Machine Learning Model Based Expert System for Pig Disease Diagnosis is thus presented is this study which is most suitable for the purpose. Preliminary functional experiments were performed to test their performance measures. The dataset in the system development of ths study is subjected to additional processing to handle the various issues of data in the multiclass dataset.

2   Literature Review

There are many different studies on disease prediction expert system in the past which are having the disadvantages of lots of manual works, time consuming, very less learning ability, limitations in complex problems. The present day model based expert systems for disease diagnosis are based on functional reasoning which have the advantages over other diagnosis approaches of more accurate explanations of the result can be provided to the user [3].

"Development of Expert System Program for Pre-Diagnosis of Important Swine Gastrointestinal Diseases in Thailand" [4] is a system developed for swine farmers in decision making on the proper initial treatment for their swine in the absence of Vet-experts. There are 17 diseases, 12 symptom groups, 38 symptoms and 31 gross or macroscopic fecal characters in this system. This system can help swine farmers make their initial diagnosis but some swine diseases must later be confirmed by a scientific laboratory. "Animal Knowledge Base Systems in Egypt" [5] presents the different methodologies use to develop the different KBS in Egypt. The different methodologies used in these systems are: the hierarchical classification generic task, the case based reasoning and the common KADS.

"A web based expert system for pig disease diagnosis" [6] analyses the user's needs and describes the architecture, main components and their functions. In their system, the production rule has been applied for knowledge representation by applying multiple knowledge acquisition approach to obtain the adequate information. Each rule has two sections—a symptom pattern section and an action section, in the form of 'IF symptom pattern A, THEN the disease B'. The system has over 300 rules and 202 images and graphics for different types of diseases and symptoms. It can diagnose 54 types of common diseases of pigs.

"An expert system for diagnosis of cattle disease" [7] is designed to address the loss of productivity from endemic and trans-boundary cattle diseases. The system is developed with hybrid reasoning which is consisting of case based reasoning (CBR) and rule based reasoning (RBR) for the diagnosis of cattle diseases. This study identified a disease by taking the symptoms provided by the user and searched a similar case from CBR. It has also been shown that the system is having a promising result to be used for cattle disease diagnosis in Ethiopia. A disease outbreak prediction model for livestock animals [8] with the application of ML techniques has been developed for controlling the disease and reducing the economic and social impact of the diseases in the society. A disease model is being designed by employing different ML techniques combined with a considerable amount of livestock disease data to provide new opportunities around the animal healthcare sector. Animal disease diagnoses expert system based on SVM [9] is designed to obtain the disease on an animal by mapping the relations between disease symptom and disease category. The model in this system highlights the functioning of one-versus-one (OVO) method, which is having the capability of good classification. In processing of knowledge acquisition, we take the set, S:$(x_1, y_1), (x_2, y_2), \ldots, (x_i, y_i)$, of disease samples provided by experts and users as the training samples of SVM, and $x_i \in R^n, y_i \in$ set of disease pattern (m). The method constructed decision functions among disease categories, which can construct $N = m(m - 1)/2$ decision hyper planes. Each decision function is trained by two categories of corresponding samples. Max-wins-voting strategy is adopted when a disease sample x is classified. It has shown the capability of SVM in disease diagnosis expert system for animal in this research work.

3 Materials and Methods

3.1 Data Description

The pig disease related dataset in this system has total 457 rows as instances and each row showing age, symptoms, diseases and decision to be made for disease prediction. The numbers of feature columns present in the dataset is 89 divided in to 9 groups, of which 6 features are Age related features, 12 features are General Observation symptoms, 20 features are skin changes symptoms, 8 features are Respiration or breathing problems symptoms, 7 features are Digestive change symptoms, 13 features are Behavioural changes symptoms, 9 features are Posture and gait changes symptoms, 6 features are Structure or Conformation changes symptoms, and 8 features are Discharges from body openings symptoms. Moreover, the dataset has also an additional column which is treated as decision label. The decision label is Probable diseases having a total of 65 probable diseases, It has been derived the decision on probable disease from the given 89 features (age related features and different symptom features).

3.2 Data Preparation

Data is very much important in any ML based predictive model because the ability of a model to learn depends on the quality of data and the useful information derived from the data. It is thus very important to pre-process the data before it is being feed into the model [10]. It has been learnt from many studies that there are different factors of a machine learning algorithm to be efficient for a particular task but still the representation and quality of the instance data attributes the success of the algorithm. Data Pre-processing for Supervised Leaning [11] addresses issues of data pre-processing that can have a significant impact on generalization performance of a ML algorithm. In the work presented here, it has been performed the best known algorithms for each step of data pre-processing so that one achieves the best performance for their data set.

There are many types of problems related to data collected from real world which can be solved using data pre-processing. In order to solve these problems, a number of different methods [12] are being used for data pre-processing such as sampling, transformations, denoising, normalization and feature extraction. Feature extraction is a data pre-processing technique which pulls out specified data that is significant in some particular context. The increase of dimensionality of data in this digital era poses a severe challenge to many existing feature selection and feature extraction methods with respect to efficiency and effectiveness [13,14]. It is thus reveals in these studies how feature selection techniques can be used to reduce the dimension of the data and how effectively these techniques can be used to achieve the performance of machine learning algorithms in improving the accuracy of prediction.

The dataset for this system initially has 106 instances of pig diseases and the instances are replicated for individual age groups so that the instances of pre-processed dataset can be seen as the instances individual pigs. It has thus been derived 457 instances of pig diseases after pre-processing the given dataset. In the work presented here, the SMOTE application ML algorithm is applied in the given dataset by encoding the decision label into numerical labels. It has been observed that the decision label in the given dataset is a multiclass and the number of instances of one class exceeds the other and the problem of class imbalance arises in the dataset. Hence the problem is a high dimensional multiclass imbalance problem with class frequencies ranges from 1 to 13 as shown in Fig. 1.

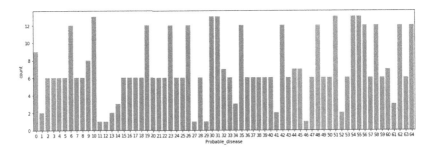

Fig. 1. Frequency count of diseased pig instances

Multiclass classification has been one of the major difficulties in machine learning problems. Random sampling methods such as over sampling and under sampling are being the most widely approaches to deal with multiclass imbalance problems [15]. As the given dataset is multiclass and imbalanced, there are many classes with single observations. Random Over-Sampling is therefore applied to duplicate the samples from minority classes so that a balance class has to be achieved. Synthetic Minority Oversampling Technique (SMOTE) is a standard pre-processing algorithm for the classification of multiclass imbalance problems. SMOTE has inspired several approaches to counter the issue of class imbalance by producing artificial samples from minority classes by interpolating the existing instances which are very close to each other [16]. After performing the necessary steps to achieve a balance class, the dataset is thus ready to apply the different ML algorithms.

3.3   Method Implementation

This system attempts to solve the problem by training and comparing the performances of 9 different ML algorithms. The algorithms are: KNN Classifier, Logistic Regression Classifier, Nearest Centroid Classifier, SVM, Bagging Trees, Random Forest Classifier, Gradient Boosting Classifier, XGBoost Classifier. A predictive model is being trained for each of 9 ML algorithms and tested it to compare the performances. Training and testing ML models on the dataset includes separating 89 features (including age group and symptom groups) and class (decision) label of probable disease. Then the dataset is divided into 90% for training and 10% for testing the algorithm using 10-fold cross validation. The performance metrics of each algorithm is being boosted by tuning the hyperparameter using Grid Search. The average accuracy, precision, recall and F1-score of the models are being measured for comparison of each iteration.

Table 1. ML classifiers and their performance measures

Model name	Average accuracy	Precision	Recall	F1-score	Model parameter
K Nearest Neighbors Classifier	0.96	0.92	0.94	0.93	K = 11
logistic Regression Classifier	0.97	0.96	0.96	0.96	C = 1
Nearest Centroid Classifier	0.95	0.92	0.93	0.92	metric = euclidean
Nearest Centroid Classifier	0.95	0.91	0.92	0.92	No hyperparameter availables
Support Vector Machine	0.98	0.97	0.98	0.97	C = 26, gamma = 0.01, kernel = rbf
Bagging Decision Trees	0.96	0.93	0.94	0.94	n_estimators = 90
Random Forest Classifier (Bagging)	0.97	0.95	0.97	0.96	n_estimators = 1800, min_sample split = 10, max features = squt, max depth = 30
GradientBoosting Classifier	0.96	0.93	0.94	0.94	learning_rate = 0.05, n_estimators = 160, subsample = 0.9, max_features = 13, min_samples_leaf = 11, max_depth = 1, min_samples_split = 14
XGBoost Classifier	0.95	0.91	0.93	0.92	0

3.4   Analysis

The above Table 1 shows the analysis of the performance of the models using the proposed classifiers for predicting probable diseases along with different scores. It is thus derived from the table that the model with Support Vector Machine (SVM) using Grid Search method gives the best performance at complexity level is 26, gamma is 0.01 and radial basis function (rbf) kernel is sigmoid which is shown in Fig. 2. SVM achieved the highest score with Accuracy at 98%, Precision at 97%, Recall at 98% and F1-Score at 97%. The next best performing models are Logistic Regression and Random Forest classifiers with both scoring the final F1-Score at 96%.

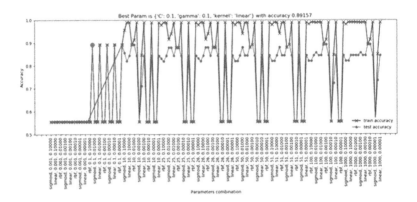

Fig. 2. SVM hyper parameter punning using grid search

4 Design and Development of Expert System

The proposed expert system is a web based application which allows the users to provide the selection of age group, symptoms and related information about the pig diseases. The system then predicts the most possible disease and remedies of the diseases based on information provided by the user. A high level Python based Django Web Framework is being used for the development of the proposed system so as to allow the integration of the ML algorithm code for disease identification which is written in Python language along with its different packages like pandas, numpy and sklearn.

4.1 Architectural Framework

The architectural framework is one of the important parts in the development of a ML based diagnosis expert system. The basic steps for developing the architectural framework in the proposed system involve the four modules and a pre-process database related to pig diseases in the form of csv files. The modules are: User Input, Disease Prediction Engine, Disease Scores, and Disease Result. A comprehensive description of each of the modules and description of database is presented here with the diagram in Fig. 3.

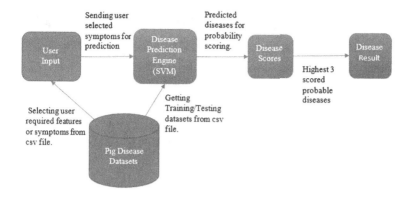

Fig. 3. Architectural framework

The pre-process database stores the pig disease related data in the form of csv files which is required for user selected data and disease prediction. The User Input module process the user selected data from the web interface which is converted in the form vector space where each selected entry is represented by 1 and 0 for others. Once the user selected data is converted in the form of vector space, it is fitted into the Disease Prediction Engine module which is

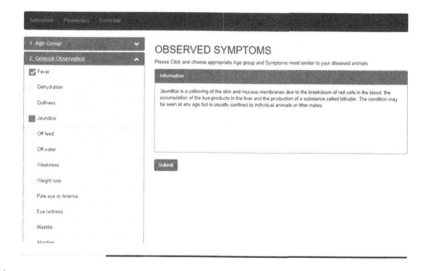

Fig. 4. User input selection page

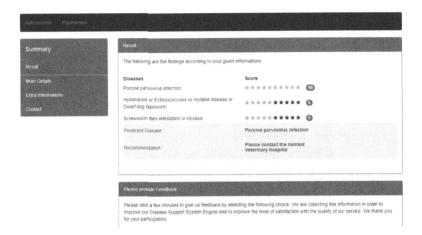

Fig. 5. Result page

implemented using the SVM classifier and is trained using the pig disease related data stored in the form of csv files. This module is of great usefulness so that a list of diseases with their probability is produced from this module. The Disease Score module is responsible for conversion of the disease scores in probability in the form of star rating ranges from 1 to 10 values and three highest score diseases are sent to the Disease Result module. The Disease Result module shows the three highest scored diseases with ratings and the highest score is recommended as predicted disease provided with their remedies.

4.2 Web Interface

The web interfaces of the system for the user are shown in the following images. The left panel of the window in the Fig. 4 is the first page which contains the drop down box with multiple headings to select the age group and groups of symptoms. The user is allowed to select a single age group at one time and multiple symptoms. Once the user selection is completed, the user can submit the details for getting the necessary information about a particular disease based on his/her input selection. The right panel of the window contains the detail and related information of particular age or symptom based on the user selection.

The Result Page shown in the Fig. 5 is the result of the developed system based on SVM classifier. An information box in the above portion of the right panel shows the three highest scored probable diseases with ratings. The highest rating is the probable disease recommended and necessary recommendation for controlling the disease are also provided in this panel. The lower portion of the right panel is for the users to provide their feedback.

5 Discussion and Conclusion

This paper has presented the design and development of ML based Expert System for the diagnosis of Pig diseases. The analysis of different machine learning algorithms in the given dataset is performed in this study and it has been observed that an efficient and accurate expert system for pig disease diagnosis can be developed using SVM classifier on the given dataset. The study has also a room for addressing the multiclass imbalance problems in the dataset and necessary steps to solve these problems. An improved version of ML based expert system will be able to develop by incorporating the user feedback data and more observations in the dataset. It has also been observed that there is still a room for improvement in the Result Page by giving more information about the resulting disease.

References

1. Haldar, A., et al.: Smallholder pig farming for rural livelihoods and food security in North East India. J. Animal Res. **7**, 471 (2017). https://doi.org/10.5958/2277-940X.2017.00070.5
2. Chauhan, A., Patel, B.H.M., Maurya, R., Kumar, S., Shukla, S., Kumar, S.: Pig production system as a source of livelihood in indian scenario: an overview. Int. J. Sci. Environ. ISSN 2278–3687 (O) and Technology, **5**(4), 2089–2096 (2016)
3. Jung-Gen, W., Yu, H.H., William, P.-C.H., David, Y.Y.Y.: A model-based expert system for digital system design. IEEE Des. Test **7**(6), 24–40 (1990). https://doi.org/10.1109/54.64955
4. Farpailin, M., Sujate, C., Neramit, S., Supaporn, T.: Development of expert system program for pre-diagnosis of important swine gastrointestinal diseases in Thailand. Kasetsart J. (Nat. Sci.) **46**, 996–1008 (2012)
5. Maryam, H., Alaa, E., Abdallah, Y.: Animal Knowledge based Systems in Egypt. Int. J. Comput. Appl. **71**(21), 0975–8887 (2013)
6. Zetian, F., Feng, X., Yun, Z., Xiaoshuan, Z.: Pig-vet: a web-based expert system for pig disease diagnosis. Expert Syst. Appl. **29**, 93–103 (2005). https://doi.org/10.1016/j.eswa.2005.01.011
7. Engidu, K.G.A., Taddesse, F.G., Assalif, A.T.: Web based expert system for diagnosis of cattle disease. In: Proceedings of the 10th International Conference on Management of Digital EcoSystems (MEDES 2018). Association for Computing Machinery, New York, NY, USA, 66–73 (2018). https://doi.org/10.1145/3281375.3281400
8. Suresh, K.P., Rashmi Kurli, D., Dheeraj, R., Roy, P.: Application of Artificial Intelligence for livestock disease prediction. Indian Farming **69**(03), 60–62 (2019)
9. Wan, L., Bao, W.: Animal disease diagnoses expert system based on SVM. In: Third IFIP TC 12 International Conference on Computer and Computing Technologies in Agriculture III (CCTA), Beijing, China. pp. 539–545 (2009). https://doi.org/10.1007/978-3-642-12220-0_78.hal-01055402
10. Shrivastava, H., Sridharan, S.: Conception of data preprocessing and partitioning procedure for machine learning algorithm. Int. J. Recent Adv. Eng. Technol. (IJRAET), **1**(3) (ISSN) 2347–2812 (2013)

11. Kotsiantis, S., Kanellopoulos, D., Pintelas, P.: Data preprocessing for supervised learning. Int. J. Comput. Sci. **1**, 111–117 (2006)
12. Malik, J.S., et al.: A Comprehensive Approach Towards Data Preprocessing Techniques & Association Rules (2010)
13. Bachu, V., Anuradha, J.: A review of feature selection and its methods. Cybern. Inf. Technol. **19**, 3 (2019). https://doi.org/10.2478/cait-2019-0001
14. Nasreen, S.: A survey of feature selection and feature extraction techniques in machine learning. In: SAI (2014)
15. Rendón, E., Alejo, R., Castorena, C., Isidro-Ortega, F.J., Granda-Gutiérrez, E.E.: Data sampling methods to deal with the big data multi-class imbalance problem. Appl. Sci. **10**, 1276 (2020)
16. Alberto, F., Salvador, G., Francisco, H., Nitesh, V.C.: SMOTE for learning from imbalanced data. progress and challenges, marking the 15-year anniversary. J. Artif. Int. Res. **61**(1), 863–905 (2018)

Combining Multiple Classifiers Using Hybrid Votes Technique with Leaf Vein Angle, CNN and Gabor Features for Plant Recognition

Pradip Salve[1]($\boxtimes$), Milind Sardesai[2], and Pravin Yannawar[1]

[1] Vision and Intelligence Laboratory, Department of Computer Science and IT,
Dr. Babasaheb Ambedkar Marathwada University, Aurangabad, Maharashtra, India
`pradipslv@gmail.com, pravinyannawar@gmail.com`
[2] Floristic Research Laboratory, Department of Botany,
Savitribai Phule Pune University, Pune, India
`sardesaimm@gmail.com`

Abstract. Modern plant identification system highly depends upon robust features and classification algorithms possibly the combination of several modalities. In this paper, we have used deep CNN features as well as we introduce venation angles present between two veins as a features to classify plants. We have employed 5 types of classifiers to classification and results have been compared using evaluation measure Mean Reciprocal Rank (MRR) and F-measure. *VISLeaf* dataset have been used to perform the experiments, the system achieved the accuracy of 96.67%, 92.76%, 80.00%, 53.89%, 96.67% for SVM, KNN and Naïve Bayes, Tree classifier, Neural Networks, respectively and 97.22% for proposed hybrid votes classifier.

Keywords: Plant classification · Leaf recognition · Vein angle · Multimodal system

1 Introduction

Various kinds of identification systems is considered as the most reliable resource to help the taxonomic people and ecologists to construct precise information of the identity of the plants, the dispersion of the medicinal plants and to study biodiversity, also the evaluation of plant over particular geographic area. Recently, Automatic plant recognition and classification provoked many researchers from computer science, ecologists, and taxonomist. Recent, shows it is very difficult to work system persistently with high accuracy by consequently adding more and more datasets from different source of collection and regions. As we add more plants we may also increase a variations between features. The similarity and dissimilarity between intra-class and inter class may also rise up. One need to add several techniques into act as a single system to maintain the state-of-the-art accuracy.In this paper we proposed 3 types of algorithms first to connect broken vein edge achieved from [1], second algorithm for calculate angles between two veins and third algorithm is for Hybrid voted classifier.

© Springer Nature Singapore Pte Ltd. 2021
K. C. Santosh and B. Gawali (Eds.): RTIP2R 2020, CCIS 1381, pp. 313–331, 2021.
https://doi.org/10.1007/978-981-16-0493-5_28

Various attempts have been proposed by many authors several authors used to employ single modal plant classification systems but [2, 3] and some others investigates the multimodal plant recognition system some of the recent work we described below.

Lee et al. [4] adopted convolutional neural networks (CNN) for feature extraction and Multilayer Perceptron (MLP), Support Vector Machine (SVM), and Radial Basis Function (RBF) employed for classification. The Malaya Kew (MK) dataset was utilized for experiments. Authors have converted RGB color space to HSV color space in order to visualize the leaf venations. The proposed system was able to obtain 99.5% of recognition result using MLP classifier.

Lee et al. [5] introduced Recurrent Neural Network (RNN) approach for plant structural learning. Authors have utilized learned attention maps for visualization of qualitative results. Authors have used 1000 plants species from PlantClef2015 dataset which was divided into 91759 and 21446 for training and testing, respectively. The dataset consist of various plant organs like fruit, branch, flower.

Ali et al. [6] combined texture features viz. Bag-of-features (BOF) and Local Binary Pattern (LBP). And Support Vector Machine used to classify Swedish leaf dataset. They got 99.4% accuracy when SVM used with Gaussian kernel.

Lee et al. [7] designed hybrid feature extraction model to classify plants. Convolutional Neural Networks (CNN) and Deconvolutional Network (DN) approaches have been utilized for feature extraction from raw images. Authors have also used venation features to produce better results.

For the pixel-based and object-based methods Dengfeng Chai et. al. [8] introduced structurally-coherent solution contrary technique. It describes the graphs that represents branches and endings of the line-networks. Authors have used Monte Carlo mechanism to find the junction points built through stochastic geometry. W. H. Rankothge et al. [9] developed Advanced Plant Identification System (APIS) based on client-server based architecture. 2D-fast furrier transform (2D-FFT) was used to deal with images rotational invariant. The system was able to produce 95% recognition rate through neural network classifier.

Rahmadhani et al. [10] proposed leaf vein patterns extraction automatically by structure searching technique. Fourier descriptor was employed shape feature extraction and segmentation was done using thresholding. The extraction of the vein was done by Canny edge detector the output was represented as b-spline.

Zhihui Sun et al. [11] proposed a method for leaf contour and veins extraction from point cloud data, leaf venation structure was extracted by curvature information of point cloud data and contour was retained by mesh algorithm. They have combined leaf vein features and boundary points to retrieve vein contour.

Amlekar et al. [12] utilized gradient information of leaf for vein extraction. After they applied Sobel, Canny, Prewitt and Roberts's operator for finding the shape features. They have employed 24 leaf samples from four kind of plant species.

Ambarwari et al. [13] addressed a biometric traits of the leaf by analyzing density of the leaf structure. Support Vector Machine (SVM) as employed to classify leaf sample based on venation type criteria. Investigation was done on 53 types of plants species including 271 leaves.

Szegedy et al. [14] performed their experiments on Inception-style networks (similar to GoogLeNet architecture). The efficiency of the network was increased by factorized aggressive regularization and convolutions. Various size of filters have been adopted and they studied to maintain the high quality networks by factorizing convolutions and aggressive dimension reductions in neural networks. The high quality networks was trained on different size of train datasets by combining lower parameter count and additional regularization with batch-normalized auxiliary classifiers and label-smoothing.

Sumathi et al. [15] employed Feed Forward Neural Network for plant classification. They have compared the system accuracy by comparing RBF, MLP and CART with their proposed model of Normalized Cubic Spline Feed Forward Neural Network (NCS –FFNN). The 10 fold cross validation was used during classification, the system achieved 94.04% of accuracy.

Goeau et al. [16] employed three types of large raw image datasets namely Trusted Training Set EoL10K, Noisy Training Set Web10K, Pl@ntNet test set for their experiments. Mean Reciprocal Rank (MRR) have been utilized for statistical measures and evaluation of the system. They have overviewed the results of LifeCLEF 2017 plant identification challenge. They concluded CNN as most effective classifier while using the raw and noisy images.

Joly et al. [17] propsed plant recognition system based on Pl@ntCLEF 2015 dataset. The dataset is consist of 1000 species herbs trees and ferns with taxonomic XML file. They have used Convolutional Neural Networks (CNN) for classification MAP for evaluation.

Elhariri et al. [18] fabricated plant classification system based on combination of leaf shape, color, texture, vein features. They have employed linear discriminant analysis (LDA) and random forests (RF) techniques for classification. The experiments were performed on UCI- machine learning repository dataset. The Dataset consisted of 340 leaf samples. The accuracy of 92.65% was observed by LDA and 88.92% by RF classifier.

2 Proposed Approach

In this section, we describe the proposed classification approach Fig. 1 depicts the proposed system architecture, using combination of CNN features, Vein angle features and Gabor features. The classification was carried out using the Support vector Machine (SVM), KNN, Naïve Bayes (NB), Tree Classifier, Neural Networks and our proposed Hybrid votestechnique for combining the output predicted by all rest of the classifiers. The system receives the input leaves followed by pre-processing [1], in successive stage we extract the features using CNN, Vein Angles and Gabor features. We have fabricated the three type of features received from each technique and produced a feature matrix described by [1]. The feature matrix fed to various classifiers for classification and the predicated labels from each classifier passed to the Hybrid votes classifier. We have evaluated the results over predicted labels obtained from each classifier and compared the results in Sect. 4.7. Each section is described briefly in following sections.

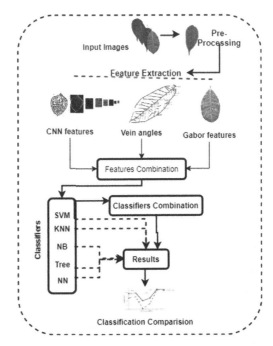

Fig. 1. Proposed system architecture

2.1   Feature Extraction

2.1.1   Deep CNN feature Extraction

We have adopted GoogLeNetimagenet-googlenet-dag [19, 20] to extract CNN feature vectors, which are pre-trained on *VISLeaf* dataset. The RGB color images of the leaves were resized to 224 * 224 of size to fitto the CNN architecture using bilinear interpolation. The images were fed through a stack of convolutional layers, with the size of the filter was set to be 3 * 3 pixels. For linear transformation of the input channels we used 1 * 1 conv. Filter. The information of the leaf images have been fed forward through pre-trained Convolutional Neural Network (CNN). The *softmax layer* have been utilized before transforming the network architecture into domain specific response to produce features from fully connected layers of CNN. The feature vectors then normalized. The final size of feature vector was 1 * 1024 features per sample (Figs. 2 and 3).

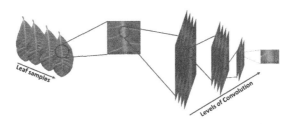

Fig. 2. Deep CNN feature extraction.

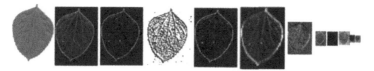

Fig. 3. Some feature maps in various layers of the CNN.

2.1.2 Leaf Vein Angle Calculation

Before calculating the angles from leaf vein structures we fill the pixels to the broken veins. Leaf received from pre-processing stage processed again in this stage due to some leaves lost their vein pixels based on threshold, we re-created the vein for obtaining better vein angles. We proposed Broken Vein Connector algorithm described below.

2.1.2.1. Broken Vein Connection

We propose algorithm for connecting or joining the vein edges, in order to produce the better angles between two veins (Fig. 4).

Algorithm 1 : broken vein connector

Input: Leaf skeleton (LS) binary image
Output: new image with connected broken parts
Begin:
1: *[Jx,Jy]←Find the Junction Points*
2: *[Ex, Ey]←Find the end points*
3: *JP_0_img←LS(Jx, Jy)==0; /* set all Junction Point to zero to isolate each vein from each other*
4: *[TVxTVy]←find the End Points from **JP_0_img***
5: *while()*
6: *if JP_0_img (TVx(i),TVy(i))==1*
7: *tmp_Img(TVx(i),TVy(i)==1)←trace Vein Edges connected to current processing JP*
8: *traced_isolated_veins{i}← tmp_Img /* store traced isolated vein edge*
9: *end if*
10: *if tmp_Img(TVx(i),TVy(i))==(Ex(i),Ey(i))*
11: *break*
12: *end if*
13: *[tmpEPx,tmpEPy]←find the End Points of tmp_Img*
14: *for k←1:length(tmpEPx)*
15: *egdeOri(θ)←atan2(tmpEPx,tmpEPy); /* calculate the orientation of the temporary vein edge*
16: *end for*
17: *new_edge_inserted_img←zeros(size(LP)); /* create empty image for storing new image pixels*
18: *for n←1:length(tmpEPx)*

19: *all_isolated_vein_edges{n}← [traced_isolated_veins(n) ,egdeOri(n)]; /* store x,*
y location of eachvein with its orientation
20: *end for*
21: *veinlength←1; /*pixel length adding with respect to its orientation*
22: *for k←1:length(all_isolated_vein_edges)*
23: *while(Jx(i),Jy(i)~= 1)*
24: *angle← egdeOri(k)+90;*
25: *x1(1)←Jx(i);*
26: *y1(1)←Jy(i);*
27: *x2←x1(1)+ veinlength*cosd(angle);*
28: *y2←y1(1)+ veinlength*sigd(angle);*
29: *new_edge_inserted_img((x1(1),y1(1)),(x2(2),y2(2))←1;/* inserting new pixel to*
the image
30:*veinlength+=1; increase the edge by 1 pixel*
31: *end while*
32: *[Jx(i),Jy(i)]=[]; remove the processed vein Junction point*
33: *end for*
/*find the similar orientation angle veins in path of the current processing vein pixels
if the orientation is similar connect it or change the path based on nearest neighbor
orientation vein (newly connected vein has similar orientation or not)

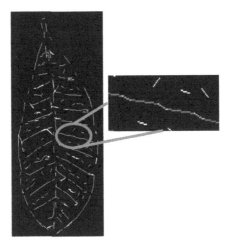

Fig. 4. Brocken gaps of veins filled by proposed algorithm showed in purple color.

We have received raw leaf skeleton images from methods defined in [1]. Further we calculate the angles of the veins. Leaf vein angles can be defined as an angles calculated between two adjacent leaf veins. Figure 5 depict the veins angle calculated. We proposed the algorithm to calculate veins which is defined below:

Algorithm 2: Vein Angle Calculator

Input : leaf venation structure binary image
Output: Angles between veins
Begin:

1:    *[EPxEPy]←find the end points*
2:    *[JPxJPy]←find the junction points*
3:    *For i←1:length(JPx)*
4:    *[JPx',JPy']←[JPx(i),JPy(i)];*   /* *current Processing Junction point*
5:    *[JPx(i),JPy(i)]←[];   /* exclude the current processing junction point*
6:    *newImg(JPx(i),JPy(i))←0; /*isolate all vein edges for calculation of orientation separately for each vein*
7:        *while(1)*
8:        *[EPx'(i),EPy'(i)]←trace_edge([JPx',JPy']) /*find end points and JP on edges connected to current processing JP*
9:            *if(JPx',JPy')==[JPx(i),JPy(i) || [EPx'(i),EPy'(i)] ==(EPx(i),EPy(i))*
10:           *Break;*
11:   *end while*
12:        *For k←1:length(EPx',1) /*run loop until end of new EP found on the (JPx',JPy')*
13:            *y←[EPy'(i)(1);JPy';EPy'(i)(2)]; /vein one*
14:            *x←[EPx'(i)(1);JPx';EPx'(i)(2)]; /vein two*
15:            *E1←atan2(y(1)-y(2),x(1)-x(2));*
16:            *E2←atan2(y(3)-y(2),x(3)-x(2));*
17:            *angle(θ)←(180/π)*0.5*(E1+E2)+90);*
18:   *end for*

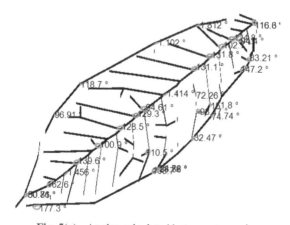

Fig. 5(a). Angles calculated between two veins

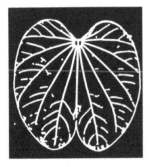 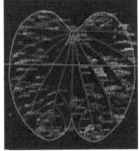 

Fig. 5(b). Leaf vein structure   **Fig. 5(c).** Vein orientation   **Fig. 5(d).** Vein angles
image

2.1.3 Gabor Feature Extraction

We have extracted Gabor features from leaves. The Gabor filters are invariant to scale, translation and rotation and they also robust to image noise. Gray images of leaves were used to extract Gabor features. A two-dimensionalGabor filter is a Gaussian kernel function moduled by complex plane defined by Eq. (1) below:

$$G(x, y) = \frac{f^2}{\pi \gamma \eta} exp\left(-\frac{x'^2 + \gamma^2 y'^2}{2\sigma^2}\right) exp\left(j2\pi fx' + \phi\right) \tag{1}$$

$$x' = x\cos\theta + y\sin\theta \tag{2}$$

$$y' = -x\sin\theta + y\cos\theta \tag{3}$$

Where, f refers to frequency of the sinusoid, θ is the orientation of strips given by a Gabor function, ϕ represents a phase offset, σ is the standard deviation of the Gaussian envelope and γ is the spatial aspect ratio which specifies the ellipticity of the support of the Gabor function [21] (Figs. 6(a), 6(b), 6(c) and 7).

Fig. 6(a). Different Gabor wavelets in 5 scales and 8 orientations.

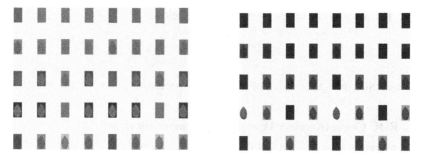

Fig. 6(b). Real parts calculated from Gabor filters.

Fig. 6(c). Magnitudes calculated from Gabor filters.

Fig. 7. Some samples of the Leaf under different orientation Filters calculated by Gabor Filter

Since leaf images are vary in shapes, we did not resized them in fixed squared size due to information loss. We get m * n * 40 size of array of features. We have downsampled the array by factoring it with 16. The downsampled array then normalized using [1].

2.2 Evaluation

Proposed system has been evaluated using Mean Reciprocal Rank (MRR). Basically it was proposed for document retrieval system for search engines. The Reciprocal Rank (RR) information retrieval measure calculates the reciprocal of the rank at which the first relevant query image was retrieved. RR is considered to be 1 if a query image retrieved at the rank 1, if query image is retrieved at rank 2 the RR is conspired to be 0.5. In simple words it is 1(rank first) divide by rank of retrieved image. 1 indicates the accuracy of 100% [22]. Therefore, the evaluated systems have to return a ranked list of possible plant species for each of the 180 test images. The used evaluation metric is the Mean Reciprocal Rank (MRR), a statistic measure for evaluating any process that produces a list of possible responses to a sample of queries ordered by probability of correctness. The reciprocal rank of a query response is the multiplicative inverse of the rank of the first correct class label. The MRR is the average of the reciprocal ranks for the whole test set. The MRR is calculated by Eq. (4) given below.

$$MRR = \frac{1}{|S|} \sum_{i=1}^{S} \frac{1}{rank_{(i)}} \qquad (4)$$

Where |S| is the total number of test samples per class.

2.2.1 ROC Curve (Receiver Operating Characteristics)

The performance visualization of different classification evaluation is calculated by ROC curve. The ROC curve illustrates the trade-off between true positive rate (TPR) and the false positive rate FPR) of the classification models. The AUC (area under the ROC) depicts the accuracy of the system. AUC of 1.0 indicated the 100% accuracy and if model is tend towards the diagonal the model is less accurate. ROC is calculate by the following proposed algorithm.

Algorithm 3: calculate MRR and Plot ROC

Input: predicated class labels (PL), test labels (test_labels)
Output: ROC curve, MRR vector array
Begin:
1: *[q1 q2]=find(~PL);*
3: *mrr=[];*
4: *for i←1:length(test_labels)*
5: *if test_labels(i)==PL(i)*
6: *mrr(i)←1/1; /* if leaf recognized in its true class (at rank 1)*
7: *else*
8:      *mrr(i)←1/PL(i)/* if leaf not recognized its true class divide rank(1) by its recognized class of rank*
9: *end if*
10: *end for*
11: *plot(sort(mrr));*

---

3 Classifiers Employed for the Experiments

3.1 Naïve Bayes Classifier

Naïve Bayes (NB) is the classification algorithm based on Bayes' theorem. The calculation of the class for each sample is based on the multinomial distribution. NB can be used on large datasets. NB used the Maximum A Posteriori decision rule for classification due to it refers to bayes rule. NB calculates the posterior probability P(c|x) from P(c), P(x) and P(x|c) the that can be derived from Eq. (5)

$$P(c|x) = \frac{P(x|c)P(c)}{P(x)} \qquad (5)$$

Where P(c|x) calculates the posterior probability of target class c based on given x data. P(c) calculates the prior probability of class.

P(x|c) represents the similarity which is the probability of predictor class. P(x) represents the prior probability of predictor.

The classification using NB is carried out using the Eq. (6)

$$y = argmax_{ci} P(c_i) \prod_{j=1}^{n} P(x_i|c_i) \tag{6}$$

Where P is the probability, x is the data and c refers to the class.

3.2 SVM Classifier

Initially SVM used for finding a hyperplane in N (number of features)-dimensional space that separate data points as per the class. In this research we have employed the one-vs-all classification strategy for SVM. It trains the single classifier for each class, the selected class treat as a positive labels and rest of the class labels consider as a negative labels and so on until it constructs N different binary classifiers. For the ith classifier, let the positive examples be all the points in class i, and let the negative examples be all the points not in class i. Let fi be the ith classifier. Classify with Eq. (7)

$$f(x) = argmax(i)f_i(x) \tag{7}$$

Where f_i is the i_{th} classifier.

3.3 KNN for Classification

The KNN classifier used distance based measures to classify M-classes into N-labels. it is primitive type of technique it used different type of distance measure viz. Euclidian distance, Hamming distance, City Block distance etc. distance function differs with respect to value of K. based on the distance between Train data and test data it categories the its labels. The highest number of occurrences from K-most identical occurrences are used by KNN for classification. Each instance in passes the votes for certain class and the class with the most votes is taken as the prediction. Following Eq. (8) has been employed for classification purpose.

$$(f_1, f_2) = \sqrt{\sum_d \left(f_1^d - f_2^d\right)^2} \tag{8}$$

Where d is the distance between feature $f1$ and $f2$.

3.4 Decision Tree Classifier

Decision Tree (DT) training model based on learning decision rules deduced from training set. We have employed ID3 (Iterative Dichotomiser 3) algorithm for making DT. DT used hierarchical tree representation for each level of comparison. Each vertex or node relates to an attribute whereas nodes connected to each vertex represents the class labels. DT uses Entropy and Information Gain to build a decision tree. The Entropy is used to find the similarity of the data by DT. Entropy is 1 if compared data is similar; whereas entropy is 0 if data distributed equally in both the classes. Basically 2 types of entropy were used to create DT viz. Entropy by frequency table of one attribute and another is by frequency table of two attributes. The Eq. (9) and Eq. (10) used to calculates the both type of entropies, respectively.

$$E(s) = \sum_{i=1}^{c} -p_i log_2 p_i \tag{9}$$

$$E(T, X) = \sum_{c \in X} P(c)E(c) \tag{10}$$

Where S represents current set of data, c is the set of classes (C = yes|no). P(c) is the proportion of the number of labels in certain class c.

In information gain DT finds the attributes that have highest similarity between two nodes that connected to the upper node. The information gain (IG) is calculated using Eq. (11). IG start with calculated entropy of the target nodes followed by slitting the different attributes. The resulted entropy subtracted from previous node entropy before it split which results in IG or decreased entropy. The node which has higher IG choose as decision node. A edge with 0 entropy consider as a node. Further divination is done on entropy higher than 0. DT make the steps recursively until all data get classified. Finally the classified nodes considered as predicted labels.

$$IG(A, S) = H(S) - \sum_{t \in T} p(t)H(t) \tag{11}$$

Where $H(S)$ is the entropy of feature set S, T is the subset built by splitting set S, $p(t)$ represents the proportion of the number of samples nodes in t to the numbers of nodes in S and $H(t)$ is the entropy of t.

3.5   Artificial Neural Network (ANN) Classifier

In this research we employed a Multilayer Feedforward Neural network with backpropagation algorithm to classify plants. ANN works as a networks of artificial neurons resembles to a human brain. Connection between neurons are used to transmit the signals. the connections are associated with weight multiplied by input net. Whereas the activation function is used to for obtaining the output which is based ion its input. Weight with the single layer connected between inputs and outputs called as Feedforward neural network. In feed forward NN the inputs are passed to its successive layer. Multilayerd NN has set of hidden layers which receives inputs from input layers and output layers fed to the output nodes. In this study our Input vector a is 1 * 1024 of size, output is 60, size of hidden layer was set to be 10.

We fed the input forward by using following Eq. (12)

$$Z - inj = \sum_{i} w_{ij}X_i + b_{oj} \tag{12}$$

Where, w represents the weights. X is input, b is the bias.
The activation function is derived from Eq. (13)

$$Z_j = f(Z - inj) \tag{13}$$

Sigmoid function is calculated from Eq. (14)

$$Y - ink = \sum_j w_{jk}Z_j + b_{ok} \tag{14}$$

Where Yk is the sum of its weighted input signal.
The errors obtained from outputs fed to hidden layer by using Eq. (15)

$$\delta_k = (T_k - Y_k)f(Y - ink) \tag{15}$$

Where δ is error.
The δ input is summed by hidden layers received from above layer using following Eq. (16)

$$\delta - inf = \sum_k \delta_k w_{jk} \tag{16}$$

Error from above δ is calculated by Eq. (17)

$$\delta_j = \delta - injf(Z - inj) \tag{17}$$

Where δ is the error at Z hidden input. Errors are minimized by updating the weight and bias using following Eq. (18) and Eq. (19) respectively.

$$\Delta w_{ij} = a\delta_j X_i \text{ and } \Delta w_{jk} = a\delta_k Z_j \tag{18}$$

Where a represents the learning rate.

$$\Delta b_{oj} = a\delta_j \text{ and } \Delta b_{ok} = a\delta_k \tag{19}$$

Finally, the network gets the goal by either the maximum epochs or minimum gradient obtained. The predicted outputs are stored in array by using its index compared to the test index for calculating the ROC and confusion matrix.

4 Experimental Results and Discussion

In this section we discussed two type of experiments. Firstly we carried out classification separately on each of the classifier viz. SVM, KNN, NB, DT and ANN and we fabricate the new classifier using the predicated results of rest of the classifiers called it as the Hybrid votes classifier. We obtained the confusion matrix from testing labels and predicted labels (classes). The MRR is used for evaluation and finally we compared the results of all the classifiers. The results from each classifier is discussed in the following subsections.

4.1 Naïve Bayes Classifier Results

We received the training and testing data from feature extraction techniques. We passed the data to naïve Bayes classifier with its labels. Before we passed the data we divide data 70:30 ratio for training and testing respectively. The overall accuracy from NB is 80.00%. The Fig. 8(a) depicts the confusion matrix produced from predictions of the labels against the 30% of test labels. The yellow diagonal pixels represents the accurate labels which are correctly classified rest of the pixels which are not diagonally fit represents the incorrect samples. The colorbar placed besides to the confusion matrix shows the accuracy of each class. Figure 8(b) illustrates the ROC curve calculated from MRR evaluation metric.

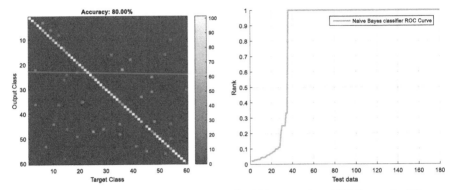

Fig. 8(a). Confusion matrix of NB classifier.   **Fig. 8(b).** MRR based ROC curve by NB classifier.

4.2 SVM Based Classification

We have adopted similar approach for SVM which is described in Sect. 4.1. SVM receives 70% of the data for training and remaining 30% data used as testing. The highest accuracy obtained using SVM is 96.67%. The Fig. 9(a) depicts the confusion matrix. The ROC curve is illustrated in Fig. 9(b). The diagonal yellow pixels symbolizes as a correctly classified data and rest of the pixels signifies the unclassified data.

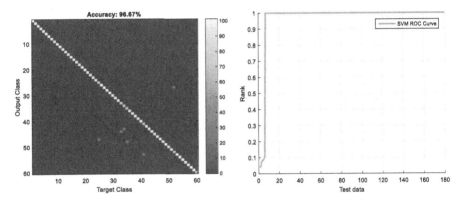

Fig. 9(a). Confusion matrix produced by SVM.   **Fig. 9(b).** MRR based ROC curve for SVM classification.

4.3 KNN Based Classification

KNN accepts the training data and testing data from combined features extracted from CNN, Vein Angles and Gabor feature extractors. The data for KNN is partitioned as exact to the other classifiers i.e. 70:30 ratio. Classifier produces the predicated labels which are used for calculating the evaluation measures confusion matrix and ROC curve. The overall accuracy produced by the KNN is 92.78%. It is observed that 12.78% of increase in the accuracy as compared to the NB classifier by KNN classifier. The confusion matrix and ROC curve are shown in the Fig. 10(a) and 10(b), respectively.

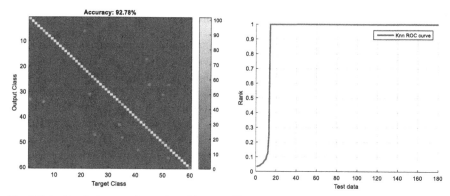

Fig. 10(a). Confusion matrix produced by KNN.

Fig. 10(b). MRR based ROC curve for KNN classification.

4.4 Decision Tree Based Classification

The decision tree accepts the feature matrix from feature extraction techniques described in Sect. 3 and calculates the decision tree in the form of tree structure, output node is a predicted class label. The predicted labels further we stored in a array vector for calculating the confusion matrix and ROC curve as a performance evaluation measure. The DT achieved the 53.89% accuracy on fused features. It is able to produce the lowest accuracy as compared to the other classifiers used in this research work. The confusion matrix and ROC curve are depicted in Fig. 11(a) and 11(b), respectively. The upper triangular and lower triangular pixels corresponds to the incorrect test labels and diagonal elements represents to the correctly recognized samples.

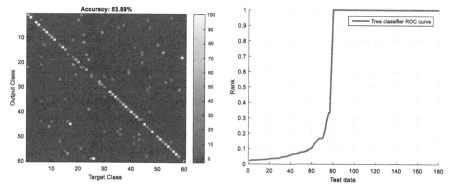

Fig. 11(a). Confusion matrix produced by DT.

Fig. 11(b). MRR based ROC curve for DT classification.

4.5 Artificial Neural Network Based Classification

We have trained the ANN by using Scaled conjugate gradient backpropagation algorithm. We randomly divide data, for training we select 70% of data, similarly we have used 15% of data for Validation and 15% of data for testing the network performance. We set bias as a 2 vectors, the size of bias vectors are first is set to 10 * 1 and second vector is 60 * 1 since the total number of classes are 60 and each class having the 10 leaves. Whereas the Layer weight is 60 * 10 and input weights set to the size of 10 * 1023 by the classifier. The overall accuracy produced by ANN is 96.67%. The confusion matrix is shown in the Fig. 12(a) and the ROC curve is depicted in Fig. 12(b) below. Colorbar indicated the accuracy of each class.

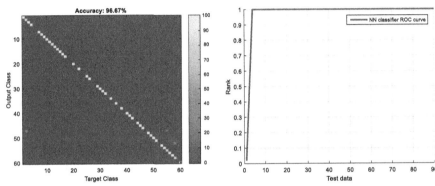

Fig. 12(a). Confusion matrix produced by ANN.

Fig. 12(b). MRR based ROC curve for ANN classification.

4.6 Hybrid Votes Classifier

We propose the Hybrid Votes Classifier for boosting the accuracy of multiple classifiers. The proposed algorithm receives the input from different classifier in the form of vector array of predicted labels. Further it combines the array into one matrix. Each predicted label is compared with actual test labels if both are equal then it assigns the 1 as vote and store it is in the form of vector. Above process repeats until it compares with all the elements of matrix. Finally it generates the similar size of matrix which contains ones and zeros. One is for correctly classified label and zero is for incorrect labels. Each row represents the class of the predicted label. Further sum the each row if the total is greater than or equal to 3 (checking for 5 classifiers) assign new predicted label as equal index to the test label index. The algorithm for Hybrid votes classifier is given below. The accuracy is calculated using predicted labels compared with actual test labels. Proposed classifier produces the accuracy of 97.22% which is significantly increased as compared to other classifiers used in this work. The confusion matrix and the ROC curve is shown in the Fig. 13(a) and 13(b).

Algorithm 4: Hybrid Votes Classifier

Input: Predicted Labels (PL) Produced by SVM, KNN, Tree classifier, NB, NN
Output: Hybrid Predicted Labels (HPL)
Begin
1: *PL←[PKNN, PSVM, PTree, PNN, PNB]*
2: *for i=1:size(PL,1)*
3:   *for j=1: size(PL,2)*
4:     *if (PL(I,j))==test_labels(i)*
5:       *votes←1*
6:     *end if*
7:   *end for*
8: *end for*
9: *for k=1:size(votes,1)*
10: *if sum(votes(k,:))>=3*
11: *HPL←test_labels(k)*
12: *end if*
13: *end for*

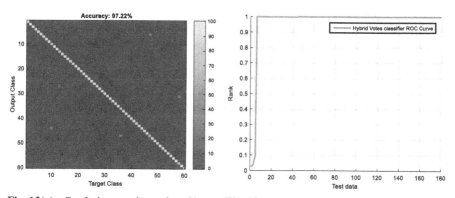

Fig. 13(a). Confusion matrix produced by Hybrid votes classifier.

Fig. 13(b). MRR based ROC curve for Hybrid votes classifier.

4.7 Comparison of All Classifiers

In this section we compare the accuracy produced over all classifiers. Figure 14 depicts the accuracy of each classifier starting from the SVM, NB, KNN, Decision Tree, NN and finally the proposed Hybrid votes classifier along with the F-measure. System generates 96.67% accuracy over the SVM classifier. It is observed that some NB and Decision Tree significantly reduced the accuracy with 80.00% and 53.89% respectively. KNN produced the accuracy of 92.76%. Similarly system obtained the 96.67% of accuracy over the ANN classifier. It is also observed the increase in the accuracy by Hybrid votes classifier with 97.67% which is highest accuracy achieved by any classifier employed in this research work. The weak classifiers can be used to optimize the classification rate by using its predictions.

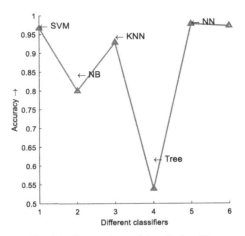

Fig. 14. Comparison of the all classifier

5 Conclusion

In this article we have proposed a technique for plant classification using Hybrid votes classifier. We carried out experiments over five types of classifiers which are significantly used in machine learning and pattern recognition. It is observed that not all classifiers able to achieve the satisfied classification accuracy but some of them produced reliable results. The weak classifiers also can be used to boost the accuracy because weak classifiers such as Decision Tree predicted the correct labels which SVM, ANN fails to classify. It is also observed in increase in the accuracy with 43.33%, 17.22% when we use Hybrid votes classifier as compared with Decision Tree and Naïve Bayes classifiers.

Acknowledgment. This research is financially supported by the Maulana Azad National fellowship, from UGC India; and we gratefully acknowledge the support of dept. of CSIT, Dr. Babasaheb Ambedkar Marathwada University, (MS), India.

References

1. Salve, P., Yannawar, P., Sardesai, M.: Multimodal plant recognition through hybrid feature fusion technique using imaging and non-imaging hyper-spectral data. J. King Saud Univ. - Comput. Inf. Sci. (2018). https://doi.org/10.1016/j.jksuci.2018.09.018. ISSN 1319-1578
2. Salve, P., Sardesai, M., Yannawar, P.: Classification of plants using GIST and LBP score level fusion. In: Thampi, S.M., Marques, O., Krishnan, S., Li, K.-C., Ciuonzo, D., Kolekar, M.H. (eds.) SIRS 2018. CCIS, vol. 968, pp. 15–29. Springer, Singapore (2019). https://doi.org/10. 1007/978-981-13-5758-9_2
3. Salve, P., Sardesai, M., Manza, R., Yannawar, P.: Identification of the plants based on leaf shape descriptors. In: Satapathy, S., Raju, K., Mandal, J., Bhateja, V. (eds.) Proceedings of the Second International Conference on Computer and Communication Technologies. Advances in Intelligent Systems and Computing AISC. IC3T 2015, vol. 379, pp. 85–101. Springer, New Delhi (2016). https://doi.org/10.1007/978-81-322-2517-1_10

4. Lee, S.H., Chan, C.S., Wilkin, P., Remagnino, P.: Deep-plant: plant identification with convolutional neural networks. In: 2015 IEEE International Conference on Image Processing (ICIP), pp. 452–456. IEEE (2015)

5. Lee, S.H., Chang, Y.L., Chan, C.S., Alexis, J., Bonnet, P., Goeau, H.: Plant classification based on gated recurrent unit. In: Bellot, P., Trabelsi, C., Mothe, J., Murtagh, F., Nie, J.Y., Soulier, L., SanJuan, E., Cappellato, L., Ferro, N. (eds.) CLEF 2018. LNCS, vol. 11018, pp. 169–180. Springer, Cham (2018). https://doi.org/10.1007/978-3-319-98932-7_16

6. Ali, R., Hardie, R., Essa, A.: A leaf recognition approach to plant classification using machine learning. In: NAECON 2018-IEEE National Aerospace and Electronics Conference, pp. 431–434. IEEE (2018)

7. Lee, S.H., Chan, C.S., Mayo, S.J., Remagnino, P.: How deep learning extracts and learns leaf features for plant classification. Pattern Recogn. **71**, 1–13 (2017)

8. Chai, D., Forstner, W., Lafarge, F.: Recovering line-networks in images by junction-point processes. In: Proceedings of the IEEE Conference on Computer Vision and Pattern Recognition, pp. 1894–1901 (2013)

9. Rankothge, W.H., et al.: Plant recognition system based on Neural Networks. In: 2013 International Conference on Advances in Technology and Engineering (ICATE), pp. 1–4. IEEE (2013)

10. Rahmadhani, M., Herdiyeni, Y.: Shape and vein extraction on plant leaf images using fourier and B-spline modeling. In: AFITA International Conference, the Quality Information for Competitive Agricultural based Production System and Commerce, pp. 306–310 (2010)

11. Sun, Z., Lu, S., Guo, X., Tian, Y.: Leaf vein and contour extraction from point cloud data. In: 2011 International Conference on Virtual Reality and Visualization (ICVRV), pp. 11–16. IEEE (2011)

12. Amlekar, M.M., Gaikwad, A.T., Manza, R.R., Yannawar, P.L.: Leaf shape extraction for plant classification. In: 2015 International Conference on Pervasive Computing (ICPC) (2015)

13. Ambarwari, A., Herdiyeni, Y., Hermadi, I.: Biometric analysis of leaf venation density based on digital image. Telkomnika **16**(4), 1735–1744 (2018)

14. Szegedy, C., Vanhoucke, V., Ioffe, S., Shlens, J., Wojna, Z.: Rethinking the inception architecture for computer vision. arXiv preprint arXiv:1512.00567 (2015)

15. Sumathi, C.S., Senthil Kumar, A.V.: Plant leaf classification using soft computing techniques. Int. J. Future Comput. Commun. **2**(3), 196 (2013)

16. Goeau, H., Bonnet, P., Joly, A.: Plant identification based on noisy web data: the amazing performance of deep learning (LifeCLEF 2017). In: CLEF 2017-Conference and Labs of the Evaluation Forum, p. 13 (2017)

17. Joly, A., et al.: LifeCLEF 2016: multimedia life species identification challenges. In: Fuhr, N. (ed.) CLEF 2016. LNCS, vol. 9822, pp. 286–310. Springer, Cham (2016). https://doi.org/10.1007/978-3-319-44564-9_26

18. Elhariri, E., El-Bendary, N., Hassanien, A.E.: Plant classification system based on leaf features. In: 2014 9th International Conference on Computer Engineering & Systems (ICCES), pp. 271–276. IEEE (2014)

19. Simonyan, K., Zisserman, A.: Very deep convolutional networks for large-scale image recognition. arXiv preprint arXiv:1409.1556 (2014)

20. Szegedy, C., et al.: Going deeper with convolutions. In: Proceedings of the IEEE Conference on Computer Vision and Pattern Recognition, pp. 1–9 (2015)

21. Haghighat, M., Zonouz, S., Abdel-Mottaleb, M.: CloudID: trustworthy cloud-based and cross-enterprise biometric identification. Expert Syst. Appl. **42**(21), 7905–7916 (2015)

22. Craswell, N.: Mean reciprocal rank. In: Liu, L., Özsu, M. (eds.) Encyclopedia of Database Systems. Springer, New York (2016). https://doi.org/10.1007/978-1-4614-8265-9_488

Hybridizing Convolution Neural Networks to Improve the Accuracy of Plant Leaf Disease Classification

Bhavana Nerkar[1]([⊠]) and Sanjay Talbar[2]

[1] NIELIT Aurangabad, Aurangabad, India
bhavananerkar7@gmail.com
[2] Department of E and TC Engineering, Shri Guru Gobind Singhji Institute of Engineering
and Technology, Nanded, India
sanjaytalbar@yahoo.com

Abstract. Plant leaf disease detection & classification is a complex image processing task, wherein proper algorithms are needed for segmentation, pre-processing, feature extraction and classification. Generally linear classification algorithms like support vector machines. (SVMs), k-nearest neighbour (kNN), Naïve Bayes (NB), Random Forest (RF), etc. do not provide high precision for classification when applied to leaf disease classification. This is due to the fact, that the features which are evaluated during the segmentation and feature extraction phases are do not vary much in terms of values, but they vary in terms of patterns of occurrence. For example, leaf images which are taken for bacterial blight and Alterneria do not show significant changes in feature values, but they show major changes in feature patterns, which is generally neglected by these linear classifiers, and thus the accuracy reduces. In order to improve the accuracy of classification, we propose a hybrid convolutional neural network (CNN) in this paper, which combines multiple methods of segmentation & feature extraction with CNN in order to improve the accuracy of the system. The developed system shows 22% higher accuracy than the existing systems, and can adapt to any type of leaf images by moderate level of training.

Keywords: Plant leaf · Disease · Detection · Classification · Convolutional · Hybrid · Neural · Network

1 Introduction

Detection and classification of plant leaf diseases is a complex image processing problem. It includes the following steps,

- Image capturing with proper scale and angular requirements
- Pre-processing of the image to remove any kind of noise
- Segmentation of the image, to detect and regions of interest
- Feature extraction to describe the image

© Springer Nature Singapore Pte Ltd. 2021
K. C. Santosh and B. Gawali (Eds.): RTIP2R 2020, CCIS 1381, pp. 332–340, 2021.
https://doi.org/10.1007/978-981-16-0493-5_29

- Feature selection, to use only non-redundant features
- Classification using the features
- Post-processing, to identify the affected areas

These steps make sure that any image which is given to the system is properly processed, and the final classification results are sufficiently accurate. Careful designing of each of these processing units is a must in order to achieve high level of accuracy. Starting with image capturing block, if the images are captured from a proper angle, and match the required scale requirements, then the quality of the input image will be high, that will allow the processing layers to get better number of pixels for processing, and thus it will help the overall system performance to be boosted. Once the images are captured properly and they satisfy the minimum criterion, then they are given to a pre-processing unit, where these images are denoised by state-of-the art image denoising algorithms like adaptive median filters, improved hexagonal denoising, etc. Upon denoising, any kind of disturbance in the image due to camera fluctuations, weather noise, etc. is removed, and the image is turned into a processable image for further evaluations.

The segmentation processing unit holds the utmost importance in the case of leaf diseases, due to the fact that this unit will be removing the pixels which are not needed by the system, and will keep only the pixels of interest for processing. Upon segmentation, only the leaf pixels with the diseases must be segmented, and in order to do so, various algorithms including but not limited to k-Means clustering, saliency maps, fuzzy C means, etc. have been proposed by researchers over the years. A correctly segmented image will always have good region of interest extraction with little or no over-segmentation or under-segmentation issues. The segmented image is then given to a feature extraction and selection unit, where the image features like grey-level co-occurrence matrix (GLCM),colour maps, edge maps, etc. are evaluated. These features and maps make sure that the image which has a large size is described using minimal number of values (called as image feature vector), and that each different image has a unique feature vector from each other. Even if the values are similar, the feature patterns must be different from each other. In order to achieve this differentiation, many techniques like variance-based feature selection, deviation-based selection, etc. have been proposed by various researchers.

Upon extraction of unique features and feature patterns, the classification process comes into picture. Using the classification process, the features are differentiated into multiple different classes, and each of these classes represent a different leaf disease type. Various classifiers can be used for this purpose, which include but are not limited to kNN, Naïve Bayes, SVM, Random Forests, Neural Networks, etc. Out of these classifiers, the neural network classifier is supposed to have good level of pattern recognition capabilities, and thus is a very good choice for the classification of leaf diseases. Once these diseases are classified, then the system can apply post processing techniques like thresholding, colour-based pixel counting, etc. in order to evaluate the severity of infection in the leaf. This makes sure that the overall process of classification is completed, and we get some conclusive information on the disease detection process. In this work, we have designed each of these blocks w.r.t. a real time self-captured dataset of leaf images, and the details of the same will follow the section next to the literature review section, which is described next. The paper also showcases the results of the given algorithm, and compares them with standard algorithms like NN & KNN in order to show its superiority of implementation. Following this, we conclude the text with some statistical observations and some further work which can be used to extend the system performance.

2 Literature Review

Over the years many techniques have been presented in order to detect plant leaf diseases. In [1], support vector machines (SVM) has been used, and compared over 5 different kinds of diseases. In our research, we found that SVMs usually have a testing accuracy of 75%, but the paper claims that the accuracy of SVM is around 91%, which might be on the training set itself. A machine learning approach is mentioned in [2], where researchers have utilized random forests (RF) in combination with histogram of gradient (HoG) features for classification. The testing set accuracy of this method is claimed to be 70%, which is a realistic measure, as random forests compare the sets with randomized features, and thus the accuracy is generally around 70 to 80%. Another learning-based approach which uses transfer learning is proposed in [3], where in researchers have developed a simple to use interface for plant leaf disease detection, but no comments are made for the accuracy of the system. They have claimed it to be fairly accurate, but the accuracy level is not mentioned in the text, so it's on the researchers to implement it and check for its performance. A second SVM based approach is mentioned in [4], wherein researchers have developed a method using grey level co-occurrence matrix (GLCM) features. These features are fairly sufficient for differentiation between various kinds of plant diseases. The vague comparison in the paper shows a 90% training level accuracy, which is quite low, as SVM implementation is based on a clustering process, which might not be accurate. Similar Matlab based implementation is done using [5], where researchers are using RF method for classification process. In this method, the researchers have found that the system is fairly accurate for detection of diseases, and also localization of the amount of disease in the given input image. No comment is done on the level of accuracy, but the output images seem to be fairly analysed by the researchers.

A brief comparison of detection algorithms is done in [6], where researchers have compared SVM, RF, k-Nearest Neighbors (kNN), and neural networks. They claim that artificial neural networks have higher level of accuracy as compared to other algorithms, and thus they must be used for real time classification process. Another research in [7], also proves that recurrent neural networks (RNNs), have better classification accuracy, for the plant disease detection process, as compared to other methods like decision tree classifier (DTC), Support Vector Machine (SVM), Naive Bayes Classifier, KNearest Neighbor (KNN), K-means clustering, Genetic algorithm, etc. It also claims that RNN requires less delay as compared to the other methods, and thus should be the method of choice for many applications. Yet another review in [8], mentions that the back propagation neural network (BPNN) has better performance when compared to other methods, which indicates that neural network should be the method of choice for the plant disease detection application. Working on this line, the research mentioned in [9], uses OTSU thresholding combined with neural networks in order to evaluate that the overall accuracy of the neural network is higher than SVMs, there is no mention of any numerical values, but from their analysis and mathematical models, it is quite inherent that neural networks are a better choice for the classification process. While in [10], the researchers have used various methods like kNN, SVM and ANN for classification, but they also claim that neural networks are a good choice for the classification process. Two more reviews published in [11, 12] also prove this point, that neural networks are the algorithm of choice for the process of plant disease classification. A similar review is published in [13], wherein researchers have claimed to compare more than 8 different methods.

3 Hybrid Convolutional Neural Network Design

The block diagram of the proposed system is shown in Fig. 1. The input images are taken from a self-prepared database, which consists of multiple images of different kinds of diseases.

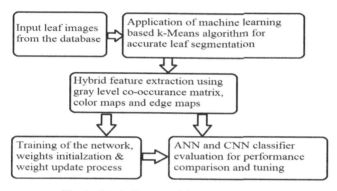

Fig. 1. Block diagram of the proposed system

All the images are given to the machine learning based k-Means algorithm, which would segment the images in order to find out the leaf regions very accurately. The machine learning based k-Means algorithm works using the following steps,

- Arrange all the pixels of the RGB image in a single 1-dimensional array
- Apply bisecting k-Means to the array, using city-block distance metric
- Re-arrange the pixels to get 2 images, one image per cluster
- Analyse the image using the following machine learning step,

  - Find all pixels which satisfy the following equation,

$$green > redANDgreen > blue \ldots \tag{1}$$

  - Find the count of pixels for both the clusters
  - The cluster which has a higher count is considered to be the leaf cluster, and the other one is a non-leaf cluster

- For the leaf cluster, remove all the pixels which do not satisfy Eq. (1)
- The obtained final image is the segmented image with the leaf regions

Once the leaf regions are obtained, then apply grey-level-co-occurrence matrix (GLCM) to the leaf image. Apply the GLCM to each of the 3-dimensions, and combine the features into a single vector array.

GLCM of an image is basically a combination of different gray level statistics, and was formulated by Haralick. It uses 2nd order statistics, to find relations between textures and the overall gray level of the image. In order to develop GLCM, Haralick analyzed that second order probabilities were sufficient for human discrimination of texture. In general, GLCM could be computed as follows. First, an original texture image D is re-arranged in image form (let's say the image name is G) with less number of gray levels, N_g. Usually the value for N_g is 16 or 32. After this, the GLCM statistics are evaluated by evaluating each pixel and its neighbor, defined by displacement d and angle ϕ. A displacement, d could take a value of 1, 2, 3... n whereas an angle, ϕ is limited 0°, 45°, 90° and 135°. The GLCM $P(i,j|d,\phi)$ is a second order joint probability density function P of grey level pairs in the image for each element in co-occurrence matrix by dividing each valueN_g. After performing all these steps, the final gray level co-occurrence matrix is found, and its internal statistical components are evaluated using the following equations,

$$\text{Energy} : \sum_{i,j} P(i,j)^2 \tag{2}$$

$$\text{Entropy:} -\sum_{i,j} P(i,j) \log P(i,j) \tag{3}$$

$$\text{Homogeneity:} \sum_{i,j} \frac{1}{1+(i-j)^2} P(i,j) \tag{4}$$

$$\text{Inertia:} \sum_{i,j} (i-j)^2 P(i,j) \tag{5}$$

$$\text{Correlation:} -\sum_{i,j} \frac{(i-\mu)(j-\mu)}{\sigma^2} P(i,j) \tag{6}$$

$$\text{Shade:} \sum_{i,j} (i+j-2\mu)^3 P(i,j) \tag{7}$$

$$\text{Prominence:} \sum_{i,j} (i+j-2\mu)^4 P(i,j) \tag{8}$$

$$\text{Variance} : \sum_{i,j} (i-\mu)^2 P(i,j) \tag{9}$$

where $\mu = \mu_x = \mu_y = \sum_i i \sum_j P(i,j) = \sum_j j \sum_i P(i,j)$

and $\sigma = \sum_i (i-\mu_x)^2 \sum_j P(i,j) = \sum_j (j-\mu_y)^2 \sum_i P(i,j)$

	1	2	3	4
1	2	3	4	1
2	2	3	3	3
4	3	3	2	1
3	3	3	4	4

	1	2	3	4
1	1	2	0	0
2	1	1	3	0
3	0	1	5	3
4	1	0	1	1

	1	2	3	4
1	0.0625	0.125	0	0
2	0.0625	0.0625	0.1875	0
3	0	0.0625	0.3125	0.1875
4	0.625	0	0.0625	0.0625

(a) Image Matrix   **(b)** Co-occurrence matrix   **(c)** Actual GLCM values

Fig. 2. **(a)** Image Matrix **(b)**. Co-occurrence matrix **(c)**. Actual GLCM values

Figures 2a showcases an input image matrix, while Fig. 2b showcases the evaluated GLCM components (as described previously). These components are given a normalization unit, wherein the final GLCM values are evaluated. These values are showcased in Fig. 2c.

A color map is defined by the variation of color levels (combined gray levels) in the input image these color levels are formed by combining the red, green and blue bands of the image pixels. Due to this combination, a final value of color is obtained, which is plotted on the X axis, and then the number of pixels of that particular value is counted which is plotted on the Y axis. This forms the color map of the image. While the edge map is simply the probability of edge occurrence (edges are found using canny edge operator) on the particular pixel value. These maps are useful in finding out the features of the input image (Fig. 3).

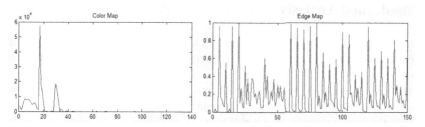

Fig. 3. Colour map and shape map of the image under test

All the 3 features combined together form a perfect descriptor for the leaf image. These features are fully capable of distinguishing between the different types of diseases in the image. Once these features have been evaluated, then the convolutional neural network (CNN) training process is performed for the features and the leaf disease types which are present in the input dataset. To recognize the plant diseases digits, a multi-layered convolutional neural network with one input layer followed by five hidden layers and one output layer is designed and illustrated in Fig. 4

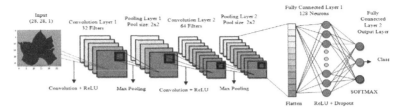

Fig. 4. A multi-layered convolutional neural network for plant disease detection

The CNN is a multi-dimensional tool for classification of any kind of data. In our case, the CNN is used to identify leaf imagery. The CNN starts by taking in the leaf images which are reduced to $28 \times 28 \times N$ in size. Each image goes through 32 convolutional filters where features like color maps, edge maps, texture maps, and GLCM are evaluated. These layers are max-pooled and the features are reduced. These features are then given to the next layer, where the feature extraction is again performed to get a better resulting feature vector. This process is repeated 2 more times, for finally getting the optimum feature vector. Finally, a ReLU layer with Softmax flat layer is used to train the CNN. The output of these layers is the final leaf disease category. Once the training and testing is completed, then the network performance is evaluated for both the CNN and simple ANN networks. The result evaluation and accuracy calculations are mentioned in the next section of this paper.

4 Results and Analysis

We tested the proposed system under various dataset sizes, and evaluated the accuracy for both ANN and CNN methods. The following table demonstrates the statistical result evaluation for ANN,

From Table 1 we can observe when the number of training images are low, then the accuracy of ANN reduces linearly with the number of evaluations for a given number of training records, but as the number of training images increase, we observe a steady increase in the number of correctly classified images, and thus the overall accuracy of the ANN increases. For real-time datasets, the accuracy is between75% to 80%, and it saturates around 80%. In contrast, as we can observe from the Table 2 as follows, the overall accuracy of CNN is much higher than that of ANN.

From Table 2, we can observe that the accuracy of the CNN also reduces linearly for low number of records, but increases steadily for a larger dataset. This is due to the fact that the proposed convolutional neural network performs pattern analysis at each layer, and thus even a small chance of matching with the provided input features is taken into consideration by the network under test.

Table 1. Accuracy analysis for ANN

Number of images in database	Number of images tested	Correct outputs ANN	Accuracy (%) ANN
10	10	9	90.00
20	15	13	86.67
20	20	16	80.00
30	25	20	80.00
30	30	22	73.33
50	35	26	74.29
50	40	30	75.00
50	50	36	72.00
0	60	42	70.00
80	70	49	70.00
80	75	53	70.67
80	80	57	71.25
100	85	61	71.76
100	90	65	72.22
100	95	70	73.68
100	100	75	75.00
125	105	80	76.19
125	110	85	77.27
125	115	90	78.26
125	120	94	78.33
125	125	99	79.20

Table 2. Accuracy comparison of CNN

Number of images in database	Number of images tested	Correct outputs CNN	Accuracy (%) CNN
10	10	10	100
20	15	14	93.33
20	20	18	90.00
30	25	24	96.00
30	30	28	93.33
50	35	33	94.29
50	40	37	92.50
50	50	47	94.00
80	60	56	93.33
80	70	66	94.29
80	75	70	93.33
80	80	75	93.75
100	85	80	94.12
100	90	85	94.44
100	95	90	94.74
100	100	94	94.00
125	105	99	94.29
125	110	103	93.64
125	115	108	93.91
125	120	113	94.17
125	125	118	94.40

5 Conclusion

The proposed hybrid CNN method has better accuracy, and goes up-to 94%, which is good for real-time purposes. Thus, CNN can be used for identification and classification of diseases for real-time datasets, and will always have a better performance than any of the state-of-the art methods. We can observe this fact from the accuracy results of ANN, and from the literature survey, which clearly indicates that neural networks out-perform all other methods for classification of plant diseases.

6 Future Work

Artificial intelligence techniques like deep nets are being utilized for segmentation and classification purposes, researchers can extend this work by adding multiple AI and deep nets based techniques, and observe their effects on the accuracy of the proposed system. The work to find and include control measures on each disease is also needed to get more convenience to farmers and users.

References

1. Pooja, V., Rahul, D.: Identification of plant leaf diseases using image processing techniques. In: IEEE International Conference on Technology Innovation in ICT. For Agriculture and Rural Development, pp. 130–133 (2017)
2. Shima, R., Hebbar, R., Vinod, P.: Plant disease detection using machine learning. In: International Conference on Design Innovations for 3Cs Compute Communicate Control, pp. 41–45 (2018)
3. Muhammad, B.: Design of plant disease detection system: a transfer learning approach work in progress. In: Proceedings of IEEE International Conference on Applied System Innovation, pp. 158–161. IEEE ICASI (2018)
4. Meena, R., Saraswathy, G., Ramalakshmi, G.: Detection of leaf diseases and classification using digital image processing. In: International Conference on Innovation in Information, Embedded and Communication Systems, ICIIECS (2017)
5. Abirami, D., Karunya, S.: Identification of Plant Disease Using Image Processing Technique. In: International Conference on Communication and Signal Processing, 749–753, India (2019)
6. Dhaware, C., Wanjale, K.: A modern approach for plant leaf disease classification which depends on leaf image processing. In: International Conference on Computer Communication and Informatics, Coimbatore, INDIA (2017)
7. Mishra, B., Lambert, M., Nema, S.: Recent technical of leaf disease detection using image processing approach – a review. In: International Conference on Innovations in Information, Embedded and Communication Systems, ICIIECS (2017)
8. Santhosh ,K., Raghavendra,B.: Diseases detection of various plant leaf using image processing techniques: a review. In: 5th International Conference on Advanced Computing & Communication Systems, ICACCS (2019)
9. Sachin, K., Patil, A.: Plant Disease Detection Using Image Processing. Int. Conf. on Computing Communication Control and Automation, pp. 768–771(2015)
10. Chaitali, G., Wanjale, K.:A Modern Approach for Plant Leaf Disease Classification which Depends on Leaf Image Processing. Int. Conf. on Computer Communication and Informatics, Coimbatore .INDIA (2017)
11. Jaskaran, S., Harpreet, K.: A review on: various techniques of plant leaf disease detection. In: Proceedings of the Second International Conference on Inventive Systems and Control, pp. 232–237 (2018)
12. Mishra, B.M., Nema, S.: Recent technologies of leaf disease detection using image processing approach – a review. In: International Conference on Innovations in Information, Embedded and Comm. System, ICIIECS (2017)
13. Vijai, S., Misra, A.: Detection of plant leaf diseases using image segmentation and soft computing techniques. In: Manuscript Information Programming in Agriculture (2016)
14. Shitala, P. Sateesh, K., Debashis, G.: Multi-resolution mobile vision system for plant leaf disease diagnosis. SIViP **10**, 379–388 (2015). https://doi.org/10.1007/s11760-015-0751-y

Signal Processing and Pattern Recognition

Automatic Speech Processing of Marathi Speaker İdentification for Isolated Words System

Pawan Kamble[1(✉)], Anupriya Kamble[1], Ramesh Manza[1], Bharati Gawali[1], Kavita Waghmare[1], Bharatratna P. Gaikwad[2], and Kavita Khobragade[3]

[1] Department of C. S. and I.T., Dr. B. A. M. University, Aurangabad 431004, Maharashtra, India
kamble.pawan@gmail.com, anupriya.k.145@gmail.com,
manzaramesh@gmail.com, drbhartirokade@gmail.com,
kavitawaghmare22@gmail.com
[2] Department of C. S., Vivekanand College, Aurangabad 431004, Maharashtra, India
bharat.gaikwad08@gmail.com
[3] Department of C. S., Fergusson, Fergusson College, Pune, Maharashtra, India
kavitanand@gmail.com

Abstract. In the prevailing period of innovation the automatic identification of speaker assumes a significant role. The application of speaker recognition is spin towards Biometric security. This paper depicted a speaker recognition framework for isolated word dataset. The feature extraction has been finished utilizing Mel Frequency Cepstral Coefficient (MFCC) techniques. The database of the research is structure and creates utilizing 25 Male and 25 Female speakers. The size of dataset is 2500 isolated words. The content for the dataset recording is chosen based on vowel letters in order. The execution of the framework is determined utilizing False Rejection Rate (FRR), False Acceptance Rate (FAR). The precision of the Speaker Recognition rate for Male is better as compare with the exactness of female. This structure is utilized for speaker recognition framework for the confined word distinguishing proof framework by applying highlight extraction methods as MFCC and arrangement is finished with Euclidian Distance. We got a normal exactness for Male rate is 85% and 81% for female. The exhibition of the haphazardly chosen subject gathering was 79%. This is the general precision pace of Speaker Recognition framework for Marathi Isolated Words.

Keywords: Marathi isolated words · Speech processing · Types of speech recognition system · MFCC

1 Introduction

Speaker recognition is the way toward distinguishing an individual based on speech alone. Speech is a speaker subordinate element that empowers us to perceive companions via cell phone. It will continue in the years ahead, it is trusted That recognition of speakers will make it conceivable to confirm the personality of people recovering

© Springer Nature Singapore Pte Ltd. 2021
K. C. Santosh and B. Gawali (Eds.): RTIP2R 2020, CCIS 1381, pp. 343–355, 2021.
https://doi.org/10.1007/978-981-16-0493-5_30

frameworks; permit computerized control of administrations by voice, for example, banking exchanges; and furthermore control the progression of private and secret data. While fingerprints and retinal CAT examines are increasingly dependable methods for distinguishing proof, communicated in language can be viewed as a non-hesitant.

Biometrics which can be assemble both with and without the individual's information or even transmitted over significant distances by means of phone [1, 2]. In contrast to different types of recognizable proof, for example, passwords or keys, an individual's voice can't be taken, overlooked or lost. Speech is a muddled sign delivered as a resultant job insect of a few changes happening at a few distinct levels: semantic, etymological, articulator, and acoustic. Contrasts in these changes show up as a contention in the properties of acoustics speech signal. Verbalize associated contrasts are an aftereffect of a mix of anatomical reference work contrasts inalienable in the vocal bundle and the picked up talking wont of various people. In verbalize recognition, every one of these distinctions can be utilized to segregate between verbalize. Speaker recognition takes into consideration a protected technique acting of verifying speakers. Throughout stage of enlistment, the speaker recognition framework creates a speaker mannequin dependent on the speaker's qualities. Speech is the most ordinarily and broadly utilized types of correspondence between people. There are different communicated in dialects which are utilized all through the world. The correspondence among the person is for the most part done by vocally, in this manner it is normal for individuals to anticipate voice communicate with a PC [3].

Perceiving an individual by his and her voice is identified as a speaker. Improvements of voice identification frameworks have achieved new statures, yet strength and commotion lenient detection frameworks are a pair of the issues which build voice identification frameworks badly designed to utilize. Several explore ventures have been finished and right now within development far and wide for the improvement of hearty voice detection frameworks.

Present are different dialects on the planet that are spoken by people for correspondence [4]. The PC framework which can comprehend the communicated in language can be exceptionally valuable in different zones like farming, social insurance and government segments and so on. Speech recognition alludes to the capacity of tuning in, expressed word and distinguishes an assortment of audio sounds present in it, and remembers them as expressions of some perceived language [5].

Semi fixed signs of speech signals. At the point when speech signals are analyzed over a brief timeframe), its attributes are fixed; be that as it may, for a more extended timeframe the sign qualities transforms; this imitate the distinctive voice pitch being spoken. Features are separated from the speech signals based on speedy time span sufficiency range (phonemes). Highlight extraction is the most significant method in voice identification framework. At this time they are a few issues is looked throughout the element pulling out method in view of the assortment of the spokesman [6]. Sorts of voice identification System is the speech recognition frameworks can be arranged into various kinds relying on various classes. The speech recognition framework can be grouped based on sort of articulations, vocabulary size and speaker reliance.

1.1   Previous Related Work

In past works certain element are removed for grouping speech influence, for example, vitality, pitch, formants frequencies, and so on. These are all prosodic features. These types of features are essential pointer of spokesperson passionate position. Sorts of Speech Recognition System can be arranged into various kinds relying on various classes. The speech recognition framework can be grouped based on kind of articulations, vocabulary size and speaker reliance.

1.2   Categorization Based on Utterances

Isolated Words: Isolated word recognizers for the most part require every expression to have very on different region of the example window. The acknowledges single words or a solitary articulation without a moment's delay. These frameworks have "Tune in/Not-hear states", where they require the speaker to hold up between articulations.

Connected Words: Connected words are allowed space in the structure of word. To be run-at the same time with an insignificant delay among the two.

Continuous Speech: Constant voice recognizers permit clients to talk normally, while the computer chooses the substance. Recognizers with consistent speech limits are some of the most hard to make since they make use of selective methods to decide the expression limits.

Spontaneous Speech: An essential stage, it very well may be consideration as a speech that is characteristic noise as well as no practiced. An ASR framework with unconstrained speech capacity ought to be able to adapt to an a selection of general speech features, as exemplar, phrases being run together, "ahs" and "um", and even slight falters [7].

1.2.1   Categorization Based on Terminology Size

Small Vocabulary: The recognition of speech in frameworks it can perceive partial and specified arrangement of vocabulary are known as restricted vocabulary speech recognition framework.

Medium Vocabulary: The voice identification framework which can perceive a colossal wide assortment of vocabularies (for example Not many from hardly any hundred up to barely any a large number of expressions or sentences) such frameworks are known as medium vocabulary speech recognition framework.

Large Vocabulary: The voice identification framework can perceive a monster scope of vocabularies (for example in excess of a couple a large numbers of expressions or sentences) such structures are perceived as enormous vocabulary speech recognition framework [8].

1.2.2 Categorization Based on Speaker Mode

Speaker Dependent: Speaker based talk affirmation structures study the unique traits of a singular person's voice, in a course for all intents and purposes indistinguishable from voice affirmation. The system is readied dependent on the readiness datasets and it possibly will utilize designs.

Speaker Independent: In without speaker talk affirmation structures, there is no planning of the machine to see a specific speaker in this manner the saved word structures must be illustrative of the collection of speakers foreseen to use the system. The word formats are derived by first getting a huge measure of test plans from a cross-section of talkers of different sex, age-assembling and tongue, and a while later gathering; these to structure an agent plan for each word.

Speaker Adaptive: In speaker flexible talk affirmation system the uses the speaker based data and acclimate to the best ideal speaker to see the talk and lessen the screw up rate with the guide of grasping [9]. Extraction method are utilize in feature extraction for incorporate extraction as Mel Frequency Cepstral Coefficient (MFCC) and Mel Energy depart LPCC, Dynamic coefficients (MEDC), LDA and PCA [10]. All more than such feature extraction systems for talk appraisal for appearing in consequent Table 1.

Table 1. Different techniques of feature extraction

Sr. no.	Techniques	Property	Methods for execution
1	Principal Component analysis (PCA)	Non-linear feature extraction method, Linear map, rapid, eigenvector-based	Traditional, eigenvector base method, as well as karhuneu-Loeve expansion; good for Gaussian data
2	Linear Discriminate Analysis (LDA)	Non-linear feature extraction method, eigenvector based, Supervised linear map, rapid	More than PCA for classification
3	Linear Predictive Coding	Static feature extraction process, 10 to 16 reduce proportion coefficient	It is used for feature Extraction at decrease order
4	Mel-frequency scale analysis	Static feature extraction process, Spectral analysis	Spectral assessment is completed with a firm resolution along a Subjective frequency Scale i.e. Mel-frequency Scale
5	Mel-frequency cepstrum (MFFCs)	Power spectrum is computed by execute Fourier Analysis	This process is to utilized for find our features
6	Dynamic feature extractions i) LPC ii) MFCCs	Acceleration and delta coefficients i.e. II and III order derivatives of normal LPC and MFCCs coefficients	It is utilized by vital or runtime Feature

2 Experimental Analysis

The formation of the human voiced part is amazing for each human being and subsequently the speech description accessible in the talk sign can be used to perceive the speaker. Seeing a person through her/his voice is perceived as a spokesperson affirmation. Since assortments in the anatomical structure are an inherent property of the speaker, voice goes under the biometric character category. Using voice for a recognizing confirmation as different focal points. The most noteworthy central focuses is far away individual approval. The component extraction performs fundamental limit in speaker affirmation. The Fig. 1 reveals the basic speaker affirmation structure.

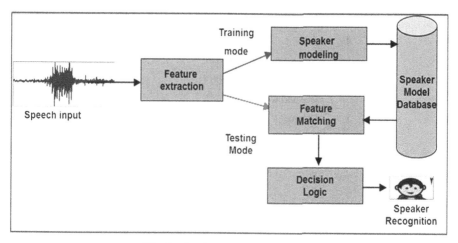

Fig. 1. Speaker recognition system.

The model affirmation systems, speaker affirmation structure similarly contain two phases exclusively planning and testing. The model affirmation structures, speaker affirmation system moreover contain two phases specifically, getting ready and testing. Testing is the most ideal affirmation task. Characteristic vectors addressing the voice properties of the speaker are removed from the training verbalization and are used to build reference models. During testing stage, comparable component vectors are expelled from the testing articulation, and the degree of their suit with the reference is obtained the usage of some organizing methodology. The period of suit shall be used to appear at the decision.

2.1 Database Creation

The database assembled from UG and PG understudies (25 Male and 25 Female) of CS and IT department of Dr. B. A. M. U, Aurangabad, Maharashtra. The age social affair of speakers is 19 years to 26 years. The 10 word close by 5 enunciations which are started from Vowel, for instance, Aadarniy, Aairani, Amrut, Antara, Aushadh, Eshanya, Ityadi, Olawa, Urja and Utapanna. Each speaker given data as 10 words with 5 explanations from different viewpoints infers 50 sound reports are recorded and by and large 2500 sound records are used for our MSRFIW investigated worked. The database is recorded in homeroom condition. Using PRAAT programming software is used for the database is recorded. According to LDCIL recording repeat is 16000 Hz. The detail specific of the database is given underneath in Table 2 (Table 3).

Table 2. Details of technical classification

Sr. no.	Parameter	Values
1	Speaker type	Students (Age 19–26 Years)
2	Gender	25 Male, 25 Female
3	Basic language	Marathi
4	Accent	Marathi
5	Mother tongue	Marathi
6	Frequency	16000 Hz
7	Isolated word	10
8	Uttrances	05
9	Region	Marathwada
10	Environment	Classroom environment

Table 3. Database of Marathi vocabulary word

Sr. No.	English Word	Pronunciation in Marathi	IPA
1.	Aadarniy	आदरणीय	/adərəɳəi:jə/
2.	Aairani	ऐरणी	/əirəɳəiː/
3.	Amrut	अमृत	/əmərutə/
4.	Antara	अंतरा	/əmtərəa/
5.	Aushadh	औषध	/əuʂədʰə/
6.	Eshanya	ईशान्य·	/iːʃəanəjə/
7.	Ityadi	इत्यादी	/itəjəadəi:/
8.	Olawa	ओलावा	/oləaʊəa/
9.	Urja	ऊर्जा	/uːrədʑəa/
10.	Utapanna	उत्पन्न	/utəpənənə/

2.2   Feature Extraction

Center of attention of anticipated research work is progress of Standardized database of speech and the utilization of that speaker recognition method to created database for improvement of. For creating Speaker Recognition System we require to remove the capacity from the obtained and recorded voice and afterward apply the acknowledgment calculation. Be that as it may, toward the beginning we have to improve the procured/recorded speech signal at some point before we can separate the element as it might contain commotion. There are particular procedures that are utilized for the Speech signal improvement and discourse include extraction. Speech is a simple sign. In the main area simple electrical markers are changed to computerized signals. This is acted in two stages sampling and quantization [11]. So a conventional portrayal of a speech signal is a progression of 8- bit numbers at the charge of 10,000 numbers for each second. When the signal change is finished, foundation clamor is separated to keep up sign to commotion proportion high. The signal is pre-stressed and afterward speech variable are extracted [12].

We utilized Mel-Frequency Cepstral Coefficient (MFCC) for separating highlights from recorded speech dataset test. The Mel-recurrence Cepstrum Coefficient (MFCC) strategy is every now and again used to make the unique fingerprint of the sound records. The MFCC depend totally on the recognized different of the human ear's major transmission capacity frequencies with channels separated straightly at lower frequencies and logarithmically at unbalanced frequencies used to catch the crucial attributes of speech [13]. The signal is segmented into imbrications frames to calculate MFCC coefficients. Let every frame consist of N samples and let adjacent frames be separated through M samples where M < N. Each frame is elevated through a Hamming window equation is given by:

$$W(n) = 0.54 - 0.46 \cos\left(\frac{2\pi n}{N-1}\right) \tag{1}$$

Then the signal is transformed from time domain to frequency area via subjecting to Fourier Transform. The signal of Discrete Fourier Transform (DFT) is described by using the subsequent:

$$X_k = \sum_{i=0} x_i e^{-\frac{j2\pi ki}{N-1}} \tag{2}$$

Then the frequency domain signal converted to the Mel Frequency Scale, which suits to better human listening to and perceptions. This is completed by way of a set of triangular filters that are used to compute a weighted sum of spectral factors so that the output of the procedure approximates a Mel scale. Each and every filter magnitude frequency response is triangular in shape and adequate to solidarity at the middle frequency and decrease linearly to zero at center frequency of two adjoining filters. The following equation is used to calculate the Mel for a given frequency [14].

$$M = 2595 \log_{10}\left(1 + \frac{j}{700}\right) \tag{3}$$

In the next step of the log Mel scale spectrum is transformed to time domain the use of Discrete Cosine Transform (DCT). DCT is defined by the following, where is a constant dependent on N:

$$X_k = \alpha \sum_{i=0} x_i \cos . \left\{ \frac{(2i + \pi k)}{2N} \right\} \tag{4}$$

The outcome of the conversion is called as Mel-Frequency Cepstral Coefficient. The lay of coefficients is known as acoustic vectors. Therefore, every input utterance is converted into an order of acoustic vectors [15–17].

Figure 2 indicates the architecture of MFCC for feature extraction of the MFCC processes. The speech wave form is cropped to eliminate silence or acoustical interference that may additionally be present in the starting or end of the sound file. The windowing block minimizes the discontinuities of the signal through tapering the beginning and end of every frame to zero. The FFT block converts every frame from the time domain to the frequency domain. In the Mel-frequency wrapping block, in opposition the signal is plotted to the Mel spectrum to mimic human hearing. In the last step, the Cepstrum, the Mel- spectrum scale is transformed again to standard frequency scale. This spectrum presents a desirable depiction of the spectral properties of the signal which is key for representing and recognizing characteristics of the speaker [18–20].

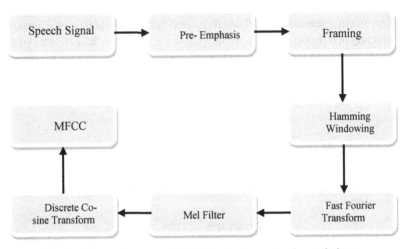

Fig. 2. The architecture of MFCC for feature extraction techniques

Table 4. MFCC feature extraction by using technical parameter

Parameter	Values
Sampling frequency	16 kHz
Window type	Hamming
Number of the coefficients (in each frame)	12
Filters in filter bank	29
The frame length	256

Outcome of the execution of speaker recognition by the number of filters of MFCC to 12, are given. The recognizer reaches the maximal performance at the filter number

K = 12. Even less or too many filters do not result in better accuracy. Here after, if not specifically stated, the number of filters is chosen to be K = 12. The step by step MFCC achievement is depict in figure. The aspect factor of feature extraction of MFCC is defined in above Table 4.

3 Classification

Lastly Cepstrum, the Mel-spectrum scale is transformed again to standard frequency scale. This spectrum presents a best representation of the spectral proper-ties of the signal which is key for representing and recognizing characteristics of the speaker [15]. According to the Euclidean distance formula, the distance be-tween two factors in the plane with coordinates (x, y) and (a, b) is given by:

$$dist((x,y),(a,b)) = \sqrt{(x-a)^2 + (y-b)^2} \qquad (5)$$

Modules has been created for the testing of the system. In it, the very first module is 10 Male subject, group second is 10 female subject and group third is random selected subject. The Table 5 is describes the performance of the speaker verification system for Male selected 10 speaker group whereas Table 6 describes the female speaker group performance (Table 7).

Table 5. The male subject group for speaker recognition system.

Speaker	Number of attempt	False acceptance	False rejection	Accuracy
SS1	10	2	1	80
SS2	10	1	1	90
SS3	10	2	2	80
SS4	10	1	2	90
SS5	10	2	2	80
SS6	10	1	1	90
SS7	10	2	1	80
SS8	10	1	2	90
SS9	10	1	1	90
SS10	10	2	1	80
Overall accuracy				**85%**

In Fig. 3 first load the audio sample and then play the sample from database after that the calculated FAR and FRR. This system is used for speaker recognition system for isolated word identification sys-tem by applied feature extraction techniques as MFCC and classification is done with Euclidian Distance.

$$FAR = \frac{\#Number\ of\ false\ acceptance}{\#Number\ of\ identification\ attempts} \times 100\%$$

Table 6. The Female subject group for speaker recognition system

Speaker	Number of attempt	False acceptance	False rejection	Accuracy
SS1	10	2	1	80
SS2	10	1	1	90
SS3	10	2	1	80
SS4	10	3	3	70
SS5	10	2	2	80
SS6	10	2	2	80
SS7	10	2	1	80
SS8	10	2	3	80
SS9	10	2	1	80
SS10	10	1	1	90
Overall accuracy				**81%**

Table 7. The speaker recognition system of random selected subject group

Speaker	Number of attempt	False acceptance	False rejection	Accuracy
SS1	10	1	1	90
SS2	10	3	3	70
SS3	10	1	2	80
SS4	10	2	1	80
SS5	10	3	4	70
SS6	10	3	5	70
SS7	10	2	2	80
SS8	10	2	3	80
SS9	10	2	1	80
SS10	10	1	1	90
Overall accuracy				**79%**

$$FRR = \frac{\#Number\ of\ false\ rejection}{\#Number\ of\ identification\ attempts} \times 100\%$$

The EER can be determined if and only if the Receiver Operating Characteristic (ROC) curve of False Acceptance Rate (FAR) against False Rejection Rate (FRR) is obtained where FAR and the FRR at this point is the equal for both. Figure 4(a) and (b) describes the ROC curve of Male subject group and Female subject group respectively.

The detail representation of the system of speaker verification for different speaker group is described in Table 8. The graphical representation of performance and error

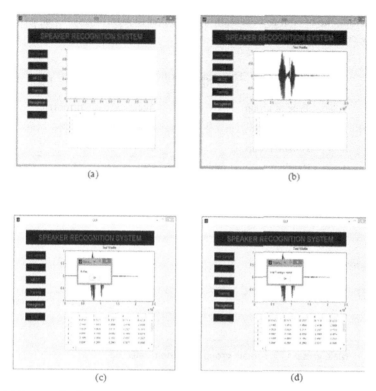

(a)   (b)

(c)   (d)

Fig. 3. (a) Open audio files, (b) Play the audio, (c) Illustrate features of MFCC, (d) Training and recognition

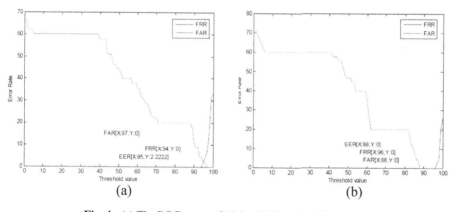

(a)   (b)

Fig. 4. (a) The ROC curve of Male, (b) Female subject group

rate of overall system is shown in Fig. 5. The mel scale is almost optimal for males, but suboptimal for females in speaker recognition, due to it quefrency quashes the cepstra containing the female formants. Anatomical characteristics of speakers to a large extent conclude the overall representation of speaker recognition systems, with females performing consistently low than males.

Table 8. The speaker verification system

Sr. no.	Type of group	Accuracy (%)	Error rate
1	Male group	85	15
2	Female group	81	19
3	Random selected subject group	79	21

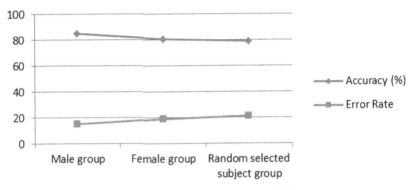

Fig. 5. Speaker verification method

4   Conclusion

We propound the outcome of proposed system of Marathi Speaker Recognition for Isolated Words (MSRFIW) system based on feature extraction. MFCC are used for feature extraction from speech utterances. The tough undertaking is to improve the recital of system. Prevail epoch of spokesperson identification is concerted to upgrade the presentation of the method. The novel intent of this exploration was to construct a MSRFIW system. We initiated our work by aiming to develop the Marathi speech database and implement the speaker recognition method using MFCC. With the help of Mel Frequency Cepstral coefficient, the system was successfully tested and Euclidian Distance was used for the classifications. The recital of the method was tested on the basis of various three groups of speaker as Male speaker, Female speaker and randomly selected speaker as gender and age group plays a significant role in speaker recognitions. This performance was calculated by using FAR and FRR. The accuracy for Male group obtained was 85% and that for female group was 81%. The recognition performance of females perform consistently low than males, when using MFCC 4%. The accuracy obtained of the random selected subject group was 79%. Hence, it is concluded by this study that the Marathi Speaker Recognition for Isolated Words (MSRFIW) system gave 85%, 81% and 79% accuracy for male, female and random selected group.

References

1. Torfi, A., Nasrabadi, N.M., Dawson, J.: Text-independent speaker verification using 3D convolutional neural networks. arXiv preprint arXiv:1705.09422 (2017)
2. Furui, S.: Speaker independent isolated word recognition using dynamic features of speech spectrum. IEEE Trans. Acoust. Speech Signal Process. ASSP **34**(1), 2–59 (1986)
3. Gaikwad, B.P., Kamble, P.S.: Speech processing for secluded Marathi words recognition using MFCC features. Int. J. Innov. Res. Comput. Commun. Eng. (IJIRCCE) **3**(8), (2015). ISSN (Online): 2320-9801, ISSN (Print): 2320-9798
4. https://en.wikipedia.org/wiki/List_of_languages_by_number_of_native_speakers.html. Accessed 1 June 2014
5. Kamble, V.V., Gaikwad, B.P., Rana, D.M.: Spontaneous emotion recognition for Marathi spoken words. In: International Conference on Communication and Signal Processing, India, 3–5 April 2014, 978-1-4799-3356-3/14/$31.00 © IEEE (2014)
6. Shrawankar, U., Thakare, V.: Techniques for feature extraction in speech recognition system: a comparative study. Int. J. Comput. Appl. Eng. Technol. Sci. (IJCAETS) 412–418 (2010). ISSN 0974-3596
7. Furui, S.: An overview of speaker recognition technology. In: ESCA Workshop on Automatic Speaker Recognition, Identification and Verification, pp. 1–9 (1994)
8. Anusuya, M.A., Katti, S.K.: Speech recognition by machine: a review. Int. J. Comput. Sci. Inf. Secur. (IJCSIS) **6**(3), 181–205 (2009)
9. Huang, X.D.: A study on speaker - adaptive speech recognition. In: Proceedings DARPA Workshop on Speech and Natural Language, pp. 278–283 (1991)
10. Rabiner, L.R., Schafer, R.W.: Digital Processing of Speech Signals, Signal Processing. Prentice-Hall, Englewood Cliffs (1978)
11. Singh, B., Kaur, R., Devgun, N., Kaur, R.: The process of feature extraction in automatic speech recognition system for computer machine interaction with humans: a review. Int. J. Adv. Res. Comput. Sci. Softw. Eng. **2**(2), 1–7 (2012)
12. Tiwari, V.: MFCC and its application in speaker recognition. Int. J. Emerg. Technol. **1**(1), 19–22 (2010)
13. Levinson, S.E.: Structural methods in automatic speech recognition. Proc. IEEE **73**(11), 1625–1650 (1985)
14. Srinivasa Kumar, Ch., Mallikarjuna Rao, P.: Int. J. Comput. Sci. Eng. **3**(8), 2942–2954 (2011)
15. Muda, L., Begam, M., Elamvazuthi, I.: Voice recognition algorithms using mel frequency cepstral coefficient (MFCC) and dynamic time warping (DTW) techniques. J. Comput. **2**(3) (2010). ISSN 2151-9617
16. Huang, C., Chang, E., Chen, T.: Accent issues in large vocabulary continuous speech recognition (LVCSR). Microsoft Research China, MSR-TR-2001-69, pp. 1–27 (2001)
17. Bachate, R.P., Sharma, A.: Automatic speech recognition systems for regional languages in India. Int. J. Recent Technol. Eng. (IJRTE) **8**(2S3) (2019). ISSN 2277-3878
18. Gawali, B.W., Gaikwad, S., Yannawar, P., Mehrotra, S.C.: marathi isolated word recognition system using MFCC and DTW features. In: Proceedings of International Conference on Advances in Computer Science, vol. 1 (2010)
19. Gaikwad, S., Gawali, B., Yannawar, P., Mehrotra, S.: Feature extraction using fusion MFCC for continuous marathi speech recognition. In: 2011 Annual IEEE India Conference. IEEE, 16 December 2011
20. Gaikwad, S.K., Gawali, B.W., Yannawar, P.: A review on speech recognition technique. Int. J. Comput. Appl. **10**(3) (2010)

Speech Recognition of Mathematical Words Using Deep Learning

Vaishali Kherdekar[1]([✉]) and Sachin Naik[2]

[1] Symbiosis International (Deemed University), Pune, Maharashtra, India
`vaishali.kherdekar@gmail.com`
[2] Symbiosis Institute of Computer Studies and Research (SICSR), Symbiosis International (Deemed University), Pune, Maharashtra, India
`sachin.naik@sicsr.ac.in`

Abstract. Speech recognition is to convert speech signal into text. It is challenging task due to natural variations present in human speech and also due to background noise. Now a day's researchers focus on deep learning due to it's effectiveness and high performance. There is need to work on recognition of speech for mathematical words; many of the currently available methods proposed by researchers are not yet adequate to recognize speech for mathematical words. In this paper we focus on CNN (Convolution Neural Network) for recognition of connected word speech recognition for mathematical words plus, minus, square and square-root. The dataset of spoken words are created using Audacity. CNN model is verified by considering Adam, Gradient descent and Adagrad optimizer with learning rate of 0.001 and 0.0001. Result shows that Adam and Gradient descent optimizer gives better result for 0.001 learning rate.

Keywords: CNN · Deep learning · Spectrogram · Speech recognition

1 Introduction

Speech recognition is nothing but to convert an audio signal into text. Speech recognition is challenging task because acoustic phonetic varies from person to person i.e. speaking style, speaking rate and emotional state may be different. Due to background noise and reverberation it is also challenging. Speech recognition systems, based on utterances are categorized into Isolated word, connected word, continuous speech and spontaneous speech. In Connected word speech recognition systems separate utterances work collectively with small pause between them. It accepts the word in minimum interval of time. Traditional speech recognition system uses GMM and HMM based models to represent sequential structure of speech signals. In traditional methods feature extraction is very critical. Now a day's deep learning plays a vital role in speech recognition, image processing etc. because of the power of deep learning to learn the features during training. Deep learning is the subset of artificial intelligence and machine learning. Now a day's performance of the speech recognition has been increased due to use of deep learning [1]. In deep learning various models are present such as DNN (Deep Neural Network),

© Springer Nature Singapore Pte Ltd. 2021
K. C. Santosh and B. Gawali (Eds.): RTIP2R 2020, CCIS 1381, pp. 356–362, 2021.
https://doi.org/10.1007/978-981-16-0493-5_31

RNN (Recurrent Neural Network), CNN (Convolution Neural Network) etc. Modern DNNs use more than one hidden layer, making them deep. As a general property, depth is an important feature for the success of modern DNNs. Several groups recently found replacing the standard sigmoidal hidden units with rectified linear units in DNNs leads to WER (Word Error Rate) gains and simpler training procedures for deep architectures [16]. The paper is organized as follows. Section 2 represents the related work. Proposed model is given in Sect. 3. Experimental details with setup, dataset, feature extraction technique, CNN and training are presented in Sect. 4 along with it's subsections respectively. Results are given in Sect. 5. Last section concludes the paper.

2 Related Work

(Honglak Lee et al. 2009) proposed model for classification of audio data. TIMIT dataset is used for training. Audio features are extracted using spectrogram and MFCC. CDBN (Convolutional Deep Belief Network) is used for classification of unlabeled audio data. Result shows that accuracy is more for second layer features than first layer. Nassif, A.B. et al. [5] have presented review of speech recognition systems using DNN (Deep Neural Network). They have considered 174 papers from 2006 to 2018. They have concluded that MFCC is most widely used feature extraction technique and use of LPC is also worth for it. Author recommended that conduct research using deep RNN specially LSTM which is very powerful for speech recognition. D. Nagajyothi and P. Siddaiah [6] have proposed speech- recognition based Airport Enquiry system using CNN for Telgu language. Dataset is prepared by considering the commonly asked question at Airport enquiry. They have used weight connectivity; local connectivity features to train the model. Model is trained and tested using CNN (Convolution Neural Network). Tanh and Relu activation function are used during training. Author have concluded that CNN gives better performance than conventional network. Sarkar et al. [7] have presented a model for speech identification, segmentation and diarization. Libri Speech dataset is used to train the model. Precision, recall and f1 scores are used for the segmentation task. Bidirectional Long Short erm Memory (LSTM) Networks, Convolutional Networks are used for classification. Relu & CTC loss function are used for activation. Gradient descents are used to minimize the loss function. Li Deng et al. [8] have reviewed the papers on use of deep learning in speech recognition. They considered the 5 parameters for technical review which include better optimization, better activation function and neural network architecture, better ways to determine the parameters of deep neural networks, more appropriate ways to preprocess speech for deep neural networks and ways of leverage multiple languages or dialects. Authors have concluded that DNN works well for filterbank output than MFCC feature. DNN also works well for noisy speech. Li Deng et al. [9] presents an overview of work done by Microsoft speech researchers. Spectrogram features of speech are superior to MFCC with DNN. In future need to perform new improvements on DNN architectures and learning. Tara N. Sainath et al. [10] proposed a combined model using CNN, LSTM and DNN named as CLDNN. Model is trained using clean as well as noisy speech and also evaluated on clean and noisy speech respectively. 40 dimensionals log mel filterbank features given as input to CNN layer. ASGD (Asynchronous Stochastic Gradient Descent) optimizer is used at the time of training.

The weights for all the layers of model are decided using Glorot-Bengio strategy. CNN is trained with 2 convolution layer and 4 fully connected layers. DNN is trained with 6 hidden layers. LSTM is trained with 2 layers of 512 dimensionals projection layer. CNN is used to reduce the spectral variation of the input feature. LSTM is used to perform temporal modeling and DNN is used to separate the features. Result shows 4–6% reduction in WER. Dong Yu et al. [13] proposed a context dependent DNN-HMM model. Authors have extended the DNN to DTNN (Deep Tensor Neural Network) which consists of tensor layers and one or more layers are double projections. Switchboard corpus is used to evaluate the model. Features are extracted using 39 dimensional HLDA and 13 dimensional PLP. Learning rate of 0.0003 is used for first 5 epoch and 0.000008 is taken for next 5 epoch. DTNN is trained using Back Propagation algorithm. WER reduces by 5% for DTNN than DNN. Navdeep Jaitly et al. [14] presented DBN network trained by using 2 datasets. VoiceSearch, Youtube dataset are used to train the model. Features are extracted using MFCC and log filter bank. Result shows WER of 16% for VoiceSearch and 52.3% for YouTube dataset. Authors have suggested that in future need to design system with fast computation without loss of accuracy. Andrew L. Maas et al. [15] empirically investigates the aspects of DNN important for improvement of speech recognition performance. Switchboard and Fisher corpus are combined for the experiment. Features are extracted using MFCC and trained on DNN, DCNN and DLUNN. Result shows that increase in the no. of parameters increases the representational capacity of DNN model. Result also shows that large DNN reduces the WER on training set. In future WER will be reduced by increasing the DNN model size [16].

3 Proposed Model

Figure 1 shows the proposed model for deep learning (CNN) for speech recognition of mathematical words. We accept the audio data in the form of.wav files. i.e. audio signals of mathematical words are given as input. Figure 2 shows signal of audio file for the word "plus". Batches are created using the audio files. Labels are created for the 4 mathematical words. MFCC (Mel Frequency Cepstral Coefficient) is used to Extracted features which are given as input to build the model. Model is trained using 2 convolution layers with ReLU(Rectified Linear Unit) activation function and 1 fully connected layer. Dropouts are used to prevent the network from over fitting the dataset.

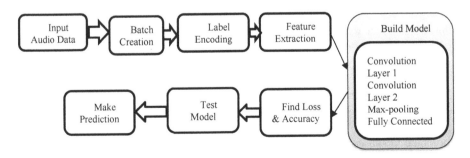

Fig. 1. Proposed model

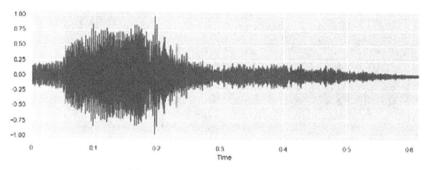

Fig. 2. Audio Signal of word "Plus"

4 Experiment

We performed the experiment to implement the model using Convolution Neural Network deep learning.

4.1 Dataset

We have recorded 4 words from 2 persons having age group between 14 to 40 using Audacity tool. Each word recorded 5 times. The dataset consists of 40 audio files of 4 words. Separate folder is created for each spoken word. The dataset is split into training and testing folders, each folder is labeled with spoken words having audio files of words "minus", "plus", "square" and "square root".

4.2 Feature Extraction Techniques

To extract features from audio data it is necessary to convert it into spectrogram. Spectrogram is visual representation of frequencies of signal varies with time. It is 2D plot between time and frequency where each point in the plot represents the amplitude of a particular frequency at a particular time in terms of intensity of color. In simple terms, it is a spectrum (broad range of colors) of frequencies as it varies with time. It is used in speech processing i.e. to identify spoken words phonetically. We first convert audio signals into spectrogram. It is given as input to CNN (Convolution Neural Network) to generate feature map or feature vector. Spectrogram is generated for the coefficients of audio file. Figure 3 shows spectrogram for audio word "plus".

4.3 Convolution Neural Network

It is type of deep learning algorithms. It is feed forward, supervised deep learning model. It accepts input as spectrogram of audio file and extracts the features through the learning phase. It is given as input to classification phase. Figure 4 shows CNN architecture. It consists of one or more convolution layer, max-pooling and one or more fully connected layer.

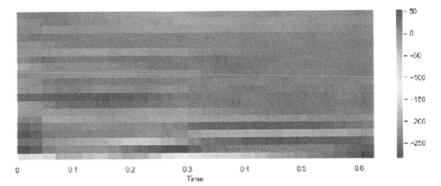

Fig. 3. Spectrogram of word "Plus"

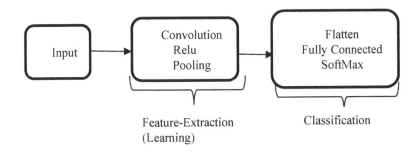

Fig. 4. CNN architecture

It consists of sparse interactions, parameter sharing, and equivariant representation as basic concepts.[6] Convolution is nothing but combining, it combines the input matrix with filter or weight matrix using dot product and gives convolved feature as output. The main concept of convolution is to use the filter for extracting the feature map from input vector. In convolution layer filters act as feature extractor to capture time-frequency spectral variations. [17] Multiscale features can be extracted using different filter size. The dimensionality of feature map is reduced by using pooling operation of the pooling layer. Fully connected layer provides a meaningful, low dimensional and invariant feature space.

4.4 Training

We trained the model using CNN for 40 audio files of mathematical words "plus", "minus", "square" and "squareroot". Spectrogram is find out for each utterance of spoken word and then it is given as input to CNN. Learning rate of 0.001 & 0.0001 is used during training. The most standard approach to DNN optimization is stochastic gradient descent (SGD) [16]. To find out the best optimizer we perform the experiment using Adam optimizer, Adagrad optimizer and Stochastic gradient descent optimizer. The selection of an appropriate activation function is important while training a model using CNN

(Convolution Neural Network). It is used to separate the data into useful and not useful data. We used the most common activation function Rectified Linear Unit (ReLU) [7] in convolution layer1 and 2 and Softmax-crossentropy is used as loss function to train the model

5 Result

Table 1 shows the accuracy result for training and test model with learning rate of 0.001, for different optimizers. It shows that final test accuracy is same for Gradient descent optimizer and Adagrad optimizer. Table 2 shows the accuracy result having learning rate of 0.0001. It shows that for Adam and Gradient descent optimizer final test accuracy remains same.

Table 1. Result with learning rate = 0.001

Optimizer	Adam optimizer	Gradient descent optimizer	Adagrad optimizer
Accuracy after 5th iteration	89.9%	50%	69.9%
Accuracy after 10th iteration	89.9%	89.9%	80%
Final test accuracy	76.6	83.33%	83.33%

Table 2. Result with learning rate = 0.0001

Optimizer	Adam optimizer	Gradient descent optimizer	Adagrad optimizer
Accuracy after 5th iteration	30%	69.9%	60%
Accuracy after 10th iteration	69.9%	80%	50%
Final test accuracy	76.66%	76.66%	63.33%

6 Conclusion

In this paper, we applied convolution neural network to audio data and evaluated it using different optimizers. We used MFCC feature extraction. Our result shows that Gradient descent is the best optimizer for 0.0001 learning rate and for 0.001 learning rate Adam optimizer is the best. In future, we want to implement the experiment on large dataset for different parameters of deep learning such as varying hidden layers, activation functions, loss functions etc.

References

1. Miao, Y., Gowayyed, M., Metze, F.: EESEN: end-to-end speech recognition using deep RNN models and WFST-based decoding. In: 2015 IEEE Workshop on Automatic Speech Recognition and Understanding (ASRU), pp. 167– 174. IEEE (2015)
2. Schmidhuber, J.: Deep learning in neural networks: an overview. Neural Netw. **61**, 85–117 (2015)
3. Trigeorgis, G., et al..: Adieu features? End-to-end speech emotion recognition using a deep convolutional recurrent network. In: 2016 IEEE International Conference on Acoustics, Speech and Signal Processing (ICASSP), pp. 5200–5204. IEEE (2016)
4. Rubi, C.R.: A review: speech recognition with deep learning methods. Int. J. Comput. Sci. Mob. Comput. **4**(8), 301–307 (2015)
5. Lee, H., Pham, P., Largman, Y., Ng, A.Y.: Unsupervised feature learning for audio classification using convolutional deep belief networks. In: Advances in Neural Information Processing Systems, pp. 1096–1104 (2009)
6. Nassif, A.B., Shahin, I., Attili, I., Azzeh, M., Shaalan, K.: Speech recognition using deep neural networks: a systematic review. IEEE Access **7**, 19143–19165 (2019)
7. Nagajyothi, D., Siddaiah, P.: Speech recognition using convolutional neural networks. Int. J. Eng. Technol. **7** (4.6), 133–137 (2018)
8. Sarkar, A., Dasgupta, S., Naskar, S.K., Bandyopadhyay, S.: Says who? Deep learning models for joint speech recognition, segmentation and diarization. In: 2018 IEEE International Conference on Acoustics, Speech and Signal Processing (ICASSP), pp. 5229–5233. IEEE (2018)
9. Deng, L., Hinton, G., Kingsbury, B.: New types of deep neural network learning for speech recognition and related applications: an overview. In: 2013 IEEE International Conference on Acoustics, Speech and Signal Processing, pp. 8599–8603. IEEE (2013)
10. Deng, L., et al.: Recent advances in deep learning for speech research at Microsoft. In: 2013 IEEE International Conference on Acoustics, Speech and Signal Processing, pp. 8604–8608. IEEE (2013)
11. Hinton, G., et al.: Deep neural networks for acoustic modeling in speech recognition. IEEE Sig. Process. Mag. **29**, 82–97 (2012)
12. Abdel-Hamid, O., Mohamed, A.R., Jiang, H., Deng, L., Penn, G., Yu, D.: Convolutional neural networks for speech recognition. IEEE/ACM Trans. Audio Speech Lang. Process. **22**(10), 1533–1545 (2014)
13. Sainath, T.N., Vinyals, O., Senior, A., Sak, H.: Convolutional, long short-term memory, fully connected deep neural networks. In: 2015 IEEE International Conference on Acoustics, Speech and Signal Processing (ICASSP), pp. 4580–4584. IEEE (2015)
14. Yu, D., Deng, L., Seide, F.: Large vocabulary speech recognition using deep tensor neural networks. In: 13th Annual Conference of the International Speech Communication Association (2012)
15. Jaitly, N., Nguyen, P., Senior, A., Vanhoucke, V.: Application of pretrained deep neural networks to large vocabulary speech recognition proceedings of interspeech (2012)
16. Maas, A.L., et al.: Building DNN acoustic models for large vocabulary speech recognition. Comput. Speech Lang. **41**, 195–213 (2017)
17. Soe, T., Maung, S.S., Oo, N.N.: Applying multi-scale features in deep convolutional neural networks for myanmar speech recognition. In: 2018 15th International Conference on Electrical Engineering/Electronics, Computer, Telecommunications and Information Technology (ECTI-CON), pp. 94–97. IEEE (2018)

Segregating Bass Grooves from Audio: A Rotation Forest-Based Approach

Himadri Mukherjee[1], Ankita Dhar[1(✉)], Sk. Md. Obaidullah[2], K. C. Santosh[3], Santanu Phadikar[4], and Kaushik Roy[1]

[1] Department of Computer Science, West Bengal State University, Kolkata, India
himadrim027@gmail.com, ankita.ankie@gmail.com, kaushik.mrg@gmail.com
[2] Department of Computer Science and Engineering, Aliah University, Kolkata, India
sk.obaidullah@gmail.com
[3] Department of Computer Science, The University of South Dakota, Vermillion, SD, USA
santosh.kc@ieee.org
[4] Department of Computer Science and Engineering, Maulana Abul Kalam Azad University of Technology, Kolkata, India
sphadikar@yahoo.com

Abstract. Notation of a music piece is an extremely important resource for musicians. It requires mastery and experience to transcribe a piece accurately which lays the path for automatic music transcription systems to help budding musicians. A piece can be divided into two parts namely the background music (BGM) and lead melody. The BGM is an extremely important aspect of a piece. It is responsible for setting the mood of a composition and at the same time makes it complete. There are different musical instruments which are used in a composition both in the BGM and lead sections one of them being the bass guitar. It bonds with the percussion instruments to form the spinal cord of a piece. It is very much important to transcribe the bass section of a composition for understanding as well as performance. Prior to identification of the notes being played, it is essential to distinguish the different patterns/grooves. In this paper, a system is presented to differentiate bass grooves. Tests were carried out with 60K clips and a best accuracy of 98.46% was obtained.

Keywords: Background music (BGM) · Automatic music transcription · Rotation forest

1 Introduction

Technological advancements have aided in simplifying things in almost all fields. The music industry has also been largely benefited by such advancements. Music production has become quite easier and user friendly in the recent times. It has also helped budding musicians as well students of music in numerous ways. The notation of a composition is a vital resource for artists and music practitioners.

© Springer Nature Singapore Pte Ltd. 2021
K. C. Santosh and B. Gawali (Eds.): RTIP2R 2020, CCIS 1381, pp. 363–372, 2021.
https://doi.org/10.1007/978-981-16-0493-5_32

It is a blueprint of a composition in terms of the notation and timing. A piece can be broadly divided into 2 sections-lead melody and background music (BGM). BGM is extremely important in a composition as it is responsible for setting the mood and ambience. It also makes a piece sound complete. Several instruments are used to put together the BGM one of them being the bass guitar. It is responsible for providing the lower frequencies in a piece thereby adding weight and depth to it. A slight change in the bass notation can alter a piece significantly thus making its notation very important. Notating a music piece is a challenging task and demands experience which is lacking in many musicians especially beginners. An automated music transcription system can help in such a scenario. It can help musicians to understand a piece for performance, practice and study.

Abeber [1] presented a system to notate bass and lead guitar clips. They worked on a dataset of 1034 isolated note clips and obtained highest f-measures of 0.90 and 0.93 for bass and lead guitar respectively. Kroher and Gomez [2] presented a system to transcribe flamenco singing clips. They extracted predominant melody followed by filtering to discard the accompaniments. Next they used volume and pitch-based attributes to distinguish the notes and reported better results than available systems. Wats and Patra [3] attempted music transcription on the disklavier dataset. They adopted a non negative matrix factorization with accelerated multiplicative update-based approach. They presented results for different instances with the highest being 68%. Wu et al. [4] reviewed different aspects involved in automatic transcription of drums. They introduced the basics of a drum kit followed by the involved challenges. They have talked about different applications of the same and different techniques for drum transcription using several approaches including deep learning techniques.

Avci and Acuner [5] presented a system to recognize open string violin notes. They used spectral difference-based approach to detect note onset on a dataset collected from 8 people whose details are presented in [5]. Puri and Mahajan [6] have discussed different aspects of automatic music transcription. They have laid ponters on different music textures in the thick of monophony, heterophony, polyphony and homophony. Different aspects of sounds in the thick of timbral features, zero crossing, centroid, roll off and flux has also been discussed. They have talked about MFCC, fourier transform-based techniques, wavelet-based techniques and several classifiers like SVM, dynamic Bayesian network, HMM, RNN, etc. Cogliati et al. [7] attempted automatic transcription of piano melody with convolutional sparse coding. They experimented with 3 datasets along with another synthetic dataset and reported better result than state-of-the art systems. Stein et al. [8] presented a system to detect audio effects in lead and bass guitar recordings. They experimented with 10 common effects and extracted over 500 harmonic, cepstral and spectral features. They further used feature selection and transformation coupled with SVM they reported nearly 100% accuracy in detecting a single effect and 98% in simultaneous detection of the entire effect range. Benetos et al. [8] have talked about different applications of automatic music transcription. They have identified some of the key challenges involved in this task and present an overview of the standard transcription methods. They have detailed about the suitability of negative matrix factorization and neural networks. Several avenues of research in this field in the thick of context-based transcription, modelling expressive pitch and timing as well as detecting and notating unpitched and percussion sounds.

Here, we differentiate bass grooves from short audio clips. This can aid in automatic transcription of BGM which is a very essential part of any composition. The used methodology is presented in Fig. 1.

In the rest of the paper, the dataset details are presented in Sect. 2 followed by the proposed methodology in Sect. 3. The results are discussed in Sect. 4 and finally we have concluded in Sect. 5.

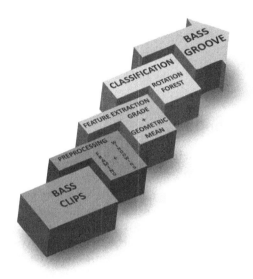

Fig. 1. The proposed system.

2 Dataset

Data is extremely important for any experiment. During data collection care needs to taken about its quality. At times the collected data may be noisy or the data collection procedure may be faulty. It is very much important to avoid such issues. In the present experiment a dataset of bass grooves was put together with the help of volunteers. They were asked to play six different grooves in the key of A. The key A was chosen because it is the first natural note in music. Audio clips of 125 ms were used in the experiment to test our system's performance for extremely short duration data. 10208 clips were obtained for each of the grooves totalling to 61248 clips. The clips were recorded different guitars which were plugged into a Steinberg ur22 mkii audio interface. The clips were recorded in .wav format at a bitrate of 2822 kbps. The constituent notes for each of the grooves is presented in Table 1 where (8va) beside a note means it is played in the higher octave. It is observed from Table 1 that different grooves had overlapping notes and some even had exactly same notes. This is often the case in real world scenario. The fret board positions of the notes are presented in Fig. 4.

Table 1. The involved notes in each of the bass grooves where (8va) denotes the higher octave of a note.

Groove	Notes
1	A, E
2	A, C, D, E
3	A, E, G, A(8va)
4	A, C, D, E
5	A, D, E, G, A(8va)
6	A, C, E, G, A(8va), C(8va)

3   Proposed Method

3.1   Framing and Windowing

Each of the clips were first split into smaller frames in order to obtain smaller spectral deviations to facilitate in analysis. The clips were partitioned into frames having 256 sample points with 100 overlapping points amidst 2 successive frames [9]. The overlapped framing technique is illustrated in Fig. 2. Post framing it is often observed that jitters are introduced, which lead to spectral leakage. In order to resolve this issue, the frames are subjected to a windowing function. Here hamming window was used due to its utility as presented in [9].

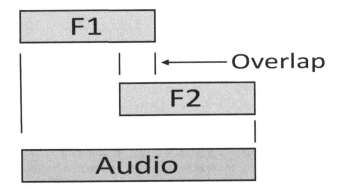

Fig. 2. The overlapped framing technique.

3.2   Feature Extraction

Standard line spectral frequency features [9] of 5, 10, 15, 20 and 25 dimensions were extracted for each of the clips. Next, the bands (frequency ranges) for the features were arranged according to their total energies. This graded band sequence formed the first half of the final feature. The second half was obtained by calculating the band wise geometric mean. The same was used to obtain the central tendency of the energy values. The geometric mean was particularly used due to its efficiency of others as presented in [10]. The same is mathematically illustrated hereafter:

$$\prod_{i=1}^{n} x_i = \sqrt[n]{x_1 x_2 ... x_n} \tag{1}$$

where, n corresponds to the number of values x_i is the ith value.

Finally, the raw LSFs yielded 10, 20, 30, 40 and 50 dimensional features. The obtained 30 dimensional features (best result) for an instance from each of the 6 classes is presented in Fig. 3.

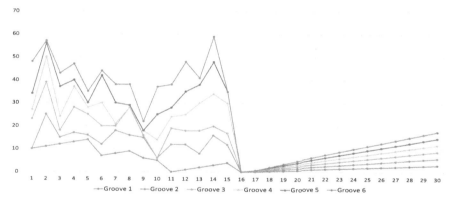

Fig. 3. Trend of the feature values for a single instance of each of the grooves.

3.3   Classification

Each of the feature sets were fed to a rotation forest [11]-based classifier. It is an ensemble learning technique, which works by splitting a feature set (F) into (k) parts. For each of these subsets, a random set of instances are chosen and 75% of this data is used for further processing. In the next stage, Principal component analysis is applied on this set and the obtained coefficients are stored in a matrix (rotation matrix). This is repeated for each of the feature subsets (F_1-F_k) and then the rotation matrix is re arranged in order to correspond the order of the features in the original undivided feature set (F). This matrix is used for training the classifier. Every classifier is accompanied with such a matrix which are later used in combination to predict the class of a test instance. In this experiment,

50% of instances were chosen for each of the feature subsets and the size of the feature groups/subsets were set to 3. The popular J48 classifier was used for the classifier ensemble. Such parameters for classification were set based on trials.

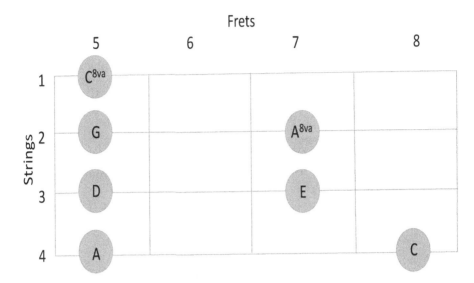

Fig. 4. The position of used notes on a bass guitar fret board.

4   Result and Discussion

Every feature set was fed to the rotation forest-based classifier with 10 epochs. 5 fold cross validation was used for evaluation. The obtained results are presented in Table 2 where it is observed that the best result was obtained for 30 dimensional features.

Table 2. Obtained accuracies for different feature sets.

Feature dimension	Accuracy (%)
10	42.66
20	48.91
30	**95.36**
40	89.19
50	94.40

The inter class confusions for the same is presented in Table 3. It is observed that groove 2 and 3 were the highest confused pair. The instances were analyzed and it was found both the grooves not only had the first same note like the rest

Table 3. Inter class confusions for the 30 dimensional features.

	Groove 1	Groove 2	Groove 3	Groove 4	Groove 5	Groove 6
Groove 1	10159	42	5	0	2	0
Groove 2	17	9332	542	103	176	38
Groove 3	7	594	9148	166	220	73
Groove 4	0	108	107	9964	27	2
Groove 5	3	225	180	48	9728	24
Groove 6	0	47	73	4	11	10073

but also the time signature of it along with some other notes were also similar. The playing style was also similar in terms of finger pressure in many instances. These are some of the reasons which perhaps led to such confusion.

The 30 dimensional feature set was further experimented with by increasing the number of training iterations whose results are presented in Table 4. It is observed from the Table that 200 iterations produced the best result showing an increase of 2.79% over the initial setup.

Table 4. Obtained accuracies for different training iterations on 30 dimensional feature set.

Training iterations	Accuracy (%)
10	95.36
50	97.71
100	97.99
150	98.04
200	**98.15**
250	98.11

The interclass confusions for 200 training iterations is shown in Table 5. It is noted that the misclassification among grooves 2 and 3 reduced to 469 from 1136 depicting a reduction of 58.71% which is very encouraging.

Finally, the cross validation folds were varied to obtain further increase in system performance with 200 training iterations whose results are tabulated in Table 6. It is observed that the best result was obtained for 20 fold cross validation whose interclass confusions are tabulated in Table 7.

It is noted that the confusion amidst grooves 2 and 3 further decreased to 391. Grooves 2 and 4 had the same notes and which were very well distinguished from each other. Only 0.19% of the total clips were confused among these 2 grooves which is extremely encouraging. Different popular classifiers from [12] in the thick of BayesNet, Simple logistic, RBF network, Decision table and multi layer perceptron were applied on this dataset whose results are presented in Table 8.

Table 5. Inter class confusions for 30 dimensional feature set with 200 training iterations.

	Groove 1	Groove 2	Groove 3	Groove 4	Groove 5	Groove 6
Groove 1	10184	23	1	0	0	0
Groove 2	12	9886	218	37	54	1
Groove 3	0	251	9847	69	27	14
Groove 4	0	80	55	10066	7	0
Groove 5	0	123	92	5	9988	0
Groove 6	0	16	46	0	1	10145

Table 6. Obtained accuracies for different folds of cross validation.

Fold	Accuracy (%)
5	98.15
10	98.30
15	98.34
20	**98.46**
25	98.39

Table 7. Inter class confusions for 20 fold cross validation with 200 training iterations on 30 dimensional feature set.

	Groove 1	Groove 2	Groove 3	Groove 4	Groove 5	Groove 6
Groove 1	10186	21	1	0	0	0
Groove 2	11	9956	180	28	33	0
Groove 3	0	211	9916	51	22	8
Groove 4	0	68	43	10091	6	0
Groove 5	0	108	83	4	10013	0
Groove 6	0	13	49	0	2	10144

Table 8. Performance of different classifiers on the 30 dimensional feature set.

Classifier	Accuracy(%)
BayesNet	78.43
Simple logistic	43.48
RBF network	59.11
Decision table	82.21
Multi layer perceptron	82.08
Rotation forest	**98.46**

5   Conclusion

This paper proposes a system for distinguishing bass grooves which can aid in automatic transcription of BGM of a piece. Experiments were performed with more 60K clips of very short duration and encouraging results have been obtained with a rotation forest-based classifier with an average precision of 0.985. In future, the dataset will be extended to grooves from other keys. Different audio capturing devices will also be used for data collection to incorporate more variability. Experiments will also be performed with other machine learning techniques including deep learning-based approaches. The system will also be subjected to clips consisting of more than a single instrument at an instance to test its performance for stereophonic audio.

Acknowledgment. The authors would like to thank Mr. Pradip Ghosh and Mr. Subho Dey for their help with the musical technicalities and data collection. They also thank WWW.PresentationGO.com for the block diagram template.

References

1. Abeßer, J.: Automatic string detection for bass guitar and electric guitar. In: Aramaki, M., Barthet, M., Kronland-Martinet, R., Ystad, S. (eds.) CMMR 2012. LNCS, vol. 7900, pp. 333–352. Springer, Heidelberg (2013). https://doi.org/10.1007/978-3-642-41248-6_18
2. Kroher, N., Gómez, E.: Automatic transcription of flamenco singing from polyphonic music recordings. IEEE/ACM Trans. Audio Speech Lang. Process. (TASLP) 24(5), 901–913 (2016)
3. Wats, N., Patra, S.: Automatic music transcription using accelerated multiplicative update for non-negative spectrogram factorization. In: 2017 International Conference on Intelligent Computing and Control (I2C2), pp. 1–5. IEEE (2017)
4. Wu, C.W., et al.: A review of automatic drum transcription. IEEE/ACM Trans. Audio Speech Lang. Process. (TASLP) 26(9), 1457–1483 (2018)
5. Avci, K., Acuner, T.Ş.: Automatic transcription of open string notes from violin recordings. In: 2017 25th Signal Processing and Communications Applications Conference (SIU), pp. 1–4. IEEE (2017)
6. Puri, S.B., Mahajan, S.P.: Review on automatic music transcription system. In: 2017 International Conference on Computing, Communication, Control and Automation (ICCUBEA), pp. 1–6. IEEE (2017)
7. Cogliati, A., Duan, Z., Wohlberg, B.: Context-dependent piano music transcription with convolutional sparse coding. IEEE/ACM Trans. Audio Speech Lang. Process. 24(12), 2218–2230 (2016)
8. Stein, M., Abeber, J., Dittmar, C., Schuller, G.: Automatic detection of audio effects in guitar and bass recordings. In: Audio Engineering Society Convention 128. Audio Engineering Society (2010)
9. Mukherjee, H., et al.: An ensemble learning-based language identification system. In: Maharatna, K., Kanjilal, M.R., Konar, S.C., Nandi, S., Das, K. (eds.) Computational Advancement in Communication Circuits and Systems. LNEE, vol. 575, pp. 129–138. Springer, Singapore (2020). https://doi.org/10.1007/978-981-13-8687-9_12

10. Thelwall, M.: The precision of the arithmetic mean, geometric mean and percentiles for citation data: an experimental simulation modelling approach. J. Informetr. **10**(1), 110–123 (2016)

11. Rodriguez, J.J., Kuncheva, L.I., Alonso, C.J.: Rotation forest: a new classifier ensemble method. IEEE Trans. Pattern Anal. Mach. Intell. **28**(10), 1619–1630 (2006)

12. Hall, M., Frank, E., Holmes, G., Pfahringer, B., Reutemann, P., Witten, I.H.: The WEKA data mining software: an update. ACM SIGKDD Explor. Newsl. **11**(1), 10–18 (2009)

Author Index